P9-DHM-844

ENGINEERING
ECONOMY

Prentice-Hall International Series in Industrial
and Systems Engineering

W. J. Fabrycky and J. H. Mize, editors

Fabrycky, Ghare, and Torgersen – *Industrial Operations
Research*
Gottfried and Weisman – *Introduction to Optimization Theory*
Mize, White, and Brooks – *Operations Planning and Control*
Thuesen, Fabrycky, and Thuesen – *Engineering Economy, 4/E*
Whitehouse – *Systems Analysis and Design Using Network
Techniques*

4th edition

ENGINEERING ECONOMY

H. G. Thuesen
Oklahoma State University

W. J. Fabrycky
Virginia Polytechnic Institute and State University

G. J. Thuesen
Georgia Institute of Technology

Prentice-Hall, Inc. / Englewood Cliffs, New Jersey

© 1971 by Prentice-Hall, Inc., Englewood Cliffs, New Jersey

*All rights reserved. No part of this book may be
reproduced in any form or by any means
without permission in writing
from the publisher.*

Printed in the United States of America

13–277442–9

Library of Congress Catalog Card Number: 72–143434

Current Printing (last digit);
10 9 8 7

Prentice-Hall International, Inc.   *London*
Prentice-Hall of Australia, Pty. Ltd.   *Sydney*
Prentice-Hall of Canada, Ltd.   *Toronto*
Prentice-Hall of India Private Limited   *New Delhi*
Prentice-Hall of Japan, Inc.   *Tokyo*

Preface

Engineering activities of analysis and design are not an end in themselves, but are a means for satisfying human wants. Thus, engineering has two aspects. One aspect concerns itself with the materials and forces of nature; the other is concerned with the needs of mankind. Because of this, engineering must be closely associated with economics.

The subject of *engineering economy* deals with the concepts and techniques of analysis useful in evaluating the worth of systems, products, and services in relation to their cost. In this text we emphasize that the essential prerequisite of successful engineering application is economic feasibility. Our objective is to help the reader grasp the significance of the economic aspects of engineering, and to become proficient in the evaluation of engineering proposals in terms of worth and cost.

The engineering approach to problem solution has advanced and broadened to the extent that success often depends upon both the ability to deal with economic aspects and physical aspects of the problem. Down through history the limiting factor has been predominantly physical, but with the development of science and technology goods and services have become physically possible that may not have utility. Fortunately, the engineer can readily extend his inherent ability of analysis to embrace economic factors. To aid him in so doing is a primary aim of this book.

A secondary aim of this book is to acquaint the engineer with operations and operational feasibility. An understanding of the organizational and personnel factors upon which the success of engineering activity depends is essential to the solution of most complex problems. Economic factors in the operation of systems and equipment can no longer be left to chance but must be considered in the design process. A basic understanding of mathematical modeling of operations is becoming more important as operational systems attract the attention of a larger number of engineers.

This text contains more than enough material for a three semester hour course. Some material will have to be skipped for a course of shorter duration. This may be easily done, since the foundation topics are concentrated in the first nine chapters.

We have had no difficulty in teaching this material to engineering sophomores as well to upper division students in management, economics, and the physical sciences who wish an introduction to engineering from the economic point of view. College algebra and elementary differential calculus is the only mathematical background required.

Those familiar with earlier editions of this text will note that we have retained the basic conceptual approach with considerable emphasis on examples. Also retained is the functional factor designation system which has been very helpful in teaching the application of compound interest to problem solution. Symbols in the system have been changed as suggested by the ad hoc standardization committee of the Engineering Economy Division of the American Society for Engineering Education.

It is our pleasure to acknowledge the very useful comments offered by many students who were exposed to this material in class. Specific thanks are due Mr. K. R. Morrison who developed the expanded set of tables for this edition and Mr. R. M. Mason who assisted in the development of problems for Chapter 15.

H. G. THUESEN
W. J. FABRYCKY
G. J. THUESEN

Contents

I
INTRODUCTION TO
ENGINEERING ECONOMY

II
INTEREST FORMULAS
AND EQUIVALENCE

III
ECONOMIC ANALYSIS
OF ALTERNATIVES

Contents

IV
ACCOUNTING, DEPRECIATION, AND INCOME TAXES

13 Income Taxes in Economy Studies 274

V
ESTIMATES, UNCERTAINTY, AND IMPLEMENTATION

14 Treatment of Estimates in Economy Studies 301

15 Dealing with Uncertainty in Economy Studies 315

16 A Plan for Engineering Economy Studies 349

VI
ECONOMIC ANALYSIS
OF OPERATIONS

APPENDICES

ENGINEERING
ECONOMY

INTRODUCTION TO
ENGINEERING ECONOMY

1

Engineering Economy
and the Engineering Process

The engineer is confronted with two important environments, the physical and the economic. His success in manipulating or altering the physical environment to produce products or services depends upon his knowledge of physical laws. However, the worth of these products and services lies in their utility measured in economic terms. There are numerous examples of structures, machines, and processes that exhibit excellent physical design but possess little economic merit. For this reason, it is essential that engineering proposals be evaluated in terms of worth and cost before they are undertaken. Therefore, in this chapter and throughout the text, we emphasize that the essential prerequisite of successful engineering application is economic feasibility.

1.1. ENGINEERING AND SCIENCE

Engineering is not a science but is rather an application of science. It is an art composed of the skill and ingenuity in adapting knowledge to the uses of the human race. As expressed in a definition adopted by the Engineers' Council for Professional Development: "Engineering is the profession in which a knowledge of the mathematical and natural sciences gained by study, experience, and practice is applied with judgment to develop ways to utilize, economically, the materials and forces of nature for the benefit of mankind." In this, as in most other accepted definitions, the applied nature of engineering activity is emphasized.

The purpose of the scientist is to add to mankind's inventory of systematic knowledge and to discover universal laws of behavior. The purpose of the engineer is to apply this knowledge to particular situations to produce products and services. To the engineer, knowledge is not an end in itself but is the raw material from which he fashions structures, machines, and processes. Thus, engineering involves the determination of the combination of materials, forces, and human factors that will yield a desired result with a reasonable degree of accuracy. Engineering activities are rarely carried out for the satisfaction that may be derived from them directly. With few exceptions, their use is confined to satisfying human wants.

Modern civilization rests to a large degree upon engineering. Most products used to facilitate work, communication, and transportation and to furnish sustenance, shelter, and even health are directly or indirectly a result of engineering activity. Engineering has also been instrumental in providing leisure time for pursuing and enjoying culture. Through the development of the printing process, television, and rapid transportation, engineering has provided the means for both cultural and economic improvement of the human race. In addition, engineering has become an essential ingredient in national survival as is evident from the considerable portion of engineering talent that is employed in national defense.

The scientist and engineer form a team that paces modern technology. In science lies the foundation upon which the engineer builds toward the advancement of mankind. With the continued development of science and the widespread application of engineering, the standard of living may be expected to improve and further increase the demand for those things that minister to man's love for the comfortable and beautiful. The fact that these human wants may be expected to engage the attention of engineers to an increasing extent is, in part, the basis for the movement to broaden the humanistic and social content of engineering curricula. An understanding of these fields is essential as engineers seek solutions to the complex social problems of our time.

1.2. THE BI-ENVIRONMENTAL NATURE OF ENGINEERING

In dealing with the physical environment the engineer has a body of physical laws upon which to base his reasoning. Such laws as Boyle's law, Ohm's law, and Newton's laws of motion were developed primarily by collecting and comparing numerous similar instances and by the use of an inductive process. These laws may then be applied by deduction to specific instances. They are supplemented by many formulas and known facts, all of which enable the engineer to come to conclusions about the physical environment that match the facts within narrow limits. Much is known with certainty about the physical environment.

Much less, particularly of a quantitative nature, is known about the economic environment. Since economics is involved with the actions of people, it is apparent that economic laws must be based upon their behavior. Economic laws, therefore, can be no more exact than the description of the behavior of people acting singly and collectively.

The usual function of engineering is to manipulate the elements of one environment, the physical, to create utility in a second environment, the economic. However, engineers are sometimes handicapped by a tendency to disregard economic feasibility and are often appalled in practice by the necessity for meeting situations in which action must be based on estimates and judgment. Yet the modern engineering graduate is increasingly finding himself in positions in which his responsibility is extended to include economic factors.

The engineer's approach to the solution of problems has broadened to such an extent that his success may depend as much upon his ability to cope with economic aspects as it does on the physical aspects of the total environment. The engineer can readily extend his inherent ability of analysis to become proficient in the analysis of the economic aspect of engineering application. Furthermore, the engineer who aspires to a creative position in engineering will find proficiency in economic analysis helpful. The large percentage of engineers who will eventually be engaged in managerial activities will find such proficiency a necessity.

Initiative for the use of engineering rests, for the most part, upon those who will concern themselves with social and economic consequences. Therefore, to maintain the initiative, engineers must operate successfully in both the physical and economic segments of the total environment. It is the objective of *engineering economy* to prepare engineers to cope effectively with the bi-environmental nature of engineering application.

1.3. PHYSICAL AND ECONOMIC EFFICIENCY

The objective of engineering application is to get the greatest end result per unit of resource input. This statement is essentially an expression of efficiency which may be stated as

$$\text{efficiency (physical)} = \frac{\text{output}}{\text{input}}.$$

If interpreted broadly enough, this statement measures the success of engineering activity in the physical environment. However, the engineer must be concerned with two levels of efficiency. On the first level is physical efficiency expressed as outputs divided by inputs of such physical units as Btu's, kilowatts, and foot pounds. When such physical units are involved, efficiency will always be less than unity or less than 100%.

On the second level are economic efficiencies. These are expressed in terms of economic units of output divided by economic units of input, each expressed in terms of a medium of exchange such as money. Economic efficiency may be stated as

$$\text{efficiency (economic)} = \frac{\text{worth}}{\text{cost}}.$$

It is well known that physical efficiencies over 100% are not possible. However, economic efficiencies can exceed 100% and must do so for economic undertakings to be successful.

Physical efficiency is related to economic efficiency. For example, a power plant may be profitable in economic terms even though its physical efficiency in converting units of energy in natural gas to electrical energy may be relatively low. As an example, in the conversion of energy in a certain plant, assume that the physical efficiency is only 14%. Assuming that output Btu's in the form of electrical energy have an economic value of $8 per million and that input Btu's in the form of natural gas have an economic value of $0.70 per million, then

$$\text{efficiency (economic)} = \frac{\text{Btu output} \times \text{value of electricity}}{\text{Btu input} \times \text{value of natural gas}}$$

$$= 0.14 \times \frac{\$8}{\$0.70} = 160\%.$$

Since physical processes are of necessity carried out at efficiencies of less than 100% and economic ventures are feasible only if they attain efficiencies of greater than 100%, it is clear that the economic worth per unit of physical output must always be greater than the economic cost per unit of physical input in feasible economic ventures. Consequently, economic efficiency must depend more upon the worth and cost per unit of physical outputs and inputs than upon physical efficiency. Physical efficiency is always significant, but only to the extent that it contributes to economic efficiency.

In the final evaluation of most ventures, even in those in which engineering plays a leading role, economic efficiencies must take precedence over physical efficiencies. This is because the function of engineering is to create utility in the economic environment by means of altering the elements of the physical environment.

1.4. THE ENGINEERING PROCESS

Man is continually seeking to satisfy his wants. In so doing, he gives up certain utilities in order to gain others that he values more. This is essentially an economic process in which the objective is the maximization of economic efficiency.

Engineering is primarily a producer activity that comes into being to satisfy human wants. Its objective is to get the greatest end result per unit of resource expenditure. This is essentially a physical process with the objective being the maximization of physical efficiency.

Want satisfaction in the economic environment and engineering proposals in the physical environment are linked by the production or the construction process. Figure 1.1 illustrates the relationship between engineering proposals, production or construction, and want satisfaction.

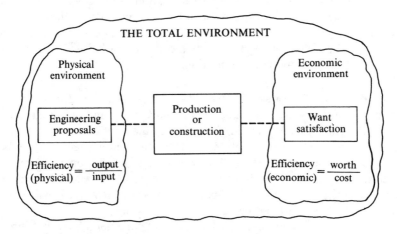

FIGURE 1.1. THE ENGINEERING ENVIRONMENT.

Engineering activity in the physical environment takes place as a result of the needs of mankind in the economic environment. The engineering process employed from the time a particular human need is recognized until the time when the need is satisfied may be divided into a number of activities. These activities will be discussed under the headings which follow and will be related to the ideas expressed above.

Determination of Objectives. One important facet of the engineering process involves the search for new objectives for engineering application; to learn what people want that can be supplied by engineering. In the field of invention, success is not a direct result of the construction of a new device; rather, it is dependent on the capability of the invention to satisfy human wants. Thus, market surveys seek to learn what the desires of people are. Automobile manufacturers make surveys to learn what mechanical, comfort, and style features people want in transportation. Highway commissions make traffic counts to learn what construction programs will be of greatest use. Considerations of technical feasibility and cost come only after what is wanted has been determined.

The things that people want may be the result of logical considerations, but more often are the result of emotional drives. There appears to be no logical reason why one prefers a certain make of car, a certain type of work, or a certain style of clothes. The bare necessities needed to maintain physical existence, in terms of calories of nourishment, clothing, and shelter are limited and may be determined with a fair degree of certainty. But the wants that stem from emotional drives seem to be unlimited.

The facet of the engineering process that seeks to learn of human wants requires not only a knowledge of the limitations of engineering capability but also a general knowledge of sociology, psychology, political science, economics, literature and other fields related to the understanding of human nature. A knowledge of these fields is recognized to be useful or essential in most branches of modern engineering.

Identification of Strategic Factors. The factors that stand in the way of attaining objectives are known as *limiting factors*. An important element of the engineering process is the identification of the limiting factors restricting accomplishment of a desired objective. Once the limiting factors have been identified they are examined to see which ones may be operated on with success. Thus, each of the limiting factors should be examined in order to locate *strategic factors;* that is those factors which, if altered, will remove limitations restricting the success of an undertaking.

Consider the following example. A shaft will not enter a hole. The limiting factors may be that the shaft is too large or that the hole is too small. The difficulty can be eliminated by reducing the diameter of the shaft or by increasing the diameter of the hole. But if decreasing the diameter of the shaft is not suitable for the total purpose, and increasing the diameter of the hole is suitable, then the diameter of the hole is the strategic factor and should be selected for alteration.

The identification of strategic factors is important, for it makes possible the concentration of effort on those areas in which success is obtainable. This may require inventive ability, or the ability to put known things together in new combinations, and is distinctly creative in character. The means that will achieve the desired objective may consist of a procedure, a technical process, or a mechanical, organizational, or administrative change. Strategic factors limiting success may be circumvented by operating on engineering, human, and economic aspects individually and jointly.

Determination of Means. The determination of means is subordinate to the identification of strategic factors, just as the identification of strategic factors is necessarily subordinate to the determination of objectives. Strategic factors may be altered in many different ways. Each possibility must be evaluated to determine which will be most successful in terms of overall economy. Engineers are well equipped by training and experience to determine means for altering the physical environment. If the means devised to overcome strategic factors come within the field of engineering they may be termed engineering proposals.

Engineers in any capacity have the opportunity to be creative by considering human and economic factors in their work. The machine designer may design tools or machines that will require a mimimum of maintenance and that can be operated with less fatigue and greater safety. The highway designer may consider durability, cost, and safety. Engineers in any capacity can see to it that both private and public projects are planned, built, and operated in accordance with good engineering practice. Engineering is an expanding profession. People have great confidence in the integrity and ability of engineers. Perhaps nothing will enhance the image of the individual engineer and the profession of engineering more than the acceptance of responsibilities which go beyond the determination of means.

Evaluation of Engineering Proposals. It is usually possible to accomplish a desired result by several means, each of which is feasible from the physical aspects of engineering application. The most desirable of the several proposals is the one that can be performed at the least cost. The evaluation of engineering proposals in terms of comparative cost is an important facet of the engineering process and an essential ingredient in the satisfaction of wants with maximum economic efficiency. Although engineering alternatives are most often evaluated to determine which is most desirable economically, exploratory evaluations are also made to determine if any likely engineering proposal can be formulated to reach a goal profitably.

A wide range of factors may be considered in evaluating the worth and cost of engineering proposals. When investment is required, the time value of money must be considered. Where machinery and plants are employed, depreciation becomes an important factor. Most proposals involve organized effort, thus making labor costs an important consideration. Material is an important ingredient which may lead to market analyses and a study of procurement policy. Risks of a physical and economic nature may be involved and must be evaluated. Where the accepted engineering proposal is successful a net income will be derived, thus making the consideration of income tax necessary. This text offers specific instruction in methods of analysis pertaining to each of these factors and general instruction regarding the engineering process as a whole.

Assistance in Decision Making. Engineering is concerned with action to be taken in the future. Therefore, an important facet of the engineering process is to improve the certainty of decision with respect to the want satisfying objective of engineering application. Correct decisions can offset many operating handicaps. On the other hand, incorrect decisions may and often do hamper all subsequent action. No matter how expertly a bad decision is carried out, results will be at best unimpressive and at worst disastrous.

To make a decision is to select a course of action from several. A correct decision is the selection of that course of action which will result in an outcome more desirable than would have resulted from any other selection. Decision rests upon the possibility of choice; that is, on the fact that there are alternatives from which to choose. The engineer acting in a creative capacity proceeds on the thought that there is a most desirable solution if he can find it.

The logical determination and evaluation of alternatives in tangible terms have long been recognized as an integral facet of the engineering process. The success of engineers in dealing with this element of application is responsible, in part, for the large percentage of engineers that are either directly or indirectly engaged in decision-making activities.

1.5. ENGINEERING ECONOMY AND THE ENGINEER

Economy, the attainment of an objective at low cost in terms of resource input, has always been associated with engineering. During much of history the limiting factor has been predominantly physical. Thus, a great innovation, the wheel, awaited invention, not because it was useless or costly, but because the mind of man could not synthesize it earlier. But, with the development of science, things have become physically possible that people are interested in only slightly or not at all. Thus, a new type of communication system may be perfectly feasible from the physical standpoint but may enjoy limited use because of its first cost or cost of operation.

There are those, and some are engineers, who feel that the engineer should restrict himself to the consideration of the physical and leave the economic and humanistic aspects of engineering to others; some would not even consider these aspects as coming under engineering. This viewpoint may arise in part because those who take pleasure in discovering and applying the well-ordered certainties of the physical environment find it difficult to adjust their thinking to consider the complexities of the economic environment.

Engineers are becoming increasingly aware of the fact that many sound proposals fail because those who might have benefited from them did not understand their significance. A prospective user of a good or service is primarily interested in its worth and cost. The person who lacks an understanding of engineering technology may find it difficult or even impossible to grasp the technical aspects of a proposal sufficiently to arrive at a measure of its economic desirability. The uncertainty so engendered may easily cause loss of confidence and a decision to discontinue consideration of the proposal.

Being accustomed to the use of facts and being proficient in computation, the engineer should accept the responsibility for making an economic interpretation of his work. It is much easier for the engineer to master the fundamental concepts of economic analysis necessary to bridge the gap between physical and economic aspects of engineering application than it is for the person who is not technically trained to acquire the necessary technical background.

Since economic factors are the strategic consideration in most engineering activities, engineering practice may be either responsive or creative. If the engineer takes the attitude that he should restrict himself to the physical he is likely to find that the initiative for the application of engineering has passed on to those who will consider economic and social factors.

The engineer who acts in a responsive manner acts on the initiative of others. The end product of his work has been envisioned by another. Although this position leaves him relatively free from criticism, he gains this freedom at the expense of professional recognition and prestige. In many ways, he is more of a technician than a professional man. Responsive engineering is, therefore, a direct hindrance to the development of the engineering profession.

The creative engineer, on the other hand, not only seeks to overcome physical limitations, but also initiates, proposes, and accepts responsibility for the success of projects involving human and economic factors. The general acceptance by engineers of the responsibility for seeing that engineering proposals are both technically and economically sound, and for interpreting proposals in terms of worth and cost, may be expected to promote confidence in engineering endeavor and enhance the worth of engineering service. In assuming this responsibility, the engineer also assumes greater control of the application of engineering. This seems desirable for it is unlikely that the future potential of engineering can be fully utilized unless engineers participate in determining its objectives.

QUESTIONS

1. Contrast the role of the engineer and the scientist.
2. What is the important difference between a physical and an economic law?
3. Give several examples of products that would be technically feasible but that would possess little economic merit.
4. Why should the engineer concern himself with both the physical and the economic aspects of the total environment?
5. Explain how physical efficiency below 100% may be converted to economic efficiency above 100%.
6. Why must economic efficiency take precedence over physical efficiency?
7. Name the essential activities in the engineering process.
8. What is involved in seeking new objectives for engineering application?
9. What is a limiting factor; a strategic factor?

10. Give reasons why engineers are particularly well equipped to determine means for the attainment of an objective.
11. Why is it essential that engineering proposals be evaluated in terms of economy?
12. How may engineers assist in decision making?
13. As compared with the economic aspect of engineering application, give reasons why the physical aspect is decreasing in relative importance.
14. Why should the economic interpretation of engineering proposals be made by engineers?
15. Describe the responsive engineer; the creative engineer.

2

Some Fundamental
Economic Concepts

Concepts are crystallized thoughts. They are usually qualitative in nature and are not necessarily universal in application. Economic concepts, if carefully related to fact, may be useful in suggesting solutions to problems in engineering economy. Those given in this chapter are by no means exhaustive. Many others will be found throughout the text to support the quantitative material presented. The ability to arrive at sound decisions through engineering economic analysis is dependent jointly upon a sound conceptual understanding and the ability to handle the quantitative aspects of the problem. In this chapter we give special attention to some basic economic concepts useful in engineering economy studies.

2.1. ECONOMICS DEALS WITH THE BEHAVIOR OF PEOPLE

The engineer is often confused by the lack of certainty associated with the economic aspect of engineering. However, it must be recognized that economic considerations embrace many of the subtleties and complexities characteristic of people. Economics deals with the behavior of people individually and collectively, particularly as their behavior relates to the satisfaction of their wants.

The wants of people are motivated largely by emotional drives and tensions and to a lesser extent by logical reasoning processes. A part of human wants can be satisfied by physical goods and services such as food, clothing, shelter, transportation, communication, entertainment, medical care, educational opportunities, and personal services; but man is rarely satisfied by physical things alone. In food, sufficient calories to meet his physical needs will rarely satisfy. He will want the food he eats to satisfy his energy needs and also his emotional needs. In consequence we find people concerned with the flavor of food, its consistency, the china and silverware with which it is served, the person or persons who serve it, the people in whose company it is eaten, and the "atmosphere" of the room in which it is served. Similarly, there are many desires associated with clothing and shelter, in addition to those required merely to meet physical needs.

Anyone who has a part in satisfying human wants must accept the uncertain action of people as a factor with which he must deal, even though he finds such action unexplainable. Much or little progress has been made in discovering knowledge on which to base predictions of human reactions, depending upon one's viewpoint. The idea that human reactions will someday be well-enough understood to be predictable is accepted by many people; but in spite of the fact that this has been the objective of the thinkers of the world since the beginning of time, it appears the progress in psychology has been meager compared to the rapid progress made in the physical sciences. But, in spite of the fact that human reactions can be neither predicted nor explained, they must be considered by those who are concerned with satisfying human wants.

2.2. VALUE AND UTILITY

The term *value*, like most other widely used terms, has a variety of meanings. In economics, value designates the worth that a person attaches to an object or a service. Thus the value of an object is not inherent in the object but is inherent in the regard that a person or people have for it. The subject of engineering valuation, for instance, is directed at assigning values to property in accordance with precepts that will be judged fair by the parties concerned. Value should not be confused with the cost or the price of an object in engineering ecomony studies. There may be little or no relation between the value a person ascribes to an article and the cost of providing it or the price that is demanded for it.

The general economic meaning of the term *utility* is the power to satisfy human wants. The utility that an object has for an individual is determined by him. Thus the utility of an object, like its value, is not inherent in the object itself but is inherent in the regard that a person has for it. Utility and value in the sense used here are closely related. The utility that an object has for a person is the satisfaction he derives from it. Value is an appraisal of utility in terms of a media of exchange.

In ordinary circumstances a large variety of goods and services is available to an individual. The utility that available items may have in the mind of a prospective user may be expected to be such that his desire for them will range from abhorrence, through indifference, to intense desire. His evaluation of the utility of various items is not ordinarily constant but may be expected to change with time. Each person also possesses either goods or services that he may render. These have the utility for the person himself that he regards them to have. These same goods and possible services may also be desired by others, who may ascribe to them very different utilities. The possibility for exchange exists when each of two persons possesses utilities desired by the other.

If the supernatural is excluded from consideration all that has utility is physically manifested. This statement is readily accepted in regard to physical objects that have utility, such as an automobile, a tractor, a house, or a steak dinner. But this statement is equally true in regard to the more intangible things. Music, which is regarded as pleasing to a person, is manifested to him as air waves which strike his ears. Pictures are manifested as light waves. Even friendship is realized only through the five senses and must, therefore, have its physical aspects. It follows that utilities must be created by changing the physical environment.

For example, the consumer utility of raw steak can be increased by altering its physical condition by an appropriate application of heat. In the area of producer utilities the machining of a bar of steel to produce a shaft for a rolling mill is an example of creating utility by manipulation of the physical environment. The purpose of much engineering effort is to determine how the physical environment may be altered to create the most utility for the least cost in terms of the utilities that must be given up.

2.3. CONSUMER AND PRODUCER GOODS

Two classes of goods are recognized by economists; consumer goods and producer goods. Consumer goods are products and services that directly satisfy human wants. Examples of consumer goods are television sets, houses, shoes, books, orchestras, and health services. Producer goods also satisfy human wants but do so indirectly as a part of the production or the construction process. Broadly speaking, the ultimate end of all engineering activity is to supply goods and services that people may consume to satisfy their desires and needs.

Producer goods are, in the long run, used as a means to an end; namely, that of producing goods and services for human consumption. Examples of this class of goods are lathes, bulldozers, ships, and railroad cars. Producer goods are an intermediate step in man's effort to supply his wants. They are not desired for themselves, but because they may be instrumental in producing something that man can consume.

Once the kind and amount of consumer goods to be produced has been determined, the kinds and amounts of facilities and producer goods necessary to produce them may be approached objectively. The energy, ash, and other contents of coal, for instance, can be determined very accurately and are the basis of evaluating the utility of the coal. The extent to which producer utility may be considered by logical processes is limited only by factual knowledge and the ability to reason.

Utility of Consumer Goods. A person will consider two kinds of utility. One kind embraces the utility of goods and services that he intends to consume personally for the satisfaction he gets out of them. Thus, it seems reasonable to believe that the utility a person ascribes to goods and services that are consumed directly is in large measure a result of subjective, nonlogical mental processes. That this is so may be inferred from the fact that sellers of consumer goods apparently find emotional appeals more effective than factual information. Early automobile advertising took the form of objective information related to design and performance, but more recent practice stresses such subjective aspects as comfort, beauty, and prestige values.

Some kinds of human wants are much more predictable than others. The demand for food, clothing, and shelter, which are needed for bare physical existence, is much more stable and predictable than the demand for those items that satisfy man's emotional needs. The amount of foodstuffs needed for existence is ascertainable within reasonable limits in terms of calories of energy, and the clothing and shelter requirement may be fairly accurately determined from climate data. But once man is assured of physical existence he reaches out for satisfactions related to his being a person rather than merely to his being a physical organism.

An analysis of advertising and sales practices used in selling consumer goods will reveal that they appeal primarily to the senses rather than to reason and perhaps rightly so. If the enjoyment of consumer goods stems almost exclusively from how one feels about them rather than what one reasons about them, it seems logical to make sales presentation on the basis on which customers ascribe utility.

Utility of Producer Goods. The second kind of utility that an object or service may have for a person is its utility as a means to an end. Producer goods are not consumed for the satisfaction that can be directly derived from them but as a means of producing consumer goods usually by facilitating the alteration of the physical environment.

Although the utility of consumer goods is primarily determined subjectively, the utility of producer goods as a means to an end may be, and usually is, in large measure considered objectively. In this connection consider the satisfaction of the human want for harmonic sounds as in music. Suppose it has been decided that the desire for music can be met by 100,000 stereophonic records. Then the organization of the artists, the technicians, and the equipment necessary to produce the records becomes predominantly objective in character. The amount of material that must be compounded and processed to form one record is calculable to a high degree of accuracy.

If a concern has been making records for some time, it will know the various operations that are to be performed and the unit times for performing them. From these data, the kind and amount of producer service, the amount and kind of labor, and the number of various types of machines are determinable within rather narrow limits. Whereas the determination of the kinds and amounts of consumer goods needed at any one time may depend upon the most subjective of human consideration, the problems associated with their production are quite objective by comparison.

2.4. THE ECONOMY OF EXCHANGE

Economy of exchange occurs when utilities are exchanged by two or more people. In this connection, a utility means anything that a person may receive in an exchange that has any value whatsoever to him; for example, a lathe, a dozen pencils, a meal, music, or a friendly gesture.

A buyer will purchase an article when he has money available and when he believes that the article has equal or greater utility for him than the amount required to purchase it. Conversely, a seller will sell an article when he believes that the amount of money to be received for the article has greater utility than the article has for him. Thus, an exchange will not be effected unless at the time of exchange both parties believe that they will benefit. Exchanges are made when they are thought to result in mutual benefit. This is possible because the objects of exchange are not valued equally by the parties to the exchange.

As an illustration of the economy of exchange consider the following example. Two workmen, upon opening their lunch boxes, discover that one contatins a piece of apple pie and the other a piece of cherry pie. Suppose further that Mr. A evaluates his apple pie to have 10 units of utility for him and the cherry pie to have 30 units. Suppose, also that Mr. B evaluates his cherry pie to have 20 units of utility for him and the apple pie 40 units. If Mr. A consumes his apple pie and Mr. B consumes his cherry pie, the utility realized in the system is 10 plus 20, or 30 units. But, if the two workmen exchange pieces of pic, the resulting utility will be increased to 30 plus 40, or 70 units. The utility in the system can be increased by 70 less 30, or 40 units by exchange.

Economy of exchange is possible because consumer utilities are evaluated by the consumer almost entirely, if not entirely, by subjective consideration. Thus, if the workmen in the example believe that the exchange has resulted in a gain of net satisfaction to them, no one can deny it. On the other hand, at the time of exchange, unless each person valued what he had to give less than what he was about to receive, an exchange could not take place. Thus, we conclude that an exchange of consumer utilities results in a gain for both parties because the utilities are subjectively evaluated by the participating parties. The reason that people can be found who will subjectively evaluate consumer utilities so that an exchange will permit each to gain is that people have different needs by virtue of their history and their environment.

Assuming that each party to an exchange of producer utilities correctly evaluates the objects of exchange in relation to his situation, what makes it possible for each person to gain? The answer is that the participants are in different economic environments. The fact may be illustrated by an example of a merchant who buys lawn mowers from a manufacturer. For example, at a certain volume of activity the manufacturer finds that he can produce and distribute mowers at a total cost of $70 per unit and that the merchant buys a number of the mowers at a price of $80 each. The merchant then finds that by expending an average of $40 per unit in selling effort, he can sell a number of the mowers to homeowners at $160 each. Both of the participants profit by the exchange. The reason that the manufacturer profits is that his environment is such that he can sell to the merchant for $80 a number of mowers that he cannot sell elsewhere at a higher price, and that he can manufacture lawn mowers for $70 each. The reason that the merchant profits is that his environment is such that he can sell mowers at $160 each by applying $40 of selling effort upon a mower of certain characteristics which he can buy for $80 each from the manufacturer in question, but not for less elsewhere.

The questions may be asked, why doesn't the manufacturer enter the merchandising field and thus increase his profit or why doesn't the merchant enter the manufacturing field? The answer to these questions is that neither the manufacturer nor the merchant can do so unless each changes his environment. The merchant, for example, lacks physical plant equipment and an organization of engineers and workmen competent to build lawn mowers. Also, he may be unable to secure credit necessary to engage in manufacturing, although he may easily secure credit in greater amounts for merchandising activities. It is quite possible that he cannot alter his environment so that he can build mowers for less than $80. Similar reasoning applies to the manufacturer. Exchange consists essentially of physical activity designed to transfer the control of things from one person to another. Thus, even in exchange, utility is created by altering the physical environment.

Mutual Benefit in Exchange. Each party in an exchange should seek to give something that has little utility for him but that will have great utility for the receiver. In this manner each exchange can result in the greatest gain for each party. Nearly everyone has been a party to such a favorable exchange. When a car becomes stuck in snow, only a slight push may be required to dislodge it. The slight effort involved in the dislodging push may have very little utility for the person giving it, so little that he expects no more compensation than a friendly nod. On the other hand it might have very great utility for the person whose car was dislodged, so great that he may offer a substantial tip. The aim of much sales and other research is to find products that not only will have great utility for the buyer but that can be supplied at a low cost; that is, have low utility for the seller. The difference between the utility that a specific good or service has for the buyer and the utility it has for the seller represents the profit or net benefit that is available for division between buyer and seller, and may be called the range of mutual benefit in exchange.

The factors that may determine a price within the range of mutual benefit at which exchange will take place are infinite in variety. They may be either subjective or objective. A person seeking to sell may be expected to make two evaluations: the minimum amount he will accept and the maximum amount a prospective buyer can be induced to pay by persuasion. The latter estimate may be based upon mere conjecture or upon a detailed analysis of buyers' subjective and objective situation. In bargaining it is usually advantageous to obscure one's situation. Thus, sellers will ordinarily refrain from revealing the costs of the things they are seeking to sell or from referring a buyer to a competitor who is willing to sell at a lower price.

Persuasion in Exchange. It is not uncommon for an equipment salesman to call on a prospective customer, describe and explain a piece of equipment, state its price, offer it for sale, and have his offer rejected. This is concrete evidence that the machine does not possess sufficient utility at the moment to induce the prospective customer to buy it. In such a situation, the salesman may be able to induce the prospect to listen to further sales talk, during which the prospect may decide to buy on the basis of the original offer. This is concrete evidence that the machine now possesses sufficient utility to induce the prospective customer to buy. Since there was no change in the machine or the price at which it was offered, there must have been a change in the customer's attitude or regard for the machine. The pertinent fact to grasp is that a proposition at first undesirable now has become desirable as a result of a change in the customer, not in the proposition.

What brought about the change? A number of reasons could be advanced. Usually it would be said that the salesman persuaded the customer to buy, in other words that the salesman induced the customer to believe something, namely that the machine had sufficient utility to warrant its purchase. There are many aspects to persuasion. It may amount only to calling attention to the availability of an item. A person cannot purchase an item he does not know exists. A part of a salesman's function is to call attention to the things he has to sell.

It is observable that persuasive ability is much in demand, is often of inestimable beneficial consequences to all concerned, and is usually richly rewarded. Persuasion as it applies to the sale of goods is of economic importance to industry. A manufacturer must dispose of the goods he produces. He can increase the salability of his products by building into them greater customer appeal in terms of greater usefulness, greater durability, or greater beauty, or he may elect to accompany his products to market with greater persuasive effort in the form of advertising and sales promotion. Either plan will require expenditure and both are subject to the law of diminishing returns. It is a study in economy to determine what levels of perfection of product and sales effort will be most profitable.

It is interesting to note, in this connection, that the costs of some things that increase the sale of a product may be more than outweighed by the economies that result. Thus qualities that make a product more desired, or sales promotion, may have two beneficial effects. One of these is that it may be feasible to increase the price asked for the product and the other is that the volume of product sold may increase to the extent that a lower manufacturing cost per unit may result.

Whatever the approach, factual or emotional, persuasion consists of taking a person on an excursion into the future in an attempt to show and convince him what will happen if he acts in accordance with a proposal. The purpose of engineering economy analysis is to estimate, on as factual a basis as possible, what the economic consequences of a decision will be. It is therefore a useful technique in persuasion.

2.5. CLASSIFICATIONS OF COST

The ultimate objective of engineering aplication is the satisfaction of human wants. But human wants are not satisfied without cost. Alternative engineering proposals will differ in regard to the costs they involve relative to the objective of want satisfaction. The engineering proposal resulting in least cost will be considered best if its end result is identical to that of competing proposals.

A number of cost classifications have come into use to serve as a basis for economic analysis. As concepts, these classifications are useful in calling to mind the source and effect of costs that will have a bearing on the end result of a proposal. This section will define and discuss these cost classifications.

First Cost. By definition, *first cost* is considered to involve the cost of getting an activity started. The chief advantage in recognizing this classification is that it calls attention to a group of costs associated with the initiation of a new activity that might not otherwise be given proper consideration. Ordinarily, this classification is limited to those costs which occur only once for any given activity.

There are degrees of action in a great many fields below which the effect of the action is insignificant. For example, a small movement of a gear in a gear train will merely take up the backlash without moving an adjacent gear. As an analogy, it must be recognized that many activities that otherwise may be profitable cannot be undertaken for the reason that their associated first cost represents a level of input above that which can be met. Many engineering proposals that are otherwise sound are not initiated because the first cost involved is beyond the reach of the controlling organization. Other engineering proposals meet with failure after initiation because it was found that the threshold input required was not successfully met due to financial limitations.

As an example, consider a proposed mining operation involving the extraction of an estimated 120,000 tons of ore from a mountain. Engineering Proposal A involves the construction of a tunnel beginning at an existing railroad spur and terminating within the mountain at the ore deposit. The ore will be removed by a gravity conveyor within the tunnel. The first cost of the tunnel and installed conveyor is estimated to be $138,000. The conveyor will cost $0.10 per ton to operate and will have a salvage value of $18,000.

Engineering Proposal B involves the removal of a quantity of overburden so the ore deposit will be exposed for loading into trucks. The first cost of removing the overburden and improving an existing road is estimated to be $22,000. Trucks with drivers will be leased to haul the ore to the spur at a cost of $1.30 per ton of ore removed. All other costs for the two proposals are equal. Neglecting interest, the cost difference is calculated as follows:

$$[\$138{,}000 - \$18{,}000 + \$0.10(120{,}000)]$$
$$- [\$22{,}000 + \$1.30(120{,}000)] = \$46{,}000.$$

Suppose that further analysis indicates that the mining operation will result in sufficient income to yield a profit regardless of the proposal chosen, but that financial limitations will not allow the successful completion of Proposal A because of its high first cost. In such a case, the mining operation might be initiated by accepting Proposal B even though it will result in a cost of $46,000 more than A. Although A appears to be best, it would result in failure because of the inability to meet its first cost.

Fixed and Variable Cost. *Fixed cost* is ordinarily defined as that group of costs involved in a going activity whose total will remain relatively constant throughout the range of operational activity. The concept of fixed cost has a wide application. For example, certain losses in the operation of an engine are in some measure independent of its output of power. Among its fixed costs, in terms of energy for a given speed and load, are those for the power to drive the fan, the valve mechanism, and the oil and fuel pumps. Almost any task involves preparation independent of its extent. Thus, to paint a small area may require as much effort for the cleaning of a brush as to paint a large area. Similarly, manufacturing involves fixed costs that are independent of the volume of output.

Variable cost is ordinarily defined as that group of costs which vary in some relationship to the level of operational activity. For example, the consumption of fuel by an engine may be expected to be proportional to its output of power and the amount of paint used may be expected to be proportional to the area painted. In manufacturing, the amount of material needed per unit of product may be expected to remain constant and, therefore, the material cost will vary directly with the number of units produced. In general, all costs such as direct labor, direct material, direct power, and the like, which can readily be allocated to each unit of product, are considered to constitute the variable costs and the balance of the costs of the enterprise are regarded as fixed costs.

Fixed costs arise from making preparation for the future. A machine is purchased now in order that labor costs may be reduced in the future. Materials which may never be needed are purchased in large quantities and stored at much expense and with some risk in order that idleness of production facilities and men may be avoided. Research is carried on with no immediate benefit in view in the hope that it will pay in the long run. The investments that give rise to fixed cost are made in the present in the hope that they will be recovered with a profit as a result of reductions in variable costs or of increases in income.

Fixed costs are made up of such cost items as depreciation, maintenance, taxes, insurance, lease rentals, and interest on invested capital, sales programs, certain administrative expenses, and research. It will be observed that these arise from the decisions of the past and in general are not subject to rapid change. Volume of operational activity, on the other hand, may fluctuate widely and rapidly. As a result, fixed costs per unit may easily get out of hand. It is probable that this is the cause of more unsuccessful activity than any other, for few have the foresight or luck to make commitments in the present which will fit requirements of the future even reasonably well. Since fixed costs cannot be changed readily, consideration must be focused upon maintaining a satisfactory volume and character of activity.

In a practical situation, fixed costs are only relatively fixed and their total may be expected to increase somewhat with increased activity. The increase will probably not follow a smooth curve but will vary in accordance with the characteristics of the enterprise.

Consider a plant of several units that has been shut down or is operating at zero volume. No heat, light, janitor, and many other services will have been required. Many of these services must be reinstated if the plant is to operate at all, and if reinstated only on a minimum basis it is probable that these services will be adequate for quite a range of activity. Further increases in activity will require expenditures for other services that cannot be provided to just the extent needed. Thus, what are termed "fixed costs" in business may be expected to increase in some stepped pattern with an increase in activity.

Variable expense may also be expected to increase in a stepped pattern. To increase production beyond a certain extent another machine may be added. Even though its full capacity may not be utilized, it may be necessary to employ a full crew of men to operate it. Also, an increase in productivity may be expected to result in the use of materials in greater quantities, and thus in their purchase at a lower cost per unit due to quantity discounts and volume handling.

The practices followed in designating fixed and variable costs are usually at variance with the strict interpretation of these terms. Analyses in which they are a factor must recognize this fact or the results may be grossly misleading.

Incremental and Marginal Cost. The terms incremental cost and marginal cost refer to essentially the same concept. The word increment means increase and an increment cost means an increase in cost. Usually, reference is made to an increase of cost in relation to some other factor, thus resulting in such expressions as increment cost per ton, increment cost per gallon, or increment cost per unit of production. The term marginal cost refers specifically to an increment of output whose cost is barely covered by the return derived from it.

Figure 2.1 illustrates the nature of fixed and variable cost as a function of output in units. The incremental cost of producing 10 units between outputs of 60 and 70 units per year is illustrated to be $8. Thus, the average incremental cost of these 10 units may be computed as Δ cost/Δ output = $8/10 = $0.80 per unit.

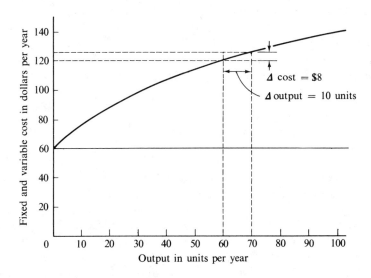

FIGURE 2.1. FIXED, VARIABLE, AND INCREMENTAL COST.

Increment costs are difficult to determine. In the example above, a curve was given which enabled the increment cost to be precisely determined as $8 for 10 units. As a practical matter in actual situations it is ordinarily difficult to determine increment cost. There is no general approach to the problem, but each case must be analyzed on the basis of the facts that apply to it at the time and the future period involved. Increment costs can be overestimated or underestimated, and either error may be costly. The overestimation of increment costs may obscure a profit possibility; on the other hand, if they are underestimated, an activity may be undertaken which will result in a loss. Thus, accurate information is necessary if sound decisions are to be made.

Sunk Cost. When making engineering economic analyses the objective of the decision maker is to choose that course of action which is expected to result in the most favorable future benefits. Because it is only the future consequences of investment alternatives that can be affected by current decisions, an important principle in economic studies is to disregard cost incurred in the past. A past cost or *sunk cost* is one that can not be altered by future action and is therefore irrelevant in engineering economy studies.

Although the principle that sunk costs should be ignored seems reasonable it is quite difficult for many people to apply. For example, suppose you had purchased 1,000 shares of stock at a price of $50 per share two years ago and now it is worth $15 per share. In all probability there are other stocks available that have a better future than the stock you presently possess. Many people react to this type of situation by holding on to their present stock until they can recover their losses. By not selling your stock you do not have to openly admit your losses and therby admit your failure in judgment. However, it should be clear that because of better opportunities elsewhere it is certainly sound decision making to acknowledge your losses by selling the present stock and to use what money remains more productively from now into the future. It is the emotional involvement with past or sunk costs that makes it difficult for such costs to be ignored in practice.

2.6. PRICE IS DETERMINED BY SUPPLY AND DEMAND

In a free enterprise system, the price of goods and services is ultimately determined by supply and demand. Typical demand and supply curves are illustrated in Figure 2.2. The demand curve shows the relationship between the quantity of a product that people are willing to buy and the price of the product. The supply curve shows the relationship between the quantity of a product that vendors will offer for sale and the price of the product. Therefore, the intersection of these curves determines the price at which exchange

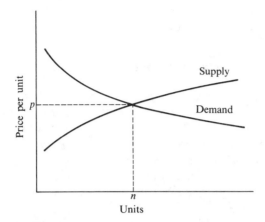

FIGURE 2.2. TYPICAL SUPPLY AND DEMAND CURVES.

will take place. The quantity exchanged is equal to both quantity of supply and quantity of demand. In the illustration, the number of units that will be exchanged is n and the price at which exchange will take place is p.

As a concept, the law of supply and demand is important in engineering economy studies since proposed ventures frequently involve action that will increase the supply of a product or influence its demand. The effect of such action upon the price at which the product can be sold is an important factor to be considered in evaluating the desirability of the venture.

The Elasticity of Demand. Consumer goods and services may be classified as being either necessities or luxuries. The classification is relative, for that which is considered to be a necessity by one person may be considered a luxury by others. The classification depends upon the individual's economic and social position.

People will decrease their consumption of luxury goods at a faster rate than their consumption of necessities for a given increase in price. The extent to which price changes affect demand is measured by a concept called the elasticity of demand. If the product of volume and price is constant, the elasticity of demand is unitary. Under this condition, the total amount which will be spent for the product will be constant regardless of the selling price. A drop in price would cause an increase in demand. Thus, the demand for a product is *elastic* when a decrease in price results in a greater than proportionate increase in sales. On the other hand, if a decrease in selling price produces a less than proportionate increase in sales, the demand is said to be *inelastic*.

Obviously, luxuries have a much greater elasticity of demand than do necessities. A change in demand will have a much greater effect on the price of necessities than on the price of luxuries. This relationship is important in evaluating the probable effect of alternate engineering proposals.

2.7. THE LAW OF DIMINISHING RETURNS

The term *law of diminishing returns* was originally used to designate the relation of input of fertilizers to land and the output of crops. It is a special application of the law of diminishing productivity, which may be stated as follows: *The amount of product obtained in a productive process varies with the way the agents of production are combined. If only one agent is varied, the product per unit of this agent may increase to a maximum amount, after which the product per unit may be expected to diminish but not necessarily proportionately.*

This concept may be illustrated by an example. A certain production line where an assembly operation is performed may be manned by a varying number of men. As the number of men on the line is increased from 15 to 19, production increases from 40.6 units per hour to 53.7 units per hour. The men are paid at the rate of $3.00 per hour. The output relationships are given in Table 2.1.

In Table 2.1 the output is given in terms of units of product and the agent of production is given in terms of man-hour input and dollar input per unit of product. It will be noticed that output per unit of input increases, reaches a maximum, and then diminishes.

The law of diminishing returns is directly applicable in a great number of engineering situations where it manifests itself in terms of providing a service at least cost. In a given situation one element may be increased and for a time costs will be decreased to a minimum point only to rise again if the element is further increased. This may be illustrated by increasing the amount of fuel

Table 2.1. AN ILLUSTRATION OF THE LAW OF DIMINISHING RETURNS

Number of men employed on line	Labor input in man hours per hour	Labor input in dollars per hour	Output in number of assemblies per hour	Units of output per man-hour of input	Units of output per dollar of labor input
15	15	45.00	40.6	2.71	0.90
16	16	48.00	46.7	2.92	0.97
17	17	51.00	52.5	3.09	1.03
18	18	54.00	53.3	2.96	0.99
19	19	57.00	53.7	2.83	0.94

fed into an engine. Below a certain fuel-air ratio the engine will not start. When fed a little more fuel, it may produce just enough power to run itself. As the fuel fed into the engine is increased, the cost of power delivered will reach a minimum value per unit of power delivered. As the fuel intake is increased, cost of power per unit will increase. The engine will sputter and stop, with the result that no power is delivered as at first.

Most situations have many elements, and where the optimum whole is sought, the optimum of elements must often be compromised. Long bridge spans reduce the cost of piers but will increase the cost of superstructure. Prices can be increased but sales may fall so that profit is less. Managements are continually confronted with finding the balance between supervisory and nonsupervisory workers.

The chief value of the concept of the law of diminishing returns in engineering economy situations is that it produces an awareness that output does not necessarily increase in a straight-line relationship with an increase in input of an agent of production. The solution of many problems in economy centers around adjusting the amounts of agents of productivity to produce maximum output per unit of input. This end is frequently expressed as an effort to find the input ratio that will result in least cost per unit of output. Such expressions as horsepower-hour output per pound of fuel, miles per gallon of fuel, sales per dollar of advertising, units of output per fatal accident, and defects per 100 units of product are associated with analyses in search of a maximum output per unit of input.

2.8. QUALITATIVE AND QUANTITATIVE KNOWLEDGE

Qualitative knowledge is knowledge of the attributes of the thing under consideration. It is descriptive and tells how a thing is constituted by naming its distinctive characteristics. In the expression of qualitative knowledge such statements are used as: Repair cost of the truck will be low; many units of this item will be sold this year; this method was less expensive than that method; the turbine had much power and used little fuel. Such expressions cannot be precisely evaluated and have comparatively little value in economy studies.

Quantitative knowledge is knowledge of the amount or extent of the attributes of a thing in terms that are capable of being counted. It is information that is capable of being expressed in numbers. Representative of quantitative knowledge are such statements as: The repair cost of the truck amounted to $358 per year; this year 6,200,000 units of this item were sold; this method costs $0.07 per piece but that method costs $0.09 per piece; the turbine developed 286 horsepower and used 0.48 pound of fuel per horsepower-hour.

It appears that qualitative knowledge is primarily evaluated by one's feeling for it and one's judgment. Different people are likely to vary widely in their interpretation of the significance of qualitative ideas. Quantitative knowledge, on the other hand, is quite precise and may be evaluated in large measure by the processes of reasoning. Also, most people may be expected to attach about the same significance to quantitative statements pertaining to fields with which they are familiar. This is an important characteristic because it facilitates communication of ideas.

A comparison of the effectiveness of the two kinds of knowledge may be gained from a consideration of two statements: (1) The turbine had much power and used little fuel; and (2) The turbine developed 286 horsepower and used 0.48 pounds of fuel per horsepower-hour. Let each statement be considered in conjunction with the knowledge that a machine for which a prime mover is sought will require a torque of 420 foot-pounds at 3,200 revolutions per minute (r. p. m.), that the cost of fuel is $0.02 per pound, and that

$$\text{h.p.} = \frac{\text{torque} \times 2\pi \times \text{r.p.m.}}{33,000}.$$

If the economy of using the turbine as the prime mover for the machine were being considered, two questions would be pertinent. These are: (1) Is the turbine powerful enough? and (2) What will be the fuel cost per hour? It will be noted that qualitative statements cannot be used to arrive at the desired answers. The quantitative information can be used as follows:

$$\text{required h.p.} = \frac{420 \times 2\pi \times 3,200}{33,000} = 256.$$

This answers the first question. The answer to the second question may be found as follows:

$$\text{fuel cost per hour} = 256 \times 0.48 \times \frac{256}{286} \times \$0.02 = \$2.20.$$

One important aspect of quantitative data is that it can often be combined with other quantitative data by logical processes to produce new quantitative knowledge as was illustrated above. This important characteristic suggests that quantitative knowledge should be exhausted in economy studies before considering qualitative knowledge.

QUESTIONS

1. Why should engineers be concerned with and seek to understand the behavior of people?
2. Define value; utility.
3. Explain how utilities are created.
4. Describe the two classes of goods recognized by economists.
5. Contrast the utility of consumer goods with the utility of producer goods.
6. Why is it that the utility of consumer goods is determined subjectively whereas the utility of producer goods is usually determined objectively?

7. Explain the economy of exchange.
8. Why is it possible for both parties to profit by an exchange?
9. How may persuasion increase the utility of an item?
10. Define first cost and explain how it may be a limiting factor in successful engineering activity.
11. Discuss the difference between fixed cost and variable cost.
12. List some difficulties associated with classifying a cost as either fixed or variable.
13. Discuss the difference between incremental and marginal cost.
14. Define sunk cost and explain why it should not be considered in engineering economic analysis.
15. Explain the relationship between price and supply; price and demand.
16. Why is the law of diminishing returns a useful concept in economy studies?
17. Contrast the attributes of qualitative and quantitative knowledge.

3

Elementary Selections in
Economic Analysis

There are many situations in which alternate designs, materials, or methods will provide identical results and will involve expenditures occurring in substantially the same short time period considered to be the present. Since the results of the alternatives will be identical, the immediate cost of each is a measure of its comparative economy. Economic evaluations that fall into this category do not require consideration of the time value of money. In this chapter we examine some simple examples of engineering economic analysis in which the time value of money is not an important factor thus making the present difference the basis for decision. Additional concepts useful in engineering economy studies are also presented.

3.1. DESIGN AND ECONOMY

The results of design effort, as manifested in plans and specifications, crystallize the final form of the product to be produced or the structure to be constructed as to its physical form, material, and production or construction requirements. For this reason design affords many opportunities for economy. Design, no matter how poorly done, is predicated on the thought that the effort devoted to it will be outweighed by the results.

To design is to project and evaluate ideas for attaining an objective. Suppose that a person employed in a machine design department has been assigned the task of designing a machine for a special purpose. His first step will be to project, literally to invent, new combinations of materials, machine elements, forces, and kinematic motions that it is believed may meet the purposes of the machine. His next step will be to evaluate these combinations.

There may be many bases of evaluation: certainty of operation, consequences of failure, operator attention required, safety of operator, rate of operation, power required, maintenance required, floor space required, and so forth. On the whole, these bases of evaluation all relate to economy in one way or another. In fact, economy is inherent in design. Consider a roller bearing in conjunction with a plain brass bearing. The friction induced by a turning load on a roller bearing is less than that of a plain bearing, by virtue of the fact that the rolling friction inherent in the former is less than the sliding friction of the latter. Where friction is economically undesirable, the roller bearing is inherently more economical than the plain bearing.

Planning is essentially a form of design and may be defined as being the projection and evaluation of ideas. In planning and designing, ways for accomplishing a contemplated end are literally invented and then evaluated on the basis of strength, fuel consumption, weight, economy, and so forth. The conception of different ways to accomplish a desired end does not appear to be a logical process, but their evaluation is largely a matter of reasoning. One important aspect of planning is that it makes it possible to try out a great many ideas on paper more or less conclusively. The cost of such trials may be negligible in comparison with actual accomplishment.

A decision to accept a design is a decision to assume the advantages and disadvantages associated with it, and to discard other designs that may have been considered. The following examples will illustrate several aspects of economy as it relates to design.

Elimination of Overdesign. In the design of a certain building it was found that 180 2-inch by 8-inch by 16-foot rafters, 16 inches center-to-center, are just sufficient to meet the contemplated load. It was also found that the sheathing to span rafters placed 24 inches center-to-center was just adequate for strength and stiffness. Consequently, the sheathing will be stronger and stiffer than necessary for the 16-inch spacing. The excess is called *overdesign* since it has no functional value and may be disadvantageous because of its extra weight.

A new design is contemplated using 2-inch by 10-inch cross-section rafters. On the principle that the bending moment resisted by a rafter is in accordance with $bd^2/6$, the ratio of the load-carrying capacity of the 2-inch by 10-inch rafter to the 2-inch by 8-inch rafter is 1.56 to 1.00.

As far as load-carrying capacity is concerned, the 2-inch by 10-inch rafters may be spaced 16 inches $\times$ 1.56, or 25 inches. In accordance with practical standards, the rafters are spaced 24 inches center-to-center. The analysis required for 2-inch by 8-inch by 16-foot rafters, 16 inches center-to-center, is as follows:

Number of rafters required	180
Overdesign of rafters..	0%
Overdesign of sheathing, $\dfrac{24 \text{ in.} - 16 \text{ in.}}{24 \text{ in.}}$	33%
Cost of rafters @ \$0.14 per board foot, $\dfrac{180 \text{ rafters} \times 2 \text{ in.} \times 8 \text{ in.} \times 16 \text{ ft.}}{12 \text{ in.}} \times \$0.14$.........................	\$538

For 2-inch by 10-inch by 16-foot rafters, 24 inches center-to-center, the required calculations are:

Number of rafters required, $\dfrac{180 \times 16 \text{ in.}}{24 \text{ in.}}$ 120

Overdesign of rafters, $\dfrac{25 \text{ in.} - 24 \text{ in.}}{24 \text{ in.}}$ 4%

Overdesign of sheathing 0%

Cost of rafters @ $0.14 per board foot,
$\dfrac{120 \text{ rafters} \times 2 \text{ in.} \times 10 \text{ in.} \times 16 \text{ ft.}}{12 \text{ in.}} \times \$0.14$ $448

In this example, the desig‑ with 2-inch by 10-inch rafters is not only adequate, but it is less expensive by $90. In addition, savings will result from having to handle less pieces.

Designing for Economy of Production. A product may exhibit excellent functional design, but may be poorly designed from the standpoint of producibility. The drawings and specifications establishing the design also establish a minimum manufacturing cost regardless of efforts to reduce labor, material, and overhead costs.

Figure 3.1 illustrates the effect of design on production cost. The difference in production cost for the most expensive and the least expensive design is $0.84. In addition, choice of the least expensive design will result in a weight saving of 3.8 pounds. Under the assumption that each foot pedal will perform equally well, and that each will contribute equally to the salability of the final product, the designer would choose the least expensive design.

Economy in the Design of Producer Goods. In the design of equipment for carrying on manufacturing, the designer may become too engrossed in the mechanical features and give insufficient attention to overall economy. As an example, assume that a jig is to be designed for adjusting 100,000 assemblies for a special production order. Two designs are presented and are evaluated as follows:

	Jig A	*Jig B*
Estimated life	100,000 pieces	100,000 pieces
Cost	$600	$300
Hourly rate of class of labor required	$2.60	$2.80
Operator hourly output rate	62	54

The cost of adjusting the 100,000 assemblies with Jig A is:

Labor cost, $\dfrac{100,000 \text{ pc.} \times \$2.60 \text{ per hr.}}{62 \text{ pc. per hr.}}$ $4,194

Cost of Jig A $600
_____
$4,794

(a) Traditional construction. Machine foot lever, 10 inches long, weighs 6 pounds. Cost with broached keyway is $1.15.

(b) Simple steel design costs 41 % less. Can be built by the shop with only saw and shears. Weighs 2.7 pounds. Cost 68 ¢ with keyway.

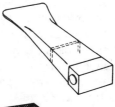

(c) Saves 53%, cast by forming lever arm and pad as integral piece from 10 gauge metal. Weighs 2.5 pounds. Cost 54 ¢.

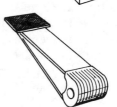

(d) Saves 73%, eliminates broaching. Hub with integral key is produced by stacking stampings in assembly. Arm is 10 gauge, brake-formed and welded to hub. Weighs 2.2 pounds. Cost 31 ¢.

FIGURE 3.1. THE EFFECT OF DESIGN ON THE ECONOMY OF PRODUCTION IS SHOWN BY THESE FOUR DESIGNS FOR A FOOT PEDAL. (*Courtesy Lincoln Electric Co.*)

The cost of adjusting the 100,000 assemblies with Jig B is:

Labor cost, $\dfrac{100{,}000 \text{ pc.} \times \$1.80 \text{ per hr.}}{54 \text{ pc. per hr.}}$ $5,185

Cost of Jig B 300

$5,485

The net advantage of selecting Jig A over Jig B is $5,485 less $4,794, or $691. This economic advantage is inherent in the design of Jig A. Its characteristics were such that it could be operated more rapidly with less operator skill.

In industrial literature, many examples may be found in which economy of production was effected through design of equipment, equipment layout, and plant. In such cases, it is apparent that the economic advantage resulted from wise choice among alternative designs.

Design for Economy of Maintenance. Drawings and specifications establishing the design of a product, structure, or complex system should exhibit consideration for maintainability. A commercial product or a weapons system is not available to serve its intended purpose if it is down for maintenance. Because of the high cost of maintenance manpower and the associated costs of test equipment, spare parts, transportation, and other logistic elements engineers must consider maintainability during design to a greater degree than at present.

The challenge is quite simple. Designers have the opportunity to reduce the need for maintenance by striving for a product or system which will not fail in use. By so doing the needed item will be available when needed and, in addition, burdensome costs associated with and resulting from maintenance will be avoided.

The ultimate goal of maintainability design is the economic reduction of the support requirements for a system after it goes into operation and the facilitation of whatever remaining maintenance will be required. This goal is logically the concern of the design engineer, for it is not likely that the disruption of service and the burden of maintenance will be reduced unless engineers assume the responsibility for this neglected area. The need is critical, for it appears that both commercial and the military products, structures, and systems may soon consume more economic resources for maintenance and related services than their first cost.

Design for Economy of Shipping. The shipping costs of products can often be reduced by proper design. Some products are designed so they can be easily assembled upon receipt by the customer. Others are designed so they may be easily packaged for shipment.

Often it may be advantageous to design so that the product can be nested to save space in shipment. In the shipment of pipe to Saudi Arabia this principle was used. The design called for pipe approximately 30 inches in diameter. To facilitate shipment by water, where bulk freight rates were in effect, half of the pipe was made 30 inches in diameter and the other half was made 31 inches in diameter to make nesting possible. As a result, shipping charges were reduced by several million dollars.

Economy of Interchangeable Design. Interchangeability in design is an extension of the mathematical axiom that things equal to the same thing are equal to each other. Components of an assembly that are to be interchangeable with components of other assemblies must possess specific tolerance limits.

As an example of the economy inherent in interchangeability, consider an engine re-manufacturing process. If the parts were not originally designed to be interchangeable with like parts of other engines of the same model, it would be necessary to keep all parts of each engine together during re-manufacturing. The needed coordination, tagging, and cost of locating stray parts is eliminated through initial design that allows interchangeability. In this example interchangeability contributes to maintainability.

3.2. ECONOMY OF MATERIAL SELECTION

A designer has a choice of specifying either aluminum or grey iron castings for an intricate housing for an instrument to be permanently mounted in a power plant. He has ascertained that either metal will serve equally well. The aluminum casting will weigh 0.8 pound and the grey iron casting 2.2 pounds. His analysis of the cost of providing each type of casting follows:

	Grey Iron	Aluminum
Cost of casting delivered to factory	$0.76	$0.92
Cost of machining	0.63	0.52
	$1.39	$1.44

On the basis that either material will serve equally well and provide an identical service, the grey iron casting will be selected because its immediate or present cost is the lower of the two.

In the above example, in which it is specified that the instrument is to be mounted on a power plant wall, differences in weight were not considered to be a factor. However, for an instrument for airplane service the difference in weight might be the deciding economic factor. In airplane service the lighter instrument casting would have an economic value in lessened fuel consumption or greater pay load and would thus have greater utility.

In determinations of economy, care must always be exercised to see that alternatives provide identical services. "Sales appeal" of one type of casting over the other may easily render their worths nonidentical. Even for power plant use it might have been recognized that the lighter metal had "sales appeal" over the heavier metal.

If concrete value can be given to a quality, this difference in service can be taken into account to render the services of the two alternatives identical. For instance, the quality of lightness might have service value to the manufacturer in that it would reduce the delivery cost of his product to the consumer. If this had applied in the example of the instrument casting, the two types of castings could have been placed on an identical service basis for comparison as follows:

	Grey Iron	*Aluminum*
Cost of casting delivered to factory	$0.76	$0.92
Cost of machining	0.63	0.52
Average additional cost to deliver grey-iron-casting-equipped instruments to customer	0.26	0.00
	$1.65	$1.44

In cases where two or more materials may serve a purpose equally well from a functional standpoint, the relationship of their cost and the cost of processing them may be the factor that determines which is chosen. Brass, for example, is often found to be less costly for parts than cold rolled steel because it can be machined at a higher rate, in spite of its greater weight per unit volume and greater cost per pound. Aluminum, which is easily machinable and in addition has a low specific weight, is being used in increasing amounts as a replacement for steel, cast iron, and other metals whose cost per pound is considerably less. Because of the ease with which they can be processed, plastics have proved to be an economy in many applications as a replacement for materials of less cost per pound. The same can be said of die castings.

In some cases the decision to substitute one material for another will result in an entirely different sequence of processing. For instance, a change from grey iron to zinc alloy castings will require marked change in equipment. To determine the comparative economic desirability of two materials, it is necessary to make a detailed study of the costs that arise when each is used. In some cases this may involve the economy of disposing of present equipment and acquiring new equipment.

3.3. PERFECTION AND ECONOMY

Perfection is often a false ideal unless perfection is taken to mean that which is most appropriate. To secure a desirable result even in the physical world, compromises from perfection must often be made. For example, increasing the compression ratio of an internal combustion engine, ideally, will increase its efficiency but at the same time necessitate use of fuel of higher anti-knock characteristics. In order to produce a smooth-running engine, a compromise between an ideal compression ratio and characteristics of available fuel will have to be made. If economy of operation is now a factor to be considered, both the compression ratio and the anti-knock characteristics of the fuel may be compromised further on the basis of the cost of fuel of different anti-knock characteristics.

But it usually is not enough that a machine shall function properly as a physical unit; it must also function properly as an economic unit. Thus, compromises of physical perfection may be suggested by cost of the materials and processes which will be needed to produce it. And still further compromises of physical perfection may be dictated by the economic climate in which the engine will be marketed.

Figure 3.2 shows an approximate general relationship between tolerance

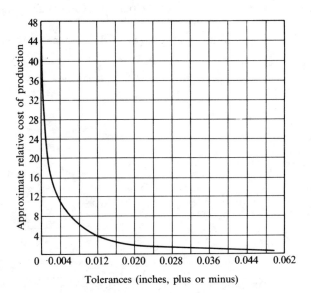

Tolerances (inches, plus or minus)

FIGURE 3.2. GENERAL COST RELATIONSHIP OF VARIOUS DEGREES OF ACCURACY.

in inches and the cost of production. According to this figure, the specification of a tolerance of 0.001 inch, when 0.003 inch is adequate, might increase the cost by

$$\frac{32 - 12}{12} = 1.67 \text{ or, } 167\%.$$

The concept that this figure exhibits has many applications. Whenever a product is to be designed or an activity planned, consideration should be given to the determination of the degree of perfection which will lead to the most desirable result in economic terms. In addition, it must be realized that there is a degree of perfection that, if specified, will be beyond the capability of the process to which it applies.

3.4. SIZE AND ECONOMY

The cost of many items varies on the basis of size. Extremely large units and extremely small units may cost relatively more than sizes that come in between. One reason for this variation in unit cost is that the number produced is often greater in the sizes that lie between the extremes. Some products, such as valves, electric motors, lathes, and engines, are essentially identical except for size. Thus the same number of parts will have to be processed. Though larger parts may be expected to require longer time for processing than smaller parts, it is unlikely that a part twice the size of another will require twice the time to process.

Figure 3.3 shows the relationship between the cost per unit and the size

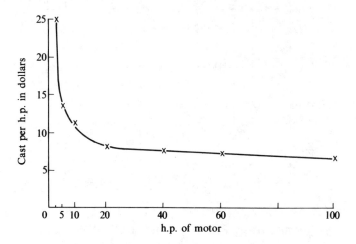

FIGURE 3.3. COST OF ELECTRIC MOTORS PER HORSEPOWER, BASED ON A MANUFACTURER'S LIST PRICE FOR SQUIRREL-CAGE INDUCTION, 3-PHASE, 60-CYCLE, 220-VOLT, 1,800-R.P.M. MOTORS.

of a certain make of electric motor. From this figure, it may be seen that a motor capacity of 20 horsepower, which could be provided for approximately $160 with a 20-horsepower motor, would cost approximately $260 with four 5-horsepower units. It is clear that the latter installation could only be justified where it would permit offsetting advantages.

Similarly, there are many activities that cannot be carried on economically above or below a certain scale. Products that require a large variety of heavy equipment for their manufacture such as locomotives, steam turbines, automobiles and steamships cannot be made economically in small scale operations. On the other hand, certain hazardous processes in the manufacture of explosives are carried on in small isolated units as a matter of economy. There are many activities that are best carried on a small scale. This fact is obscured in many cases because of an extreme regard for size by many people.

3.5. ECONOMY AND LOCATION

There are many situations in which geographical location is a factor in economy. Situations in which location is important over a short period of time, say a year or less, come properly under the heading of present economy. Small differences of location, as for example the location of part bins and tools in a work place, may have considerable effect upon economy and are often given very detailed consideration.

The selection of a location for a plant is a long-time commitment. A new enterprise may be hampered throughout its life by an unfavorable location. Once a plant has been built, the expense and disruption of activities necessary to move it to a more favorable location is generally so great as to be impractical, even though failure may result from the unfavorable characteristics of the original location. Therefore, the search for and evaluation of plant sites justify very careful consideration.

When profit is the measure of success, the best location is the one in which production and marketing effort will result in the greatest profit. Location may affect the cost at which raw materials are gathered, the cost of production, the cost of marketing, and the volume of product that can be sold.

Evaluation of a location begins with research to determine the volume of sales and income promised by the given location. Then research is directed along the lines suggested by the factors listed above to gain data with which to calculate the cost of gathering raw material, processing it, and delivering finished products to the consumer in the volume that can be sold. Evaluation of plant locations consists essentially of operating the enterprise under consideration "on paper" at each location studied. The result of this ap-

Table 3.1. COMPARATIVE EVALUATION OF THREE PLANT LOCATIONS

	Location A	Location B	Location C
Market:			
Annual sales ....(A)	$260,000	$260,000	$260,000
Selling expense ....(B)	44,000	43,000	46,000
Net income from sales	$216,000	$217,000	$214,000
Production costs:			
Supplies and raw materials	$ 69,000	$ 71,000	$ 62,000
Transportation—in and out	27,000	26,000	23,000
Fuel, power, and water	13,000	17,000	18,000
Wages and salaries	64,000	63,000	61,000
Miscellaneous items	8,000	8,000	10,000
Fixed costs other than interest	11,000	11,000	11,000
Net production costs ....(C)	$192,000	$196,000	$185,000
Selling plus production costs,			
$B + C$ ....(D)	236,000	239,000	231,000
Annual profit, $A - D$ ....(E)	24,000	21,000	29,000
Rate of return, $E \div \$180,000$	13.3%	11.7%	16.1%

proach in evaluating three locations for a small glassware plant requiring an investment of approximately $180,000 are given in Table 3.1.

Consider the competitive advantage of a plant at Location C over one at Location B. A plant at Location C has an advantage in net income of over 4% by virtue of its location to offset price reductions and operating and selling efficiency of a competing plant at Location B.

It is not uncommon for concerns to set up temporary service facilities in an area in which they have unusual temporary activities. They are then confronted with determining the economy of setting up such facilities and the economy of various locations within the area.

A firm having a contract to build a dam requiring 300,000 cubic yards of gravel found two feasible sources whose characteristics are summarized as follows:

	Source A	Source B
Distance, pit to dam site	1.8 miles	0.6 mile
Cost of gravel per cubic yard at pit	....	$0.08
Purchase price of pit	$7,200	....
Road construction necessary	$2,400	None
Overburden to be removed at $0.24 per cu. yd	....	60,000
Hauling cost per cubic yd. per mile as bid by hauling contractor	$0.072	$0.084

In order to make a selection on the basis of economy, the cost of securing the required gravel from either source should be determined. The cost of 300,000 cubic yards of gravel from Source A is calculated as follows:

Purchase price of pit	$ 7,200
Road construction	2,400
Hauling cost, 300,000 cu. yd. × 1.8 miles @ $0.072 per cu. yd. per mile ..	38,900
	$48,500

The cost of 300,000 cubic yards of gravel from Source B is:

Cost of gravel at pit, 300,000 yd. @ $0.08	$24,000
Removal of overburden, 60,000 yd. @ $0.24	14,400
Hauling cost, 300,000 cu. yd. × 0.6 mile @ $0.084 per cu. yd. per mile ..	15,100
	$53,500

The cost of making analyses similar to the one covering the situation above is often a small percentage of the saving that may result from a current decision. A cursory analysis or "hunch" cannot often be defended for decision making on the basis of its lesser cost.

3.6. ECONOMY OF STANDARDIZATION
AND SIMPLIFICATION

A standard is a specification. Products designated as *standardized* conform to a previously accepted specification. Standardization, the conduct of activities in accordance with previously determined specifications, appears to be a modern development, particularly as it relates to the production of goods. Both products and the procedures by which they are made are subject to standardization.

Simplification is a name for the practice of examining a line of products for the purpose of eliminating useless variety in style, color, dimensions, size range, and the like. Its practice results in setting up what might be termed *most useful specifications* for a line of products. Thus simplification is closely related to standardization. Simplification may be thought of as being the practice of eliminating undesirable standards.

There are a number of reasons, related to economy, why standardization has become a widespread practice. One factor is undoubtedly that human actions are characterized by variation and individualism whereas machine action is characterized by repetition of identical patterns. Thus it is relatively easy to build machines that operate and produce goods in accordance with predetermined specifications. As the burden of production has been shifted from human effort and skills to machines, it has been necessary, at least to the extent of machine processes, to standardize.

The chief drive to standardization is probably specialization. It is recognized as characteristic that the skill and speed of performance is increased and the effort of performance is decreased as a person repetitiously performs a specified task. It is also known that persons have greater aptitude for doing some things than others; more people may be found with one skill than several. It is rarity that places triple-threat men in demand in football.

Specialization is not economically practical unless it can be engaged in for some time. It is rarely profitable to build a machine for a total lifetime production of a single part. Nor would it be profitable to set up an assembly line of a number of specialized people to assemble a single unit of product.

Standardization is a method of increasing the number of units of one kind for a given total production. Suppose that a pottery has had an annual output of 1,000 nonstandardized vases made in accordance with 40 different patterns, each in 5 different sizes. Thus the output involves 200 different specifications and an average annual output of 5 units per specification. If it were decided to standardize on one vase of one size, the annual output per specification would be increased from 5 to 1,000 units. Thus it appears that the economy of standardization lies, in large measure, in the increased opportunity for specialization afforded by increased volume of one kind of product or activity.

Since standards ordinarily have extended usefulness in terms of time and volume, it is often economical to take great pains to perfect them. A standard is in reality a predetermined decision to act in accordance with a certain specification in certain situations. Even though a standard may be developed only after a considerable expenditure, it will probably result in a lower cost per decision than if each specification problem were decided as it arose without benefit of a standard. Suppose that a certain situation necessitating decision arises in a concern 1,000 times per year and that the cost of making each decision individually, owing to confusion, error, and loss of time, amounts to $1. A standard procedure that could reduce this cost to $0.10 per decision, for instance, would justify an expenditure of $900 (or more, if applicable for more than one year).

One valuable feature of standards is that they greatly facilitate communication. Involved directions, for instance, may be given simply by pointing out the standards that should be applied. This is of tremendous value in coordinating the activities of large groups of people as in modern industry.

3.7. ECONOMY IN THE APPROPRIATE COMBINATION OF ELEMENTS

There are many instances in which economy rests on the appropriateness of a combination of several elements. Consider the machining of a metal shaft on a lathe. The job will be accomplished economically if the feed and speed of the tool is appropriate in the light of the metal being machined, the character of the tool, the rigidity of the tool holder, the power available, and the strength of the machine components. Each of these elements must be appropriate for the total purpose if the total purpose is to be pursued economically.

A great many machines can be built from such common machine elements as levers, cams, columns, beams, ratchets, gears, and pulleys. However, the economic worth of the resulting machines depends largely upon the appropriateness of the combination of the several machine elements, rather than upon the excellence of the individual elements. In a similar manner, the success of any economic venture results from the effectiveness of the combination of elements employed rather than upon the success of the separate elements.

For example, a well-designed office building might be constructed on a well-located site for the purpose of economic gain by lease of office space. But if the building is too small to yield a return sufficient for the purpose as a whole, the income derived might only cover the cost of the building and not the cost of the expensive site. In this case, the building was not matched to the site in a manner that would contribute to the economic success of the entire undertaking.

Much excellent work produces little result because it is directed to perfecting details rather than to perfecting their joint effect. One objective of engineering economy is to prepare the engineer to cope with social and economic elements as well as with physical factors so that his ability will be improved to consider the joint effect of these elements in situations involving engineering application.

3.8. ADVANTAGES AND DISADVANTAGES

Any contemplated action has advantages and disadvantages. A person who has enthusiasm for a proposal, particularly its originator, is likely to give undue favorable weight to its advantages and to minimize its disadvantages. But experience has taught most people to be skeptical of new things. The opposition to a new proposal may tend to stress its disadvantages and disregard its advantages.

Certain advantages favor a new activity. Since it is independent of previous commitments, it may be implemented in large measure on the basis of present and future consideration. Little consideration need be taken of the past except to profit by previous mistakes. Contrast this with the situation of railroads, whose future development is for all practical purposes limited by the present standard distance between rails. This gauge is not the best for either economy or comfort for higher-speed trains, but to change it is not feasible.

A new activity may enjoy a tremendous temporary advantage from being first in the field. Its launching is analogous to the surprise attack in war, where traditionally a small force overwhelms a much larger force. Where the new enterprise is based upon a patented device, initiative of discovery results in the advantage of monopoly to a greater or lesser extent. But even without patents the discovery and the vigorous prosecution of an economic opportunity often result in a tremendous initial gain before rivals can come into the field.

A serious disadavantage faced in launching many new enterprises is that they often require an input of money and effort for long periods of time prior to realization of income from them. Many new ventures of ultimate promise fail because funds and enthusiasm are exhausted before adequate returns can be established. Unless strong financial backing without returns for long periods of time is available, many otherwise profitable undertakings are not feasible. Inadequate financing is recognized as one of the most important causes of failure of new enterprises.

In any pioneering activity lack of experience often results in the arising of unforeseen contingencies for which provision has not been made. Lack of experience or ability in regard to financing, organization, marketing, production, and other important activities is an important cause of failure in otherwise promising opportunities. Thus in launching a new enterprise competent personnel is of great importance. Incompetence of management ranks along with inadequate financing as a cause of failure of new enterprises.

Many new enterprises fail because the opportunities they are designed to exploit are inadequate for success. Although the opportunities for profit should be ascertained prior to deciding upon a proposal if possible, many opportunities cannot be evaluated even reasonably well except by trial. When this is the case, the uncertainty of the enterprise should be realized so that steps may be taken to meet the outcome, whatever it may be. If the outlook for the new activity seems promising, immediate steps should be taken to exploit the opportunity with vigor. If it seems unpromising, steps should be taken to withdraw from the venture with a minimum loss of money and effort.

The point to realize is that every proposal has its advantages and disadvantages whether these exist in the proposal itself or in people's regard for it. The sponsor of a proposal should school himself to consider advantages and disadvantages as impersonally as possible, because the economic advantage of a proposal depends upon its advantages and disadvantages in comparison with those of other alternatives.

PROBLEMS

1. Two alternate designs are under consideration for a tapered fastening pin. Either design will serve equally well and will involve the same material and manufacturing cost except for the lathe and grinder operations.

   Design A will require 16 hours of lathe time and 4.5 hours of grinder time per 1,000 units. Design B will require 7 hours of lathe time and 12 hours of grinder time per 1,000 units. The operating cost of the lathe including labor is $8.60 per hour. The operating cost of the grinder including labor is $6.80 per hour. Which design should be adopted if 95,000 units are required per year and what is the saving over the alternate design?

2. The flooring for an area 12 feet by 15 feet has been designed for a floor joist spacing of 18 inches center-to-center. The joists may span either the 12-foot or the 15-foot dimension. The joists are to support a uniform loading of 120 pounds per square foot including the weight of the flooring. Space restrictions limit the depth of the joists to 10 inches. Joists are to be selected from standard size lumber; that is, the width is available in increments of 1 inch and the length in increments of 2 feet. The allowable stress in the joist is 1,400 psi and the cost of the lumber is $120 per 1,000 board feet.

   The stress in a beam of rectangular cross section is given by the expression $S = 6l^2w/8bd^2$, where S = stress in psi; l = span of joist in inches; w = load on beam in pounds per inch of length; b = width of beam in inches; d = depth of beam in inches. Determine the most economical span for the joists and find the saving over the alternate design.

3. It has been decided by a previous design that 4 by 4-inch rough pine shores will be used to support certain concrete forms. The contractor will need 175 shores, each

10 feet long and each capable of supporting a load of 5,000 pounds induced by the forms and concrete.

The maximum safe load on a shore is given by the expression $P = 1,000$ $(1 - g/80b)bh$, where P = maximum safe load in pounds; g = height of shore in inches; b = width of shore in inches; h = depth of shore in inches. It has been decided that the 4-inch width of the shores will be maintained in any redesign consideration. If rough pine lumber costs $85 per 1,000 board feet, what total amount can be saved by the elimination of overdesign?

4. The chief engineer in charge of refinery operations is not satisfied with the preliminary design for storage tanks to be used as part of a plant expansion program. The engineer who submitted the design was called in and asked to reconsider the overall dimensions in the light of an article in *The Chemical Engineer*, titled "How to Size Future Process Vessels."

The original design submitted called for 4 tanks 15 feet in diameter and 20 feet high. From a graph in the article, the engineer found that the present ratio of height to diameter of 1.33 is 111% of the minimum cost and that the minimum cost for a tank occurred when the ratio of height to diameter was 4 to 1. The cost for the tank design as originally submitted was estimated to be $28,000. What are the optimum tank dimensions if the volume remains the same as for the original design? What total saving may be expected through redesign?

5. In the design of buildings to be constructed in Northern Greenland, the designer is considering the type of window frame to specify. Either steel or aluminum window frames will satisfy design criteria. Due to the remote location of the building site and the lack of building materials in Greenland, the window frames will be purchased in the United States and transported a distance of 3,500 miles to the site. The price of window frames of the type required is $42 each for steel frames and $56 each for aluminum frames. The weight of steel window frames is 150 pounds each and the weight of aluminum window frames is 50 pounds each. The transportation shipping rate is $0.008 per pound per 100 miles. Which design should be specified and what is the economic advantage of the selection?

6. An architectural engineer is considering means for saving design and construction cost for a new suburban development involving 100 new homes. The approach of using a limited number of house designs repeated throughout the area is eliminated because of the resulting reduction in sales appeal produced by an area of "look alike" houses. Individual designs for each house would increase design and construction costs to such an extent that sales would be difficult.

The engineer decides that one means of decreasing costs without making houses look alike is to use interchangeable foundation and floor structure designs. Estimates show that the average cost of design for a foundation is $260 and that floor structure design averages $180 per design; the average cost of construction of a foundation is $2,400 and the average cost of floor structure construction is $1,600. The use of basic designs repeated throughout the housing area will save the total costs of design after the first 10 houses and will save 6% of the costs of construction of foundations and floor structures after the first 10 houses. How much can the sale price of each house be reduced without a loss of profit by using this system of interchangeable design for foundations and floor structure?

7. In the design of a jet engine part, the designer has a choice of specifying either an aluminum alloy casting or a steel casting. Either material will provide equal service, but the aluminum casting will weigh 1.20 pounds as compared with 1.35 pounds for the steel casting.

The aluminum can be cast for $2.40 per pound and the steel can be cast for $1.30 per pound. The cost of machining per unit is $4.80 for the aluminum and $5.35 for the steel. Every pound of excess weight is assessed a penalty of $40 due to increased fuel consumption. Which material should be specified and what is the economic advantage of the selection per unit?

8. The volume of the raw material required for a metal part is 0.123 cubic inch. Its finished volume is 0.064 cubic inch. The machining time per piece is 0.246 minute for steel and 0.144 minute for brass. The cost of the specified steel is $0.18 per pound and the value of steel scrap is negligible. The cost of the specified brass is $0.46 per pound and the value of brass scrap is $0.08 per pound. The hourly cost of the required machine and operator is $8.90. The weight of brass and steel is 0.309 and 0.283 pounds per cubic inch, respectively. Determine the comparative costs per piece for steel and brass parts.

9. Either aluminum alloy or stainless steel will serve equally well in a certain corrosive environment. Aluminum alloy has a yield strength of 20,000 pounds per square inch and stainless steel has a yield strength of 33,000 pounds per square inch. The aluminum alloy will cost $0.68 per pound and the stainless steel will cost $0.94 per pound. The specific gravities of aluminum alloy and stainless steel are respectively 2.79 and 7.77. If selection is based upon yield strength which material will be more economical?

10. A chemical processing company is currently producing 10,000,000 pounds of Class B sulphuric acid per year at a cost of $2.40 per 100 pounds. At the present time, the process is controlled so that the product is 94.6% pure.

Upon checking the specifications for Class B sulphuric acid, the engineer in charge of the process finds that the purity must fall between 90 and 95%. The decision is made to adjust the process so that the degree of purity obtained drops by 4.2%, thus resulting in a product purity of 90.4%. After one year of operation, cost data indicates that production costs fell to $2.25 per 100 pounds of product. As a result of this reduction in perfection, how much was the company able to save for the year's operation?

11. Two locations are under consideration for a new brick plant. A cost analysis applicable to each location is given below.

	Location A	Location B
Capital cost per 1,000 bricks	$ 2.17	$ 1.56
Material cost per 1,000 bricks	2.92	3.36
Labor cost per 1,000 bricks	6.73	5.28
Distribution cost per 1,000 bricks	6.35	6.41
Total cost per 1,000 bricks	$18.17	$16.61

(a) Calculate the ratio of the profit per 1,000 bricks produced at Location A to the profit per 1,000 bricks produced at Location *B* for selling prices of $17.50, $18.50, $19.50, and $20.50 per 1,000 bricks.

(b) What is the minimum selling price per 1,000 bricks at each location if the profit is at least 10%?

12. A contractor has been awarded the contract to construct an earth fill dam requiring 500,000 cubic yards of fill. Two possible sources of fill have been found: Source *A*, above the dam site at a distance of 4 miles at which the pit will be operated and maintained by the contractor, and Source *B* which is below the dam site at a distance of 2.6 miles from which the contractor must purchase the fill by the cubic yard. The haul cost per cubic yard will be different because in one instance the loaded trucks will travel down slope, in the other they will travel up slope. The following estimates are available:

	A	*B*
Purchase price of pit	$10,000	....
Cost of fill per cu. yd. at pit	....	$0.12
Construction of haul road	$ 3,500	....
Maintenance of haul road	$ 1,800	....
Hauling cost per cu. yd. per mile	$ 0.095	$0.13

Which source should the contractor choose? What savings will result from choosing this source?

13. A manufacturer of heavy earth-moving equipment estimates the rolling resistance of different qualities of roadbeds as follows:

Unstabilized, rutted, dirt roadway; soft under travel	150 lb. per ton
Maintained, firm, smooth roadway; dirt surfacing flexing slightly under load	65 lb. per ton
Maintained, hard-surfaced roadway; little penetration under load	40 lb. per ton

The outputs of earth haulers over a 5,000-foot roadway for various values of rolling resistance are as follows:

Rolling resistance of roadway, lbs. per ton	150	65	40
Output, cubic yards per hour	30	60	80

Twelve thousand cubic yards of earth are to be moved; the hauler and driver cost $18 per hour; and a roadway with a rolling resistance of 150 pounds per ton costs $260. What expenditure is justified for the construction of a roadway whose rolling resistance is (a) 65 and (b) 40 pounds per ton?

14. An automobile parts manufacturer located in City *A* dispatches a shipment of finished parts to its warehouse in City *B* each day. The distance is 200 miles and the cost per mile is estimated to be $0.15 plus $4.00 per hour for the driver. The truck makes the return trip without payload. A trucking company located in City *B* has a contract to haul finished castings to City *A* and has established its cost per mile to be $0.18 plus $3.80 per hour for the driver. The return trip is made without payload. The round-trip time for the trucking company and the manufacturer is 9 hours.

The parts manufacturer has offered to subcontract the hauling of the machined cast-ings. It is estimated that the task will cause an increase in expenses equivalent to an additional 20 miles and an increased roundtrip time of 1 hour and 45 minutes.
(a) What is the minimum that the manufacturing company can charge for the casting haul?
(b) What is the maximum that the trucking company can pay for subcontracting the haul?

15. A manufacturer purchases ball bearings in cases containing 1,000 bearings each. Some bearings from each case are inspected as part of the acceptance procedure. The cost of inspecting one bearing is $0.20 and the loss resulting from the acceptance of a defective bearing is $2. From past experience, the number of defective bearings that will be accepted if 100, 200, 300, and 400 bearings are inspected from each case is, respectively, 45, 30, 25, and 20. Determine the most economical number of bear-ings to inspect from each case.

16. The cost of manufacturing a certain tool is $2.75 per unit of which $0.85 arises from material cost and the balance is processing cost. A new design is under considera-tion which will require an initial cost of $19,500. If the new design is adopted, the cost of material will be reduced by 6% and the cost of processing will be reduced by 3.5%. If 800,000 units are to be manufactured per year, and if the benefits from the improved design are to be considered for one year only, what amount can be spent for the new design if a return of $3 is required for each $1 spent?

II

INTEREST FORMULAS
AND EQUIVALENCE

4

Interest and
Interest Formulas

The term *interest* is used to designate a rental for the use of money. Fundamentally, the rental paid for the use of equipment is essentially the same as interest paid for the use of capital. Charging a rental for the use of money is a practice dating back to the time of man's earliest recorded history. The ethics and economics of interest have been a subject of discussion for philosophers, theologians, statesmen, and economists throughout the ages. The individual interested in these aspects of interest will find extensive literature for study, but the fact that money has time value must be considered by those who make engineering economy studies. In this chapter we present interest formulas useful in the evaluation of engineering proposals.

4.1. INTEREST RATE AND INTEREST

An *interest rate* is the ratio of the gain received from an investment and the investment over a period of time, usually one year. Also, an interest rate may be expressed as a ratio between the amount paid for the use of funds and the amount of funds used.

In one aspect, interest is an amount of money *received* as a result of investing funds, either by loaning it or by using it in the purchase of materials, labor, or facilities. Interest received in this connection is gain or profit. In another aspect, interest is an amount of money *paid out* as a result of borrowing funds. Interest paid in this connection is a *cost*.

Interest Rate from the Lender's Viewpoint. A person who has a sum of money is faced with several alternatives regarding its use:

1. He may exchange the money for goods and services that will satisfy his

personal wants. Such an exchange would involve the purchase of consumer goods.

2. He may exchange the money for productive goods or instruments. Such an exchange would involve the purchase of producer goods.

3. He may hoard the money, either for the satisfaction of gloating over it, or in awaiting an opportunity for its subsequent use.

4. He may lend the money asking only that the original sum be returned at some future date.

5. He may lend the money on the condition that the borrower will repay the initial sum plus interest at some future date.

If the decision is to lend the money with the expectation of its return plus interest, the lender must consider a number of factors in deciding on the interest rate. The following are perhaps the most important:

1. What is the probability that the borrower will not repay the loan? The answer to this question may be derived from the integrity of the borrower, his wealth, his potential earnings, and the value of any security granted the lender. If the chances are one in fifty that the loan will not be repaid, the lender is justified in charging 2% of the sum to compensate him for the risk of loss.

2. What expense will be incurred in investigating the borrower, drawing up the loan agreement, transferring the funds to the borrower, and collecting the loan? If the sum of the loan is $1,000 for a period of one year and the lender values his efforts at $10, then he is justified in charging 1% of the sum to compensate for the expense involved.

3. What net amount will compensate for being deprived of electing other alternatives for disposing of the money? Assume that $3 per hundred or 3% is considered as adequate return considering the investment opportunities foregone.

On the basis of the reasoning above, the interest rate arrived at will be 2% plus 1% plus 3%, or 6%. Therefore, an interest rate may be thought of, for convenience, as being made up of percentages for (1) risk of loss, (2) administrative expenses, and (3) pure gain or profit.

Interest Rate from the Borrower's Viewpoint. In many, if not most, cases the alternatives open to the borrower for the use of borrowed funds are limited by the lender, who may grant the loan only on condition that it be used for a specific purpose. Except as limited by the conditions of a loan, the borrower has open to him essentially the same alternatives for the use of money as a person who has ownership of money, but the borrower is faced with the necessity of repaying the amount borrowed and the interest on it in accordance with the conditions of the loan agreement or suffering the consequences. The consequences may be loss of reputation, seizure of

property or of other moneys, or the placing of a lien on his future earnings. Organized society provides many pressures, legal and social, to induce a borrower to repay a loan. Default may have serious and even disastrous consequences to the borrower.

The prospective borrower's viewpoint on the rate of interest will be influenced by the use he intends to make of funds he may borrow. If he borrows the funds for personal use, the interest rate he is willing to pay will be a measure of the amount he is willing to pay for the privilege of having satisfactions immediately instead of in the future.

If funds are borrowed to finance operations expected to result in a gain, the interest to be paid must be less than the expected gain. An example of this is the common practice of banks and similar enterprises of borrowing funds to lend to others. In this case it is evident that the amount paid out as interest, plus risks incurred, plus administrative expenses must be less than the interest received on the money reloaned, if the practice is to be profitable. A borrower may be expected to seek to borrow funds at the lowest interest rate possible.

4.2. THE EARNING POWER OF MONEY

Funds borrowed for the prospect of gain are commonly exchanged for goods, services, or instruments of production. This leads us to the consideration of the earning power of money that may make it profitable to borrow money. Consider the following example:

Mr. Digg manually digs ditches for underground cable. For this he is paid $0.10 per linear foot and averages 200 linear feet per day. Weather conditions limit this kind of work to 180 days per year. Thus, he has an income of $20 per day worked or $3,600 per year.

An advertisement brings to his attention a power ditcher that can be purchased for $1,200. He buys the ditcher after borrowing $1,200 at 8% interest. The machine will dig an average of 800 linear feet per day. By reducing the price to $0.06 per linear foot he can get sufficient work to keep the machine busy when the weather will permit.

At the end of the year the ditching machine is abandoned because it is worn out. A summary of the venture follows:

Receipts		
Amount of loan	$1,200	
Payment for ditches dug, 180 days × 800 lin. ft. × $0.06	$8,640	$9,840
Disbursements		
Purchase of ditcher	$1,200	
Fuel and repairs for machine	700	
Interest on loan, $1,200 × 0.08	96	
Repayment of loan	1,200	$3,196
Receipts Less Disbursements		$6,644

An increase in net earnings for the year over the previous year of $6,644 — $3,600 = $3,044 is enjoyed by Mr. Digg.

The above example is an illustration of what is commonly spoken of as the "earning power of money." It should be noted that the money borrowed was converted into an instrument of production. It was the instrument of production, the ditcher, which enabled Mr. Digg to increase his earnings. If Mr. Digg had held the money throughout the year it could have earned him nothing; also, if he had exchanged it for an instrument of production that turned out to be unprofitable, he might have lost money. Indirectly money has earning power when exchanged for profitable instruments of production.

In many engineering economy studies only small elements of a whole enterprise are considered. For example, studies are often made to evaluate the consequences of the purchase of a single tool or machine in a complex of many facilities. In such cases it would be desirable to isolate the element from the whole by some means analogous to the "free body" diagram in mechanics. To do this, for example, with respect to a machine being considered for purchase, it would be necessary to learn all the receipts and all the disbursements that would arise from the machine. If this could be done, the disbursements could be subtracted from the receipts. This difference would represent profit or gain, from which a rate of return could be calculated.

It is usually difficult and often impossible to learn what receipts result from a small part of a much larger whole. Therefore, numerous engineering economy studies are made on the basis of costs. For example, it may be assumed that two machines will perform a necessary service, i.e., result in equal income for the enterprise as a whole. Then the two machines can be compared on the basis of a summation of their costs.

In a cost analysis of an element, the element is isolated from the whole, even to the point of considering that the funds necessary to put it into effect are borrowed by the element. The interest on the funds assumed to be borrowed will then become a charge against the element and thus a cost. This is merely a method of taking cognizance of the amount of funds required to put into effect alternatives to be compared.

4.3. THE TIME VALUE OF MONEY

Because money can earn at a certain interest rate through its investment for a period of time, usually one year, it is important to recognize that a dollar received at some future date is not worth as much as a dollar in hand at present. It is this relationship between interest and time that leads to the concept of "the time value of money." For example, a dollar in hand now can accumulate interest for two years while a dollar received two years from now will not yield any return. Thus, the time value of money means that equal dollar amounts at different points in time do not have equal value if the interest rate is greater than zero.

Engineering economic analysis is concerned with the evaluation of alternatives. These alternatives are often described by indicating the amount and timing of estimated future receipts and disbursements that will result from each decision. Since the time value of money is concerned with the effect of time and interest on monetary amounts, it is essential that this topic be given primary attention in engineering economy. In this connection it is useful to think in terms of simple interest and then to expand this to consider the more common case of compound interest.

Simple Interest. The rental rate for a sum of money is usually expressed as the percent of the sum that is to be paid for the use of the sum for a period of one year. Interest rates are also quoted for periods other than one year, known as interest periods. In order to simplify the following discussion, consideration of interest rates for periods of other than one year will be deferred until later.

In simple interest the interest to be paid on repayment of a loan is proportional to the length of time the principal sum has been borrowed. The interest that will be earned may be found in the following manner. Let P represent the principal, n the interest period, and i the interest rate. Then

$$I = Pni.$$

Suppose that \$1,000 is borrowed at simple interest at a rate of 6% per annum. At the end of one year, the interest would be

$$I = \$1,000(1)(0.06) = \$60.$$

The principal plus interest would be \$1,060 and would be due at the end of the year.

A simple interest loan may be made for any period of time. Interest and principal become due only at the end of the loan period. When it is necessary to calculate the interest due for a fraction of a year, it is common to consider the year as composed of twelve months of thirty days each, or 360 days. For example, on a loan of \$100 at an interest rate of 7% per annum for the period February 1 to April 20, the interest due on April 20 along with the principal sum of \$100 would be $0.07(\$100)(80 \div 360) = \$1.55$.

Compound Interest. When a loan is made for a length of time equal to several interest periods, provision is made that the earned interest is *due at the end of each interest period.* For example, the payments on a loan of \$1,000 at 6% interest per annum for a period of four years would be calculated as shown in Table 4.1.

Table 4.1. APPLICATION OF COMPOUND INTEREST WHEN INTEREST IS PAID
ANNUALLY

Year	Amount Owed at Beginning of Year	Interest to Be Paid at End of Year	Amount Owed at End of Year	Amount to Be Paid by Borrower at End of Year
1	$1,000	$60	$1,060	$ 60
2	1,000	60	1,060	60
3	1,000	60	1,060	60
4	1,000	60	1,060	1,060

If the borrower is allowed to keep the earned interest until the entire loan
becomes due, the loan will be increased by an amount equal to the interest
due at the end of each year. In this case, no yearly interest payments are required and interest is said to be compounded. On this basis, a loan of $1,000
at 6% interest compounded annually for a period of four years will be as
shown in Table 4.2.

Table 4.2. APPLICATION OF COMPOUND INTEREST WHEN INTEREST IS PERMITTED
TO COMPOUND

Year	Amount Owed at Beginning of year (A)	Interest to Be Added to Loan at End of Year (B)	Amount Owed at End of Year (A + B)	Amount to Be Paid by Borrower at End of Year
1	$1,000.00	$1,000.00 × 0.06 = $60.00	$1,000(1.06) = $1,060.00	$ 00.00
2	1,060.00	1,060.00 × 0.06 = 63.60	$1,000(1.06)^2 = 1,123.60	00.00
3	1,123.60	1,123.60 × 0.06 = 67.42	$1,000(1.06)^3 = 1,191.02	00.00
4	1,191.02	1,191.02 × 0.06 = 71.46	$1,000(1.06)^4 = 1,262.48	1,262.48

Where the interest earned each year is added to the amount of the loan,
as in the example above, it is said to be *compounded annually*. The section
which follows will present interest formulas useful in dealing with the case
where annual payments and annual compounding interest are incurred.

4.4. INTEREST FORMULAS, ANNUAL COMPOUNDING INTEREST-ANNUAL PAYMENTS

The derivations which follow will explore the common situation of annual
compounding interest and annual payments. The following symbols will
be used. Let

i = the annual interest rate;
n = the number of annual interest periods;
P = a present principal sum;

A = a single payment, in a series of n equal payments, made at the end of each annual interest period;

F = a future sum, n annual interest periods hence, equal to the compound amount of a present principal sum P, or equal to the sum of the compound amounts of payments, A in a series.

Single-Payment Compound-Amount Factor. When interest is permitted to compound, as in Table 4.2, the interest earned is added to the principal at the end of each annual interest period. By substituting general terms in place of numerical values in Table 4.2, the results shown in Table 4.3 are

Table 4.3. DEVELOPMENT OF SINGLE-PAYMENT COMPOUND-AMOUNT FACTOR

Year	Amount at Beginning of Year	Interest Earned During Year	Compound Amount at End of Year		
1	P	Pi	$P + Pi$		$= P(1 + i)$
2	$P(1 + i)$	$P(1 + i)i$	$P(1 + i)$	$+ P(1 + i)i$	$= P(1 + i)^2$
3	$P(1 + i)^2$	$P(1 + i)^2i$	$P(1 + i)^2$	$+ P(1 + i)^2i$	$= P(1 + i)^3$
n	$P(1 + i)^{n-1}$	$P(1 + i)^{n-1}i$	$P(1 + i)^{n-1}$	$+ P(1 + i)^{n-1}i$	$= P(1 + i)^n$
					$= F$

developed. The resulting factor, $(1 + i)^n$, is known as the *single-payment compound-amount factor*[1] and is designated $(\overset{F/P\ i,\ n}{\quad})$. This factor may be used to find the compound-amount, F, of a present principal amount, P. The relationship is

$$F = P(1 + i)^n$$

or

$$F = P(\overset{F/P\ i,\ n}{\quad}).$$

Referring to the example of Table 4.2, if \$1,000 is invested at 6% interest compounded annually at the beginning of year one, the compound amount at the end of the fourth year will be

$$F = \$1,000(1 + 0.06)^4 = \$1,000(1.262)$$
$$= \$1,262.$$

Or, by use of the factor designation and its associated tabular value,

$$F = \$1,000(\overset{F/P\ 6,\ 4}{1.262}) = \$1,262.$$

---

[1] Values for interest factors for annual compounding interest-annual payments are given in Appendix D, Tables D.1 through D.18.

Single-Payment Present-Worth Factor. The single-payment compound-amount relationship may be solved for P as follows:

$$P = F\left[\frac{1}{(1+i)^n}\right].$$

The resulting factor, $1/(1+i)^n$, is known as the *single-payment present-worth factor* and is designated ($\overset{P/F\,i,\,n}{}$). This factor may be used to find the present worth, P, of a future amount, F. As an example, if \$1,262 is to be received four years hence, its present worth at 6% compounded annually is

$$P = \$1,262\left[\frac{1}{(1+0.06)^4}\right] = \$1,262(0.7921) = \$1,000.$$

Or, by using the factor designation and the interest tables

$$P = \$1,262(\overset{P/F\,6,\,4}{0.7921}) = \$1,000.$$

Note that the single-payment compound-amount factor and the single-payment present-worth factor are reciprocals.

Equal-Payment-Series Compound-Amount Factor. In many engineering economy studies, a series of equal payments occurring at the end of succeeding annual interest periods is encountered. The sum of the compound amounts of the several payments may be calculated by use of the single-payment compound-amount factor. For example, the calculation of the compound amount of a series of five \$100 payments made at the end of each year at 6% interest compounded annually is illustrated in Table 4.4.

Table 4.4. THE COMPOUND AMOUNT OF A SERIES OF YEAR-END PAYMENTS

End of Year	Year-end Payment Times Compound-Amount Factor	Compound Amount at End of 5 Year Period	Total Compound Amount
1	$\$100(1.06)^4$	\$126	
2	$\$100(1.06)^3$	\$119	
3	$\$100(1.06)^2$	\$113	
4	$\$100(1.06)^1$	\$106	
5	$\$100(1.06)^0$	\$100	\$564

It is apparent that the method illustrated will be cumbersome for calculating the compound amount of an extensive series. Therefore, it is desirable that a compact solution for this type of problem be available. If A represents a series of n equal payments such as the \$100 series in Table 4.4

$$F = A(1) + A(1+i) + \ldots + A(1+i)^{n-2} + A(1+i)^{n-1}$$

since the total future amount F is equal to the sum of individual future amounts calculated for each payment A. Multiplying this equation by $(1 + i)$ results in

$$F(1 + i) = A(1 + i) + A(1 + i)^2 + \ldots + A(1 + i)^{n-1} + A(1 + i)^n.$$

Subtracting the first equation from the second gives

$$F(1 + i) - F = -A + A(1 + i)^n$$

$$F = A\left[\frac{(1 + i)^n - 1}{i}\right].$$

The resulting factor, $[(1 + i)^n - 1]/i$, is known as the *equal-payment-series compound-amount factor* and is designated ($\overset{F/A\ i,\ n}{}$). This factor may be used to find the compound amount, F, of an equal-payment-series, A. For example, the compound amount of the five \$100 payments mentioned above will be

$$F = \$100\left[\frac{(1 + 0.06)^5 - 1}{0.06}\right] = \$100(5.637) = \$563.70$$

which agrees with the result found in Table 4.4. Using the factor designation and the interest tables

$$F = \$100(\overset{F/A\ 6,\ 5}{5.637}) = \$563.70.$$

Equal-Payment-Series Sinking-Fund Factor. The equal-payment-series compound-amount relationship may be solved for A as follows:

$$A = F\left[\frac{i}{(1 + i)^n - 1}\right].$$

The resulting factor, $i/[(1 + i)^n - 1]$, is known as the *equal-payment series sinking-fund factor* and is designated ($\overset{A/F\ i,\ n}{}$). This factor may be used to find the required year-end payments, A, to accumulate a future amount, F. If, for example, it is desired to accumulate \$563.70 by making a series of five annual payments at 6% interest compounded annually, the required amount of each payment will be

$$A = \$563.70\left[\frac{0.06}{(1 + 0.06)^5 - 1}\right]$$
$$= \$563.70(0.1774) = \$100.$$

or

$$A = \$563.70(\overset{A/F\ 6,\ 5}{0.1774}) = \$100.$$

The derivation of this factor and this example illustrate that the equal-payment-series compound-amount factor and the equal-payment-series sinking-fund factor are reciprocals.

Equal-Payment-Series Capital-Recovery Factor. The substitution of $P(1 + i)^n$ for F in the equal-payment-series sinking-fund relationship results in

$$A = P(1 + i)^n \left[\frac{i}{(1 + i)^n - 1} \right]$$

$$= P \left[\frac{i(1 + i)^n}{(1 + i)^n - 1} \right].$$

The resulting factor, $i(1 + i)^n/[(1 + i)^n - 1]$ is known as the *equal-payment-series capital-recovery factor* and is designated ($\overset{A/P\,i,\,n}{\quad}$). This factor may be used to find the year end payments, A, that will be provided by a present amount, P. For example, $1,000 invested at 5% interest compounded annually will provide for eight equal year end payments of

$$A = \$1,000 \left[\frac{0.05(1 + 0.05)^8}{(1 + 0.05)^8 - 1} \right]$$

$$= \$1,000(0.1547) = \$154.72.$$

or

$$A = \$1,000 \overset{A/P\,5,\,8}{(0.1547)} = \$154.72.$$

Equal-Payment-Series Present-Worth Factor. The equal-payment-series capital-recovery factor may be solved for P as follows:

$$P = A \left[\frac{(1 + i)^n - 1}{i(1 + i)^n} \right].$$

The resulting factor, $[(1 + i)^n - 1]/i(1 + i)^n$, is known as the *equal-payment-series present-worth factor* and is designated ($\overset{P/A\,i,\,n}{\quad}$). This factor may be used to find the present worth, P, of a series of equal annual payments, A. For example, the present worth of a series of eight equal annual payments of $154.72 at an interest rate of 5% compounded annually will be

$$P = \$154.72 \left[\frac{(1 + 0.05)^8 - 1}{0.05(1 + 0.05)^8} \right]$$

$$= \$154.72(6.4632) = \$1,000.$$

or

$$P = \$154.72 \overset{P/A\,5,\,8}{(6.4632)} = \$1,000.$$

This example and the derivation illustrate that the equal-payment-series capital-recovery factor and the equal-payment-series present-worth factor are reciprocals.

Uniform Gradient-Series Factor. In many cases, annual payments do not occur in an equal-payment series. For example, a series of payments that would be uniformly increasing is $100, $125, $150 and $175 occurring at the end of the first, second, third, and fourth year. Similarly, a uniformly decreasing series would be $100, $90, $80, and $70 occurring at the end of the first, second, third, and fourth year. In general, a uniformly increasing series of payments for n interest periods may be expressed as $A_1, A_1 + g, A_1 + 2g,$ $\ldots, A_1 + (n - 1)g$ as shown in Figure 4.1 where A_1 denotes the first year-

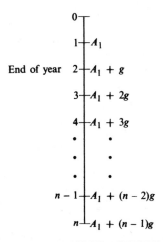

FIGURE 4.1. A UNIFORM GRADIENT SERIES.

end payment in the series and g the annual change in the magnitude of the payments.

One way of evaluating such a series is to apply the interest formulas developed previously to each payment in the series. This method will yield good results but will be very time consuming. Another approach is to reduce the uniformly increasing series of payments to an equivalent equal-payment series so that the equal-payment series factor can be used. Let

A_1 = payment at the end of the first year;
g = annual change or gradient;
n = the number of years;
A = the equivalent equal annual payment.

A uniform gradient series may be considered to be made up of two separate series, an equal-payment series with equal annual payment A_1, and a gradient series $0, g, 2g, \ldots, (n - 1)g$ at the end of successive years. Each payment in an equal-payment series equivalent to this series can be represented as

$$A = A_1 + A_2$$

where

$$A_2 = F(\overset{A/F\,i,\,n}{}) = F\left[\frac{i}{(1+i)^n - 1}\right]$$

and F is the future amount equivalent to the gradient series. The gradient series can be separated into $(n-1)$ distinct equal-payment series with annual payments of g as shown in Table 4.5. The future amount equivalent to the gradient series can be derived from the table as follows:

Table 4.5. GRADIENT SERIES AND AN EQUIVALENT SET OF SERIES

End of Year	Gradient Series	Set of Series Equivalent to Gradient Series
0	0	0
1	0	0
2	g	g
3	$2g$	$g + g$
4	$3g$	$g + g + g$
.	.	. . .
.	.	. . .
.	.	. . .
$n-1$	$(n-2)g$	$g + g + g + \cdots + g$
n	$(n-1)g$	$g + g + g + \cdots + g + g$

$$
\begin{aligned}
F &= g(\overset{F/A\,i,\,n-1}{}) + g(\overset{F/A\,i,\,n-2}{}) + \ldots + g(\overset{F/A\,i,\,2}{}) + g(\overset{F/A\,i,\,1}{}) \\
&= g\left[\frac{(1+i)^{n-1}-1}{i}\right] + g\left[\frac{(1+i)^{n-2}-1}{i}\right] + \ldots + g\left[\frac{(1+i)^2-1}{i}\right] \\
&\quad + g\left[\frac{(1+i)^1-1}{i}\right] \\
&= \frac{g}{i}[(1+i)^{n-1} + (1+i)^{n-2} + \ldots + (1+i)^2 + (1+i) - (n-1)] \\
&= \frac{g}{i}[(1+i)^{n-1} + (1+i)^{n-2} + \ldots + (1+i)^2 + (1+i) + 1] - \frac{ng}{i}.
\end{aligned}
$$

The terms in the brackets define the equal-payment-series compound-amount factor for n years. Therefore

$$F = \frac{g}{i}\left[\frac{(1+i)^n - 1}{i}\right] - \frac{ng}{i}$$

and

$$A_2 = F\left[\frac{i}{(1+i)^n - 1}\right] = \frac{g}{i}\left[\frac{(1+i)^n - 1}{i}\right]\left[\frac{i}{(1+i)^n - 1}\right]$$
$$- \frac{ng}{i}\left[\frac{i}{(1+i)^n - 1}\right]$$

$$A_2 = \frac{g}{i} - \frac{ng}{i}\left[\frac{i}{(1+i)^n - 1}\right]$$

$$A_2 = \frac{g}{i} - \frac{ng}{i}(\overset{A/Fi,\,n}{}) = g\left[\frac{1}{i} - \frac{n}{i}(\overset{A/Fi,\,n}{})\right].$$

The resulting factor, $\left[\frac{1}{i} - \frac{n}{i}(\overset{A/Fi,\,n}{})\right]$, is called the *gradient factor* for annual compounding interest and will be designated ($\overset{A/Gi,\,n}{}$).

As an example of the use of the gradient factor assume that a man receives an annual salary of \$10,000 increasing at the rate of \$700, a year. What is his equivalent uniform salary for a period of 10 years if the interest rate is 8% compounded annually?

$$A = A_1 + g(\overset{A/Gi,\,n}{})$$
$$= \$10,000 + \$700(\overset{A/G\,8,\,10}{3.8713})$$
$$= \$12,709.91$$

The gradient factor may also be used for a uniformly decreasing gradient. In this case the formulation would be

$$A = A_1 - g(\overset{A/Gi,\,n}{}).$$

Format for the Use of Interest Factors. In engineering economy studies, disbursements made to initiate an alternative are considered to take place at the beginning of the period embraced by the alternative. Payments occurring during the period of the alternative are usually assumed to occur at the end of the year or interest period in which they occur. To use the several interest factors that have been developed, it is necessary that the monetary transactions conform to the format for which the factors are applicable. The schematic arrangement of the factors in Table 4.6 should be helpful in this connection.

Five important points should be noted in the use of interest factors for annual payments:

1. The end of one year is the beginning of the next year.
2. P is at the beginning of a year at a time regarded as being the present.
3. F is at the end of the nth year from a time regarded as being the present.
4. An A occurs at the *end* of each year of the period under consideration. When P and A are involved, the first A of the series occurs one year after P. When F and A are involved, the last A of the series occurs at the same time as F.
5. In the solution of problems, the quantities P, F, and A must be set up to conform with the pattern applicable to the factors used.

Table 4.6. SCHEMATIC ILLUSTRATION OF THE USE OF FACTORS

End of Year	Single Payment		Equal Payment Series				Gradient Series
	Use of Compound-Amount Factor	Use of Present-Worth Factor	Use of Compound-Amount Factor	Use of Sinking-Fund Factor	Use of Present-Worth Factor	Use of Capital-Recovery Factor	Use of Gradient Series Factor
0	P	P	—	—	P	P	—
1			A	A	A	A	—
2			A	A	A	A	g
3			A	A	A	A	$2g$
r			A	A	A	A	$(r-1)g$
n	F	F	$A\ \ F$	$A\ \ F$	A	A	$(n-1)g$
	$F = P(^{F/P\,i,\,n})$	$P = F(^{P/F\,i,\,n})$	$F = A(^{F/A\,i,\,n})$	$A = F(^{A/F\,i,\,n})$	$P = A(^{P/A\,i,\,n})$	$A = P(^{A/P\,i,\,n})$	$A = g(^{A/G\,i,\,n})$

There are two important advantages of using the factor designations in place of the algebraic expressions for the factors. These advantages are: (1) the equations for solving problems may be set up prior to looking up any values of factors from the tables and inserting them in the parentheses, and (2) the source and the identity of values taken from the tables are maintained throughout the solution. For example, in solving a problem where it is required to find the present worth of $800 six years hence at 4% interest compounded annually, the following format may be used:

$$P = F(\overset{P/F\,i,\,n}{\hspace{1em}})$$

$$P = \$800(\overset{P/F\,4,\,6}{0.7903}) = \$632.24.$$

4.5. NOMINAL AND EFFECTIVE INTEREST RATES

For simplicity, the discussion to this point has involved interest periods of only one year. However, agreements may specify that interest shall be paid more frequently, such as each half year, each quarter, or each month. Such agreements result in interest periods of one-half year, one-quarter year, or one-twelfth year, and the compounding of interest twice, four times, or twelve times a year, respectively.

The interest rates associated with this more frequent compounding are normally quoted on an annual basis according to the following convention. Where the actual or *effective* rate of interest is 3% interest compounded each six-month period the annual or *nominal* interest is quoted as "6% per year compounded semiannually." For an effective rate of interest of 1.5% compounded at the end of each three-month period the nominal interest is quoted as "6% per year compounded quarterly." Thus the nominal rate of interest is expressed on an annual basis and it is determined by multiplying the actual or effective interest rate per interest period times the number of interest periods per year.

The effect of this more frequent compounding is that the actual interest rate per year or effective interest rate per year is higher than the nominal interest rate. For example, consider a nominal interest rate of 6% compounded semiannually. The value of $1 at the end of one year when $1 is compounded at 3% for each half year period is

$$F = \$1(1.03)(1.03)$$
$$= \$1(1.03)^2 = \$1.0609.$$

The actual interest earned on the dollar for one year is 1.0609 minus 1.0000 = 0.0609. Therefore, the effective interest rate is 6.09%.

An expression for the effective annual interest rate may be derived from the above reasoning. If r is the nominal interest rate and if c is the number of interest periods per year, then the

$$\text{effective annual interest rate} = \left(1 + \frac{r}{c}\right)^c - 1.$$

As a limit, interest may be considered to be compounded an infinite number of times per year; that is, *continuously*. Under these conditions, the effective interest rate may be derived as follows:

$$\left(1 + \frac{r}{c}\right)^c - 1 = \left[\left(1 + \frac{r}{c}\right)^{c/r}\right]^r - 1,$$

but

$$\lim_{c \to \infty} \left(1 + \frac{r}{c}\right)^{c/r} = e,$$

hence

$$\lim_{c \to \infty} \left[\left(1 + \frac{r}{c}\right)^{c/r}\right]^r - 1 = e^r - 1.$$

Therefore, where interest is compounded continuously, the effective annual interest rate $= e^r - 1$.

The effective interest rates corresponding to a nominal annual interest rate of 6% compounded annually, semiannually, quarterly, monthly, weekly, daily, and continuously are respectively 6.000%, 6.090%, 6.136%, 6.168%, 6.180%, 6.183%, and 6.184%. Note that the effective interest rate is equal to the nominal rate when compounding occurs annually.

Since the effective interest rate represents the actual interest earned, it is this rate that should be used to compare the benefits of various nominal rates of interest. For example one might be confronted with the problem of determining whether it is more desirable to receive 16% compounded annually or 15% compounded monthly. The effective rate of interest per year for 16% compounded annually is of course 16%, while for 15% compounded monthly the effective annual interest rate is

$$\left(1 + \frac{0.15}{12}\right)^{12} - 1 = 16.1\%.$$

The interest formulas for annual compounding interest-annual payments were derived on the basis of an effective interest rate for an interest period; specifically, for an annual interest rate compounded annually. However, they may be used when compounding occurs more frequently than once a year. This may be done in one of two ways (1) find the effective interest rate from the relationships derived above or from the effective rates tabulated[2] and use

---

[2]Effective interest rates corresponding to nominal annual rates for various compounding frequencies are given in Appendix D, Table D.19.

this rate in place of the nominal annual rate, or (2) match the interest rate to the interest period and use the formula directly or its corresponding tabulated value. Consider the following example in which it is desired to find the compound amount of $1,000 four years from now at a nominal annual interest rate of 6% compounded semiannually. The effective interest rate is 6.09% and may be used with the single-payment compound-amount factor as follows:

$$F = \$1,000(1 + 0.0609)^4$$
$$= \$1,000(\overset{F/P\ 6.09,\ 4}{1.267}) = \$1,267.$$

Or, since the nominal annual interest rate is 6% compounded semiannually, the interest rate is 3% for an interest period of one-half year. The required calculation is as follows:

$$F = \$1,000(1 + 0.030)^8$$
$$= \$1,000(\overset{F/P\ 3,\ 8}{1.267}) = \$1,267.$$

This analysis may be used for nominal annual interest rates compounded with any frequency up to and including continuous compounding. Note, however, that compounding frequencies in excess of 52 times per year differ only slightly from the assumption of continuous compounding.

To help distinguish between effective and nominal rates of interest in this book, the letter i is used to represent effective rates of interest while the letter r is used for nominal rates of interest. The derivations of the interest formulas in Section 4.4 were based on an interest rate per period or an effective interest rate. The letter i was used in these derivations to indicate that these formulas require an effective rate of interest rather than the nominal rate of interest. Of course when compounding is on a yearly basis the nominal interest rate can be used in those formulas since it is equal to the effective interest rate.

4.6. INTEREST FORMULAS, CONTINUOUS COMPOUNDING INTEREST-ANNUAL PAYMENTS

In certain economic evaluations, it is reasonable to assume that continuous compounding interest more nearly represents the true situation than does annual compounding. Also, the assumption of continuous compounding may be more convenient from a computational standpoint in some applications. Therefore, this section will present interest formulas that may be used in those cases where annual payments and continuous compounding interest seem appropriate. The following symbols will be used. Let

r = the nominal annual interest rate;

$n =$ the number of annual periods;
$P =$ a present principal sum;
$A =$ a single payment, in a series of n equal payments, made at the end of each annual period;
$F =$ a future sum, n annual periods hence, equal to the compound amount of a present principal sum, P, or equal to the sum of the compound amounts of payments, A in a series.

Single-Payment Compound-Amount Factor. When finding a future amount in years hence that will result from a present amount it is necessary to consider the frequency of compounding. The single-payment compound-amount factor depends on the number of compounding periods in the following way.

$$\text{Annual compounding} \qquad F = P(1 + r)^n$$

$$\text{Semiannual compounding} \quad F = P\left(1 + \frac{r}{2}\right)^{2n}$$

$$\text{Monthly compounding} \qquad F = P\left(1 + \frac{r}{12}\right)^{12n}$$

In general, if there are c compounding periods per year

$$F = P\left(1 + \frac{r}{c}\right)^{cn}.$$

When interest is permitted to compound continuously, the interest earned is instantaneously added to the principal at the end of each infinitesimal interest period. For continuous compounding, the number of compounding periods per year is considered to be infinite. As a result

$$F = P\left[\lim_{c \to \infty}\left(1 + \frac{r}{c}\right)^{cn}\right].$$

By rearranging terms

$$F = P\left\{\lim_{c \to \infty}\left[\left(1 + \frac{r}{c}\right)^{c/r}\right]^{rn}\right\}.$$

But,

$$\lim_{c \to \infty}\left(1 + \frac{r}{c}\right)^{c/r} = e = 2.7182.$$

Therefore,

$$F = Pe^{rn}.$$

The resulting factor, e^{rn}, is the *single-payment compound-amount factor*[3] for continuous compounding interest and is designated $\left(\overset{F/P\ r,\ n}{}\right)$.

---

[3] Values for interest factors for continuous compounding interest-annual payments are given in Appendix D, Tables D.20 through D.34.

Single-Payment Present-Worth Factor. The single-payment compound-amount relationship may be solved for P as follows:

$$P = F\left[\frac{1}{e^{rn}}\right].$$

The resulting factor, e^{-rn}, is the *single-payment present-worth factor* for continuous compounding interest and is designated ($\overset{P/F\,r,\,n}{\quad}$).

Equal-Payment-Series Present-Worth Factor. By considering each payment in the series individually, the total present worth of the series is a sum of the individual present-worth amounts as follows:

$$P = A(e^{-r}) + A(e^{-r2}) + \ldots + A(e^{-rn})$$
$$= Ae^{-r}(1 + e^{-r} + e^{-r2} + \ldots + e^{-r(n-1)})$$

which is Ae^{-r} times the geometric series $\sum_{j=0}^{n-1}\left(\frac{1}{e^r}\right)^j$. Therefore,

$$P = Ae^{-r}\left[\frac{1 - e^{-rn}}{1 - e^{-r}}\right]$$
$$= A\left[\frac{1 - e^{-rn}}{e^r - 1}\right].$$

The resulting factor, $(1 - e^{-rn})/(e^r - 1)$, is the *equal-payment-series present-worth factor* for continuous compounding interest and is designated ($\overset{P/A\,r,\,n}{\quad}$).

Equal-Payment-Series Capital-Recovery Factor. The equal-payment-series present-worth relationship may be solved for A as follows:

$$A = P\left[\frac{e^r - 1}{1 - e^{-rn}}\right].$$

The resulting factor, $(e^r - 1)/(1 - e^{-rn})$, is the *equal-payment-series capital-recovery factor* for continuous compounding interest and is designated ($\overset{A/P\,r,\,n}{\quad}$).

Equal-Payment-Series Sinking-Fund Factor. The substitution of Fe^{-rn} for P in the equal-payment-series capital-recovery relationship results in

$$A = Fe^{-rn}\left[\frac{e^r - 1}{1 - e^{-rn}}\right]$$
$$= F\left[\frac{e^r - 1}{e^{rn} - 1}\right].$$

The resulting factor, $(e^r - 1)/(e^{rn} - 1)$, is the *equal-payment-series sinking-fund factor* for continuous compounding interest and is designated ($\overset{A/F\,r,\,n}{\quad}$).

Equal-Payment-Series Compound-Amount Factor. The equal-payment-series sinking-fund relationship may be solved for F as follows:

$$F = A\left[\frac{e^{rn} - 1}{e^r - 1}\right].$$

The resulting factor, $(e^{rn} - 1)/(e^r - 1)$, is the *equal-payment-series compound-amount factor* for continuous compounding interest and is designated $\left(\overset{F/A\, r,\, n}{}\right)$.

Gradient Series, Continuous Compounding. The equivalent annual payment corresponding to an initial payment A_1, linear gradient, g, number of years, n, and interest rate, r, may be found in a similar manner as for annual compounding. It can be shown that

$$A = A_1 + g\left[\frac{1}{e^r - 1} - \frac{n}{e^{rn} - 1}\right].$$

The resulting factor, $\left[\dfrac{1}{e^r - 1} - \dfrac{n}{e^{rn} - 1}\right]$, is called the *gradient factor* for continuous compounding interest and is designated $\left(\overset{A/G\, r,\, n}{}\right)$.

More Frequent Payments. The interest formulas for continuous compounding interest developed above can be modified to accomodate equal payments that occur more frequently than annually. When there are c payments per year let

$$n = c \text{ (number of years)}$$

and

$$r = \frac{\substack{\text{nominal interest rate} \\ \text{per year}}}{c}.$$

Suppose it is desired to find a future amount at the end of five years that would result from end of month deposits of $1,000 made throughout the entire five year period. Assume that the interest earned on these deposits is 15% compounded continuously.

$$n = (12 \text{ periods per year})(5 \text{ years}) = 60 \text{ periods}$$

$$r = \frac{15\%}{12 \text{ periods}} = 0.0125$$

$$F = A\left[\frac{e^{rn} - 1}{e^r - 1}\right]$$

$$= \$1,000\left[\frac{e^{(0.0125)(60)} - 1}{e^{0.0125} - 1}\right]$$

$$= \$1,000\left[\frac{1.1170}{0.0126}\right] = \$88,650.$$

4.7. INTEREST FORMULAS, CONTINUOUS COMPOUNDING INTEREST-CONTINUOUS PAYMENTS

In the previous derivations, payments were considered to be concentrated at discrete points in time. However, in many instances, it is reasonable to assume that monetary transactions occur on a relatively uniform basis throughout the year. In this case, a uniform flow of money best describes the nature of the transaction. Situations such as this involve a *funds-flow process* which may be described in terms of an annual flow rate. The following symbols will be used. Let

r = the nominal annual interest rate;
n = the time expressed in years;
P = a present principal sum;
$\bar{A}$ = the uniform flow rate of money per year;
F = a future amount equal to the compound amount of a uniform flow of money at time n.

Where there is no flow of payments, as in the case with annual payments, the compound amount and the present-worth factors are identical to those for continuous compounding interest-annual payments. Thus,

$$F = Pe^{rn}$$

as was shown in Section 4.6. Its reciprocal

$$P = Fe^{-rn}$$

was also developed previously.

Funds-Flow Compound-Amount Factor. The following symbols will be used to develop interest formulas for the funds-flow process. Let

ΔF = a future amount equal to the compound amount ΔP. This future amount occurs t years from time n as shown in Figure 4.2.
$\bar{A}$ = a uniform rate of flow of money per year.

Since it has been shown that $F = Pe^{rn}$

$$\Delta F = \Delta Pe^{rn}.$$

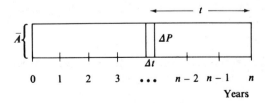

FIGURE 4.2. UNIFORM CONTINUOUS FUNDS FLOW.

But,

$$\Delta P = \bar{A}\Delta t$$

so that

$$\Delta F = \bar{A}e^{rt}\Delta t.$$

By letting Δt approach zero

$$dF = \bar{A}e^{rt}dt.$$

And, for the entire interval 0 to n

$$F = \int_0^n dF = \int_0^n \bar{A}e^{rt}dt$$

$$F = \frac{\bar{A}e^{rt}}{r}\Big]_0^n = \bar{A}\left[\frac{e^{rn}}{r} - \frac{e^0}{r}\right]$$

$$F = \bar{A}\left[\frac{e^{rn} - 1}{r}\right].$$

The resulting factor, $(e^{rn} - 1)/r$, is called the *funds-flow compound-amount factor* and is designated ($\overset{F/\bar{A}\,r,\,n}{}$).

Funds-Flow Sinking-Fund Factor. The funds-flow compound-amount relationship may be solved for $\bar{A}$ as follows:

$$\bar{A} = F\left[\frac{r}{e^{rn} - 1}\right].$$

The resulting factor, $r/(e^{rn} - 1)$, is the *funds flow sinking-fund factor* and is designated ($\overset{\bar{A}/F\,r,\,n}{}$).

Funds-Flow Capital Recovery Factor. By using the single-payment compound-amount relationship for continuous compounding, $F = Pe^{rn}$, and the funds-flow sinking-fund relationship just derived it is seen that

$$\bar{A} = Pe^{rn}\left[\frac{r}{e^{rn} - 1}\right]$$

$$\bar{A} = P\left[\frac{re^{rn}}{e^{rn} - 1}\right].$$

The resulting factor $(re^{rn})/(e^{rn} - 1)$, is the *funds-flow capital-recovery factor* and is designated ($\overset{\bar{A}/P\,r,\,n}{}$).

Funds-Flow Present-Worth Factor. The funds-flow capital recovery relationship may be solved for P as follows:

$$P = \bar{A}\left[\frac{e^{rn} - 1}{re^{rn}}\right].$$

The resulting factor, $(e^{rn} - 1)/(re^{rn})$, is the *funds-flow present-worth factor* and is designated $\left(\overset{P/\bar{A}\,r,\,n}{}\right)$.

Funds-Flow Conversion Factor. Tabulated values for the interest factors for continuous compounding interest-annual payments may be modified and used for the funds-flow factors. The required conversion factor may be derived by finding the year end equivalent of a summation of an infinite number of payments occurring during the year. The equal payment-series compound-amount factor

$$F = A\left[\frac{e^{rn} - 1}{e^r - 1}\right]$$

may be modified to reflect c interest periods per year as follows:

$$F = \frac{A}{c}\left[\frac{e^{(r/c)c} - 1}{e^{r/c} - 1}\right] = \frac{A}{c}\left[\frac{e^r - 1}{e^{r/c} - 1}\right].$$

But,

$$\lim_{c \to \infty} \frac{A}{c}\left[\frac{e^r - 1}{e^{r/c} - 1}\right] = \bar{A}\left[\frac{e^r - 1}{r}\right]$$

$$F = \bar{A}\left[\frac{e^r - 1}{r}\right].$$

The above factor expresses the equivalence between a uniform continuous flow of funds for one year, $\bar{A}$, and a future amount at the end of the year, F. For a time span greater than one year the same factor also calculates the equivalence between a uniform flow of funds occurring at the rate of $\bar{A}$, a year and equal annual amounts A at the end of each year. Thus for time spans greater than one year

$$A = \bar{A}\left[\frac{e^r - 1}{r}\right].$$

The resulting factor, $(e^r - 1)/r$, is called the *funds-flow conversion factor*[4] and is designated $\left(\overset{A/\bar{A}\,r}{}\right)$. This conversion factor may be used with the interest factors for continuous compounding interest-annual payments to yield values for the funds-flow factors in the following manner:

$$\left(\overset{\bar{A}/P\,r,\,n}{}\right) = \left(\overset{A/P\,r,\,n}{}\right) \div \left(\overset{A/\bar{A}\,r}{}\right)$$

$$\left(\overset{P/\bar{A}\,r,\,n}{}\right) = \left(\overset{P/A\,r,\,n}{}\right)\left(\overset{A/\bar{A}\,r}{}\right)$$

$$\left(\overset{\bar{A}/F\,r,\,n}{}\right) = \left(\overset{A/F\,r,\,n}{}\right) \div \left(\overset{A/\bar{A}\,r}{}\right)$$

---

[4] Values for the funds-flow conversion factor for various interest rates are given in Appendix D, Table D. 35.

Table 4.7. SUMMARY OF COMPOUNDING FACTORS AND FACTOR DESIGNATIONS

To Find	Given	Discrete Payments — Discrete Compounding	Discrete Payments — Continuous Compounding	Continuous Payments — Continuous Compounding
F	P	$F = P(1+i)^n = P(\quad)^{F/P\,i,n}$	$F = Pe^{rn} = P(\quad)^{F/P\,r,n}$	$F = Pe^{rn} = P(\quad)^{F/P\,r,n}$
P	F	$P = F\dfrac{1}{(1+i)^n} = F(\quad)^{P/F\,i,n}$	$P = F\dfrac{1}{e^{rn}} = F(\quad)^{P/F\,r,n}$	$P = F\dfrac{1}{e^{rn}} = F(\quad)^{P/F\,r,n}$
F	A	$F = A\left[\dfrac{(1+i)^n - 1}{i}\right] = A(\quad)^{F/A\,i,n}$	$F = A\left[\dfrac{e^{rn}-1}{e^r - 1}\right] = A(\quad)^{F/A\,r,n}$	$F = \bar{A}\left[\dfrac{e^{rn}-1}{r}\right] = \bar{A}(\quad)^{F/\bar{A}\,r,n}$
A	F	$A = F\left[\dfrac{i}{(1+i)^n - 1}\right] = F(\quad)^{A/F\,i,n}$	$A = F\left[\dfrac{e^r - 1}{e^{rn}-1}\right] = F(\quad)^{A/F\,r,n}$	$\bar{A} = F\left[\dfrac{r}{e^{rn}-1}\right] = F(\quad)^{\bar{A}/F\,r,n}$
P	A	$P = A\left[\dfrac{(1+i)^n - 1}{i(1+i)^n}\right] = A(\quad)^{P/A\,i,n}$	$P = A\left[\dfrac{1-e^{-rn}}{e^r - 1}\right] = A(\quad)^{P/A\,r,n}$	$P = \bar{A}\left[\dfrac{e^{rn}-1}{re^{rn}}\right] = \bar{A}(\quad)^{P/\bar{A}\,r,n}$
A	P	$A = P\left[\dfrac{i(1+i)^n}{(1+i)^n - 1}\right] = P(\quad)^{A/P\,i,n}$	$A = P\left[\dfrac{e^r - 1}{1-e^{-rn}}\right] = P(\quad)^{A/P\,r,n}$	$\bar{A} = P\left[\dfrac{re^{rn}}{e^{rn}-1}\right] = P(\quad)^{\bar{A}/P\,r,n}$
A	g	$A = g\left[\dfrac{1}{i} - \dfrac{n}{(1+i)^n - 1}\right] = g(\quad)^{A/G\,i,n}$	$A = g\left[\dfrac{1}{e^r - 1} - \dfrac{n}{e^{rn}-1}\right] = g(\quad)^{A/G\,r,n}$	

$$\begin{pmatrix} F/\bar{A}\,r, n \\ \ \end{pmatrix} = \begin{pmatrix} F/A\,r, n \\ \ \end{pmatrix}\begin{pmatrix} A/\bar{A}\,r \\ \ \end{pmatrix}$$

As an example of the use of the funds-flow conversion factor consider the following example. Find the present amount of $800 per year flowing uniformly for a period of 6 years at an interest rate of 6% compounded continuously. The required calculations are

$$P = \bar{A}\begin{pmatrix} P/\bar{A}\,r, n \\ \ \end{pmatrix} = \bar{A}\begin{pmatrix} P/A\,r, n \\ \ \end{pmatrix}\begin{pmatrix} A/\bar{A}\,r \\ \ \end{pmatrix}$$

$$= \$800\underset{P/A\,6,6}{(4.8891)}\underset{A/\bar{A}\,6}{(1.030608)} = \$4,031.$$

4.8. SUMMARY OF INTEREST FORMULAS

The three groups of interest formulas derived in this chapter are summarized in Table 4.7. Each group is based on assumptions about the nature of payments and the compounding of interest. In engineering economic analysis that group which most accurately represents the situation under study should be used.

PROBLEMS

1. For what period of time will $5,000 have to be invested to amount to $6,400 if it earns 8% simple interest per annum?

2. If $210 in interest is earned in three months on an investment of $12,000, what is the annual rate of simple interest?

3. What is the principal amount if the principal plus interest at the end of $1\frac{3}{4}$ years is $3,315 for a simple interest rate of 6% per annum?

4. A man lends $1,200 at 8% simple interest for four years. At the end of this time he invests the entire amount (principal plus interest) at 6% compounded annually for 10 years. How much will he have at the end of the 14-year period?

5. Compare the interest earned by $100 for 15 years at 8% simple interest with that earned by the same amount for 15 years at 8% compounded annually.

6. From the interest tables given in the text, determine the value of the following factors by interpolation:
   (a) The single-payment compound-amount factor for 12 periods at $5\frac{1}{2}$% interest.
   (b) The equal-payment-series sinking-fund factor for 44 periods at 6% interest.
   (c) The equal-payment-series present-worth factor for 12 periods at $3\frac{1}{4}$% interest.
   (d) The equal-payment-series compound-amount factor for 39 periods at 8% interest.

7. From the interest tables in the text determine the following value of the factors by interpolation:
   (a) The single-payment present-worth factor for 37 periods at $5\frac{1}{2}$% interest.
   (b) The equal-payment-series capital-recovery factor for 48 periods at $6\frac{1}{4}$% interest.

8. What will be the amount accumulated by each of these present investments?
    (a) $7,200 in 8 years at 9% compounded annually.
    (b) $675 in 11 years at 4% compounded semiannually.
    (c) $3,500 in 41 years at 7% compounded annually.
    (d) $11,000 in 7 years at 8% compounded quarterly.
9. What is the present value of these future payments?
    (a) $4,300 6 years from now at 9% compounded annually.
    (b) $1,700 12 years from now at 6% compounded monthly.
    (c) $6,200 15 years from now at 12% compounded annually.
    (d) $6,200 15 years from now at 12% compounded monthly.
10. What is the present value of the following series of prospective payments?
    (a) $3,500 a year for 12 years at 7% compounded annually.
    (b) $300 a year for 38 years at 9% compounded annually.
    (c) $1,000 a month for 4 years at 9% compounded monthly.
    (d) $5,000 every six months for 6 years at 10% compounded semiannually.
11. What is the accumulated value of each of the following series of payments?
    (a) $700 at the end of each year for 12 years at 5% compounded annually.
    (b) $1,400 at the end of each quarter for 10 years at 8% compounded quarterly.
    (c) $3,800 at the end of every year for 47 years at 9% compounded annually.
    (d) $500 at the end of each month for 2 years at 10% compounded monthly.
12. What equal series of payments must be paid into a sinking fund to accumulate the
    following amounts?
    (a) $7,000 in 5 years at 7% compounded annually when payments are annual.
    (b) $15,000 in 8 years at 12% compounded quarterly when payments are quarterly.
    (c) $4,000 in 11 years at 9% compounded semiannually when payments are semian-
    nual.
    (d) $17,000 in 15 years at 8% compounded quarterly when payments are monthly.
13. What equal series of payments are necessary to repay the following present amounts?
    (a) $3,000 in 5 years at 4% compounded annually with annual payments.
    (b) $16,000 in 8 years at 10% compounded semiannually with semiannual payments.
    (c) $37,000 in 5 years at 9% compounded monthly with monthly payments.
    (d) $8,000 in 3 years at 12% compounded quarterly with quarterly payments.
14. How many years will it take for an investment to double itself if interest is compounded
    annually for an interest rate of 5%? 10%?
15. At what rate of interest compounded annually will an investment triple itself in 8
    years?
16. What rate of interest compounded annually is involved if:
    (a) An investment of $10,000 made now will result in a recept of $15,010, 6 years
    from now?
    (b) An investment of $1,000 made 21 years ago is increased in value to $7,400.
17. What nominal interest rate is paid if:
    (a) Payments of $7,000 per year for 4 years will repay an original loan of $22,000?
    (b) Twenty-four monthly deposits of $100 will result in $2,800 at the end of 2 years.
18. How many years will be required for:
    (a) An investment of $3,000 to increase to $5,000 if interest is 8% compounded
    annually.

(b) An investment of $1,000 to increase to $10,641 if interest is 12% compounded quarterly.

19. For an interest rate of 8% compounded annually, find:

(a) How much can be loaned now if $2,000 will be repaid at the end of 7 years?

(b) How much will be required 4 years hence to repay a $1,500 loan made now?

20. For interest at 6% compounded semiannually, find:

(a) What payment can be made now to prevent an expense of $450 every 6 months for the next 7 years?

(b) What semiannual deposit into a fund is required to total $10,000 in 8 years?

21. Graphically illustrate the function of each of the six interest factors for annual compounding.

22. Graphically illustrate the function of the uniform gradient series factor for annual compounding; for continuous compounding.

23. Develop a formula for finding the accumulated amount F at the end of n interest periods which will result from a series of beginning-of-period payments each equal to B, if the latter are placed in a sinking fund for which the interest rate per period is i, compounded each period.

24. How would you determine a desired equal-payment-series sinking-fund factor if you only had a table of:

(a) Single-payment compound-amount factors?

(b) Single-payment present-worth factors?

(c) Equal-payment-series compound-amount factors?

(d) Equal-payment-series capital-recovery factors?

25. How would you determine a desired equal-payment-series capital-recovery factor if you only had a table of:

(a) Single-payment present-worth factors?

(b) Equal-payment-series present-worth factors?

(c) Equal-payment-series sinking-fund factors?

(d) Equal-payment-series compound-amount factors?

(e) Single-payment compound-amount factors?

26. Rewrite the formula given for the single-payment compound-amount factor to apply to the compounding of interest at the end of each period, where p represents the number of compounding periods per year, y represents the number of years, and r represents the nominal annual rate of interest. Use P as the present sum and F as the compound amount and express F in terms of P, r, p, and y.

27. What equal annual amount must be deposited for 10 years in order to provide withdrawals of $100 at the end of the second year, $200 at the end of the third year, $300 at the end of the fourth year, and so on, up to $900 at the end of the tenth year? The interest rate is 9% compounded annually.

28. What is the equal payment series that is equivalent to a payment series of $10,000 at the end of the first year decreasing by $200 each year over 15 years. Interest is 7% compounded annually.

29. Find the interest factor ($\overset{P/G\,i,\,n}{\quad}$) that will convert a gradient series as defined in Table 4.5 to its equivalent value at the present.

30. What effective annual interest rate corresponds to the following:
    (a) Nominal interest rate of 12% compounded semiannually.
    (b) Nominal interest rate of 12% compounded monthly.
    (c) Nominal interest rate of 12% compounded quarterly.
    (d) Nominal interest rate of 12% compounded weekly.

31. The Square Deal Loan Company offers money at $\frac{1}{2}$% interest per week compounded weekly. What is the effective annual interest rate? What is the nominal interest rate?

32. An annual effective interest rate of 10% is desired:
    (a) What nominal rate should be asked if compounding is to be semiannually?
    (b) What nominal rate should be asked if compounding is to be quarterly?

33. How much more desirable is 15% compounded monthly than 15% compounded yearly?

34. What is the effective interest rate if a nominal rate of 7% is compounded continuously? If an effective interest rate of 7% is desired, what must the nominal rate be if compounding is continuous?

35. What is the present worth of a uniform series of year-end payments of $350 each for 8 years if the interest rate is 6% compounded continuously?

36. What is the present worth of the following prospective payments?
    (a) $2,200 in 35 years at an interest rate of 8% compounded continuously.
    (b) $1,200 in 5 years at an interest rate of 9% compounded weekly.

37. How many years will it take an investment to triple itself if the interest rate is 7% compounded annually; compounded continuously?

38. An interest rate of 9% compounded continuously is desired on an investment of $14,500. How many years will be required to recover the capital with the desired return if $3,500 is received each year?

39. What will be the required quarterly payment to repay a loan of $5,000 in 3 years if the interest rate is 6% compounded continuously?

40. What equal semiannual payment must be deposited into a sinking fund to accumulate $25,000 in 18 years at 8% interest compounded continuously?

41. What is the accumulated value of each of the following series of payments?
    (a) $500 at the end of each month for 9 years at 6% interest compounded continuously.
    (b) $400 at the end of each quarter for 37 years at 8% interest compounded continuously.

42. Find the annual uniform payment series which would be equivalent to the following increasing series of payments if the interest rate is 8% compounded annually; compounded continuously.

$400 at the end of the first year
$450 at the end of the second year
$500 at the end of the third year
$550 at the end of the fourth year
$600 at the end of the fifth year
$650 at the end of the sixth year
$700 at the end of the seventh year

43. Find the equal quarterly series that would be exchanged for the following decreasing series if the interest rate is 8% compounded quarterly; compounded annually.

$3,000 at the end of the first quarter
$2,750 at the end of the second quarter
$2,500 at the end of the third quarter
$2,250 at the end of the fourth quarter
$2,000 at the end of the fifth quarter
$1,750 at the end of the sixth quarter
$1,500 at the end of the seventh quarter
$1,250 at the end of the eighth quarter

44. What is the present value of the following continuous funds-flow?
   (a) $5,000 per year for 7 years at 6% compounded continously.
   (b) $800 per month for 10 years at 7% compounded continuously.
   (c) $1,500 per quarter for 8.3 years at 8% compounded continuously.
   (d) $900 per year for 13.7 years at 5% compounded continuously.
45. What amount will be accumulated by each of these continuous funds flow?
   (a) $3,000 per year in 6 years at 9% compounded continuously.
   (b) $1,700 per month in 11.9 years at 7% compounded continuously.
46. What annual interest rate compounded continuously will be earned on an investment of $13,000 that provides a continuous flow of funds at the rate of $250
   monthly for 5.6 years?
47. For how many years must an investment of $32,000 provide a continuous flow of
   funds at the rate of $10,000 per year so that an annual interest rate of 15% com-
   pounded continuously is earned?

5

Calculations of Equivalence
Involving Interest

Many calculations in engineering economy require that prospective receipts and disbursements of two or more alternative proposals be placed on an equivalent basis for comparison. This may be accomplished by the proper use of the interest formulas developed in the previous chapter. Also, it is essential that the economic meaning of equivalence be understood. This chapter will illustrate the concept of equivalence and will present computational methods required when interest formulas are used in engineering economy studies.

5.1. THE MEANING OF EQUIVALENCE

If two or more situations are to be compared, their characteristics must be placed on an equivalent basis. Which is worth more, 4 ounces of Product A or 1,800 grains of Product A? In order to answer this question, it is necessary to place the two amounts on an equivalent basis by use of the proper conversion factor. After conversion of ounces to grains, the question becomes: Which is worth more, 1,750 grains of Product A or 1,800 grains of Product A? The answer is now obvious.

Two things are said to be equivalent when they have the same effect. For instance, the torques produced by applying forces of 100 pounds and 200 pounds 2 feet and 1 foot, respectively, from the fulcrum of a lever are equivalent since each produces a torque of 200 foot-pounds.

Equivalence of Value in Exchange. In engineering economy the meaning of equivalence pertaining to value in exchange is of primary importance; for example, a present amount of $300.00 is equivalent to $478.20 if the amounts are separated by 8 years and if the interest rate is 6% per annum.

This is so because a person who considers 6% to be a satisfactory rate of interest would be indifferent to receiving $300 now or $478.20, 8 years from now.

This equivalence may be illustrated by use of the single payment formulas for annual compounding interest. A sum of $300 in the present is equivalent to

$$\$300(1 + 0.06)^8 = \$300(\overset{F/P\ 6,8}{1.594}) = \$478.20$$

8 years from now. Similarly, $478.20 to be received 8 years from now is equivalent to

$$\$478.20\left[\frac{1}{(1 + 0.06)^8}\right] = \$478.20(\overset{P/F\ 6,8}{0.6274}) = \$300.00$$

at the present.

Three factors are involved in the equivalence of sums of money. These are (1) the amounts of the sums, (2) the time of occurrence of the sums, and (3) the interest rate. The interest formulas developed consider time and the interest rate. Thus, they constitute a convenient way of taking the time value of money into consideration when calculating the equivalence of monetary amounts occurring at different points in time.

Equivalence is Not Directly Apparent. The relative value of several alternatives is usually not apparent from a simple statement of their future receipts and disbursements until these amounts have been placed on an equivalent basis. Consider the following example: An engineer sold his patent to a corporation and is offered a choice of $12,500 now or $1,650 per year for the next 10 years, the estimated beneficial life of the patent to the corporation. The engineer is paying 6% interest on his home mortgage and will use this rate in his evaluation. The patterns of receipts are shown in Table 5.1.

Table 5.1. PATTERN OF RECEIPTS FOR TWO ALTERNATIVES

End of Year Number	Receipts Alternative A	Receipts Alternative B
0	$12,500	0
1	0	$1,650
2	0	1,650
3	0	1,650
4	0	1,650
5	0	1,650
6	0	1,650
7	0	1,650
8	0	1,650
9	0	1,650
10	0	1,650
Total Receipts	$12,500	$16,500

Since money has a time value it is not apparent from a cursory examination of the receipts of the two alternatives which is economically the most desirable. For instance it is incorrect to say alternative B is more desirable than alternative A because the sum of receipts from those alternatives are $16,500 and $12,500, respectively. Such a statement would be correct only if the interest rate is considered to be zero.

The equivalence values for these two alternatives for an interest rate of 6% must be found by the use of interest formulas. One way to determine an equivalent value for alternative B is to calculate an amount at the present which is equivalent to 10 receipts of $1,650 each as

$$P = \$1,650(\overset{P/A\ 6,10}{7.3601}) = \$12,144.$$

This amount is equivalent to 10 future payments of $1,650 each and is directly comparable with $12,500. This is because both figures represent amounts of money at the same point in time, the present. Thus, the engineer can now see that on an equivalent basis the $12,500 lump sum is most desirable.

It should be noted that the $12,144 is only an equivalent amount determined from an anticipated series of cash receipts. An actual receipt of $12,144 would not occur, even if this alternative had been chosen. The actual receipts would be $1,650 per year for 10 years.

5.2. EQUIVALENCE CALCULATIONS REQUIRING A SINGLE FACTOR

The interest formulas derived in Chapter 4 express relationships that exist between the several elements making up the formulas. These formulas exhibit relationships between $P, A, F, i,$ and n for annual compounding and between $P, A, F, r,$ and n for continuous compounding. For the case where continuous funds flow is assumed the formulas exhibit the relationships between $P, \bar{A}, F, r,$ and n.

The paragraphs which follow will illustrate methods for calculating equivalence where these interest formulas are involved. In the examples, the quantities $P, A,$ and F will be set up to conform to the pattern applicable to the particular factor used, as was illustrated in Table 4.6, Chapter 4.

Single-Payment Compound-Amount Factor Calculations. The single-payment compound-amount factors yield a sum F, at a given time in the future, that is equivalent to a principal amount P for a specified interest rate i compounded annually, or r compounded continuously. For example, the solution for finding the compound amount on January 1, 1979 that is equivalent to a principal sum of $200 on January 1, 1971 for an interest rate of 5% compounded annually is

$$n = 1979 - 1971 = 8$$

$$F = P(\overset{F/P\,i,\,n}{})$$

$$= \$200(\overset{F/P\,5,\,8}{1.477}) = \$295.40.$$

If interest is compounded continuously, the solution is

$$F = P(\overset{F/P\,r,\,n}{})$$

$$= \$200(\overset{F/P\,5,\,8}{1.492}) = \$298.40.$$

If the principal P, the compound amount F, and the number of years n are known, the interest rate i, may be determined by interpolation in the interest tables. For example, if $P = \$300$, $F = \$525$, and $n = 9$, the solution for i is

$$F = P(\overset{F/P\,i,\,n}{})$$

$$\$525 = \$300(\overset{F/P\,i,\,9}{})$$

$$(\overset{F/P\,i,\,9}{1.750}) = \frac{\$525}{\$300}.$$

A search of the interest tables for annual compounding interest reveals that 1.750 falls between the single payment compound-amount factors in the 6% and 7% tables for $n = 9$. The value from the 6% table is 1.689 and the value from the 7% table is 1.838. By linear proportion

$$i = 6 + (1)\frac{1.689 - 1.750}{1.689 - 1.838}$$

$$= 6 + \frac{0.061}{0.149} = 6.41\%.$$

The linear interpolation used for i is illustrated by Figure 5.1.

Soultion for i might have been done without the use of tables as follows:

$$F = P(1 + i)^n$$

$$\$525 = \$300(1 + i)^9$$

$$(1 + i)^9 = \frac{525}{300}$$

$$i = \sqrt[9]{1.750} - 1$$

$$i = 1.0641 - 1 = 0.0641, \text{ or } 6.41\%.$$

Solutions accomplished without the use of tables are usually more time-consuming than solutions by interpolation. Since engineering economy studies usually require estimates of the future, the error introduced by

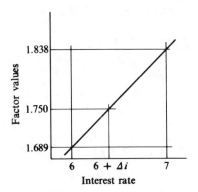

FIGURE 5.1.
INTERPOLATION FOR *i*.

interpolation will rarely be of significance. Therefore it is recommended that the interpolation method be used unless tables of factors are not available or the calculations are being done by computer where direct calculation is faster than a table look-up operation.

If the principal sum, P, its compound amount, F, and the interest rate, i, are known, the number of years n may be determined by interpolation of the interest tables. For example, if $P = \$400$, $F = \$704.40$, and $i = 0.07$, the solution for n is

$$F = P(\overset{F/P\,i,\,n}{\qquad})$$

$$\$704.40 = \$400(\overset{F/P\,7,\,n}{\qquad})$$

$$(\overset{F/P\,7,\,n}{1.761}) = \frac{\$704.40}{\$400}.$$

A search of the 7% interest table reveals that 1.761 falls between the single-payment compound-amount factors for $n = 8$ and $n = 9$. For $n = 8$, the factor is 1.718 and for $n = 9$, the factor is 1.838. By linear proportion

$$n = 8 + (1)\frac{1.718 - 1.761}{1.718 - 1.838}$$

$$= 8 + \frac{0.043}{0.120} = 8.36 \text{ years.}$$

The linear interpolation used for n is illustrated by Figure 5.2.

Solution for n might have been done without the use of tables as follows:

$$F = P(1 + i)^n$$

$$\$704.40 = \$400(1 + 0.07)^n$$

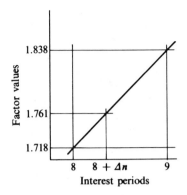

FIGURE 5.2.
INTERPOLATION FOR n.

$$(1.07)^n = \frac{\$704.40}{\$400} = 1.761$$

$$n = 8.37 \text{ years.}$$

The interpretation of n in this problem is that 8.37 years are required for $400 to earn enough interest so that the total amount available after 8.37 years equals $740.40.

Single-Payment Present-Worth Factor Calculations. The single-payment present-worth factors yield a principal sum P, at a time regarded as being the present, which is equivalent to a future sum F. For example, the solution for finding the present worth of a sum equal to $400 received 12 years hence, for an interest rate of 6% compounded annually, is

$$P = F(\overset{P/F\,i,\,n}{})$$
$$= \$400(\overset{P/F\,6,\,12}{0.4970}) = \$198.80.$$

If the future amount F, its present worth P, and the number of years n are known, the interest rate i may be determined by interpolation of the interest tables. For example, if $F = \$400$, $P = \$294.80$, $n = 6$ years, and if it is desired to find r, the nominal interest rate, when compounding occurs semiannually, then the approach described in Section 4.5 may be used. This method requires that i, the interest rate per compounding period, be used in the interest formula where $i = r/2$ and the number of interest periods is 12.

$$P = F(\overset{P/F\,i,\,12}{\quad})$$

$$\$294.80 = \$400(\overset{P/F\,i,\,12}{\quad})$$

$$\overset{P/F\,i,\,12}{(0.7370)} = \frac{\$294.80}{\$400}.$$

A search of the interest tables reveals that 0.7370 falls between the single-payment present-worth factors in the 2% and the 3% tables for $n = 12$. The value from the 2% table is 0.7885 and the value from the 3% table is 0.7014. By linear proportion

$$i = 2 + (1)\frac{0.7885 - 0.7370}{0.7885 - 0.7014}$$

$$= 2 + \frac{0.0515}{0.0871} = 2.59\% \text{ per period}$$

$$r = 2 \times 2.59\%$$

$$= 5.18\% \text{ per year compounded semiannually.}$$

If the interest rate to be found for the preceding problem is compounded continuously rather than semiannually, the table for continuous compounding interest is used as follows:

$$P = F(\overset{P/F\,r,\,6}{\quad})$$

$$\$294.80 = \$400(\overset{P/F\,r,\,6}{\quad})$$

$$\overset{P/F\,r,\,6}{(0.7370)} = \frac{\$294.80}{\$400}.$$

A search of the continuous compounding interest tables for $n = 6$ reveals that 0.7370 falls between the single-payment present-worth factors for $r = 5\%$ and $r = 6\%$. The value of the factor for $r = 5\%$ is 0.7408 and the value for the factor for $r = 6\%$ is 0.6977. By the linear interpolation

$$r = 5 + (1)\left[\frac{0.7408 - 0.7370}{0.7408 - 0.6977}\right]$$

$$= 5 + \left[\frac{0.0038}{0.0431}\right] = 5.09\%.$$

Equal-Payment-Series Compound-Amount Factor Calculations. The equal-payment-series compound-amount factors yield a sum F at a given time in the future, which is equivalent to a series of payments A, occurring at the end of successive years such that the last A concurs with F. The solution for finding the equivalent amount, seven years from now, of a series of seven $40 year-end payments whose final payment occurs simultaneously with the compound amount being determined, for an interest rate of 6% is

$$F = A(\overset{F/A\,i,\,n}{\quad})$$

$$= \$40(\overset{F/A\,6,\,7}{8.394}) = \$335.76.$$

If the compound amount F, the annual payments A, and the number of years n are known, the interest rate i may be determined by interpolation of the interest tables. For example, if $F = \$441.10$, $A = \$100$, and $n = 4$ the solution for i is

$$F = A(\overset{F/A\,i,\,n}{\quad})$$

$$\$441.10 = \$100(\overset{F/A\,i,\,4}{\quad})$$

$$(\overset{F/A\,i,\,4}{4.411}) = \frac{\$441.10}{\$100}.$$

This value falls between the equal-payment-series compound-amount factors in the 6% and 7% tables for $n = 4$. By linear interpolation

$$i = 6 + (1)\frac{4.375 - 4.411}{4.375 - 4.440}$$

$$= 6 + \frac{0.036}{0.065} = 6.55\%.$$

Suppose for an equal annual cash flow of $100 each year it is necessary to calculate how long it would take to accumulate $2,000 if the interest rate is 8% compounded continuously.

$$F = A(\overset{F/A\,r,\,n}{\quad})$$

$$\$2,000 = \$100(\overset{F/A\,8,\,n}{\quad})$$

$$(\overset{F/A\,8,\,n}{20.00}) = \frac{\$2,000}{\$100}$$

$$n = 12 + (1)\frac{19.351 - 20.000}{19.351 - 21.963}$$

$$n = 12 + \frac{0.649}{2.612} = 12.25 \text{ years}$$

This problem can be also be solved directly by using the interest formula. This direct approach requires solution for n in terms of F, A, and r as follows:

$$F = A\left[\frac{e^{rn} - 1}{e^r - 1}\right]$$

$$e^{rn} - 1 = \frac{F}{A}(e^r - 1)$$

$$rn = \log_e\left[\frac{F}{A}(e^r - 1) + 1\right]$$

$$n = \frac{\log_e\left[\frac{F}{A}(e^r - 1) + 1\right]}{r}$$

For $F = \$2,000$, $A = \$100$, $r = 8\%$

$$n = \frac{\log_e\left[\frac{\$2,000}{100}(e^{0.08} - 1) + 1\right]}{0.08} = \frac{\log_e[20(0.0834) + 1]}{0.08}$$

$$= \frac{\log_e[2.668]}{0.08} = \frac{0.981}{0.08} = 12.26 \text{ years.}$$

Where tables of interest factors are available it is usually less time consuming to calculate the value of the factor and then interpolate from the tables rather than using the approach just shown.

Equal-Payment-Series Sinking-Fund Factor Calculations. The equal-payment-series sinking-fund factors are used to determine the amount A of each payment of a series of payments, occurring at the end of successive years, which are equivalent to a future sum F. The solution for finding the amount of annual sinking-fund deposits A for the period June 1, 1971 to June 1, 1978, that are equivalent to a sinking fund F equal to $\$400$ on June 1, 1978 at 5% interest is

$$A = F(\overset{A/F\,i,\,n}{})$$

$$= \$400(\overset{A/F\,5,\,7}{0.1228}) = \$49.13.$$

Solution for i and n when F, A, and n or i are known may be done by interpolation in the interest tables as was illustrated for the single payment compound-amount factor.

If in the preceding problem the interest rate is 5% compounded quarterly, the equal-payment-series sinking fund factor can still be used to find A by using the effective interest rate for a year

$$i = \left(1 + \frac{r}{c}\right)^c - 1 = \left(1 + \frac{0.05}{4}\right)^4 - 1 = (1.0125)^4 - 1$$

$$= 1.050944 - 1 = 5.1\%$$

$$A = \$400(\overset{A/F\,5.1,\,7}{})$$

By interpolation

$$(\overset{A/F\,5.1,\,7}{}) = 0.12282 - (0.12282 - 0.11914)\left(\frac{5.0 - 5.1}{5.0 - 6.0}\right) = 0.12245$$

Thus,

$$A = \$400(\overset{A/F\,5.1,\,7}{0.12245}) = \$48.98.$$

This value of A is less than the value of \$49.13 previously calculated for annual compounding. This is because less deposits are required due to the increase in interest earned by more frequent compounding.

Equal-Payment-Series Present-Worth Factor Calculations. The equal-payment-series present-worth factors are used to find the present worth P, of an equal-payment-series A, occurring at the end of successive years following the time taken to be the present. For example, the solution for finding the present worth P which is equivalent to a series of five \$60 year-end payments beginning at the end of the first interest period after the present for an interest rate of 10%, is

$$P = A(\overset{P/A\ i,\ n}{})$$
$$= \$60(\overset{P/A\ 10,\ 5}{3.7908}) = \$227.45.$$

If payments of \$100 occur semiannually at the end of each six-month period for 3 years and the nominal interest rate is 12% compounded semiannually, the present worth P is determined as follows:

$$i = \frac{12\%}{2\ \text{periods}} = 6\%\ \text{per semiannual period}$$

$$n = (3\ \text{years})(2\ \text{periods per year}) = 6\ \text{periods}$$

$$P = A(\overset{P/A\ i,\ n}{})$$
$$= \$100(\overset{P/A\ 6,\ 6}{4.9173}) = \$491.73.$$

If the interest rate for the previous problem had been 12% compounded monthly then it is necessary to find the effective interest rate for a six-month period so that the equal-payment present-worth factor can be used.

The interest rate per month is

$$\frac{12\%}{12\ \text{periods}} = 1\%$$

The effective interest rate for a six-month period

$$= (1 + .01)^6 - 1 = \overset{.}{1}.062 - 1 = 6.2\%.$$

Therefore,

$$P = \$100(\overset{P/A\ 6.2,\ 6}{4.8872}) = \$488.72.$$

Equal-Payment-Series Capital-Recovery Factor Calculations. The equal-payment-series capital-recovery factors are used to determine the amount A of each payment of a series of payments occurring at the end of successive years which is equivalent to a present sum P. For example, the solution for finding the amount A of annual year-end payments for a five-year period which is equivalent to an amount P of \$300 at the present for an interest rate of 6% is

$$A = P(\overset{A/P\ i,\ n}{})$$
$$= \$300(\underset{A/P\ 6,\ 5}{0.2374}) = \$71.22.$$

For the situation where funds are flowing continuously and the interest rate is compounded continuously it is necessary to use the funds-flow factors developed in Chapter 4. To find the funds-flow equivalent of a present sum of \$600 when the time period is 12 years and the interest rate is 10%, calculate

$$\bar{A} = P(\overset{\bar{A}/P\ r,\ n}{}) = P(\overset{A/P\ r,\ n}{}) \div (\overset{A/\bar{A}\ 10}{})$$
$$\bar{A} = \$600(\overset{\bar{A}/P\ 10,\ 12}{}) = \$600(\underset{A/P\ 10,\ 12}{0.15050}) \div (\underset{A/\bar{A}\ 10}{1.0517}) = \$85.86\ \text{per year.}$$

Direct calculation gives

$$\bar{A} = P\left[\frac{re^{rn}}{e^{rn} - 1}\right]$$
$$\bar{A} = \$600\left[\frac{(0.10)e^{(0.10)(12)}}{e^{(0.10)(12)} - 1}\right] = \$600\left[\frac{0.332}{2.32}\right] = \$85.86\ \text{per year.}$$

5.3. EQUIVALENCE CALCULATIONS REQUIRING SEVERAL FACTORS

Where a number of calculations of equivalence involving several interest factors are to be made, some difficulty may be experienced in laying out a plan of attack. Also, until considerable experience has been gained with this type of calculation it may be difficult to keep track of the lapse of time.

For complex problems, the speed and accuracy can usually be improved by a schematic representation. For example, suppose that it is desired to determine what amount at the present is equivalent to the following cash flow for an interest rate of 5%: \$300 end of year 6; \$60 end of years 9, 10, 11, and 12; \$210 end of year 13; \$80 end of years 15, 16 and 17. These payments may be represented schematically as illustrated in Figure 5.3.

The plan of attack is to determine the amount at the beginning of year 1 equivalent to the single payments and groups of payments that make up the cash flow described above. By converting the various payments to their equivalents at the same point in time it then is possible to determine the total equivalent amount by direct addition. Remember that it is the dollars at

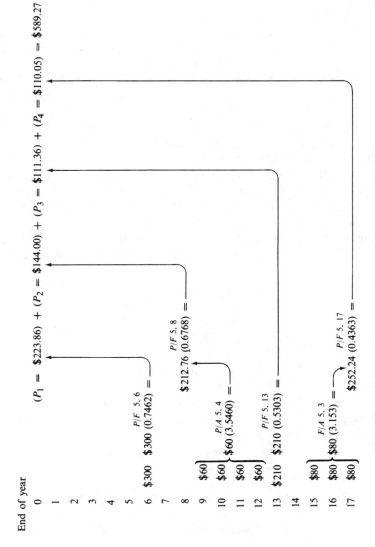

FIGURE 5.3. SCHEMATIC ILLUSTRATION OF EQUIVALENCE.

different points in time that cannot be directly added when the interest rate is positive.

To use the interest formulas properly recall that P occurs at the beginning of an interest period and that F and A payments occur at the end of interest periods. For instance the group of four $60 payments are converted to a single equivalent amount of $212.76 at the end of year 8 which is one interest period before the first $60 payment. This is in accordance with the convention of the conversion formula which requires that the amount P occur one interest period prior to the first A payment. In Figure 5.3, the $252.24 as of the end of year 17 represents the equivalent future worth of the three $80 payments. Note that the $252.24 amount concurs with the last $80. Again this is in accordance with the convention adopted for the derivation of the factor that finds F when given a series of payments A.

The sequence of calculations in the solution of this problem is clearly indicated in the diagram. The position of the arrowhead following each multiplication represents the position of the result with respect to time. The intermediate quantities $212.76 and $252.24 need not have been found. Much time may be saved if all calculations to be made in solving a problem are indicated prior to looking up factor values from the tables and making calculations. In the above example this might have been done as follows:

$$P_1 = \$300(\overset{P/F\,5,\,6}{}) \quad\quad =$$

$$P_2 = \$\,60(\overset{P/A\,5,\,4}{})(\overset{P/F\,5,\,8}{}) \;=$$

$$P_3 = \$210(\overset{P/F\,5,\,13}{}) \quad\quad =$$

$$P_4 = \$\,80(\overset{F/A\,5,\,3}{})(\overset{P/F\,5,\,17}{}) =$$

Next, all factor values are found from the tables and inserted in the parentheses. Calculations are then made as follows to obtain the results printed in **bold face type:**

$$P_1 = \$300(\overset{P/F\,5,\,6}{0.7462}) \quad\quad\quad = \textbf{\$223.86}$$

$$P_2 = \$\,60(\overset{P/A\,5,\,4}{3.5456})(\overset{P/F\,5,\,8}{0.6768}) \;= \textbf{144.00}$$

$$P_3 = \$210(\overset{P/F\,5,\,13}{0.5303}) \quad\quad\;\; = \textbf{111.36}$$

$$P_4 = \$\,80(\overset{F/A\,5,\,3}{3.153})(\overset{P/F\,5,\,17}{0.4363}) = \textbf{110.05}$$

$$P = P_1 + P_2 + P_3 + P_4 \quad = \textbf{\$589.27}$$

If interest had been 5% compounded continuously instead of 5% compounded annually, values would have been taken from the table for continuous compounding interest. All other calculations would have remained the same.

The amount of calculation required in a given situation can be kept to a minimum by the proper selection of interest factors. Consider the example shown in Figure 5.4. By recognizing that the cash flow beginning at the end of year 4 is a gradient series with a gradient of $20 per year, it is possible to convert that series into an equivalent equal annual series by the following calculation

$$A = A_1 + g(\overset{A/G\,i,\,n}{})$$

$$A = \$0 + \$20(\overset{A/G\,10,\,8}{3.0045}) = \$60.09.$$

Note that the equal payments of $60.09 begin at the end of year 3 although the gradient series begins at the end of year 4. This situation arises because of the cash flow convention in the derivation of the gradient series factor.

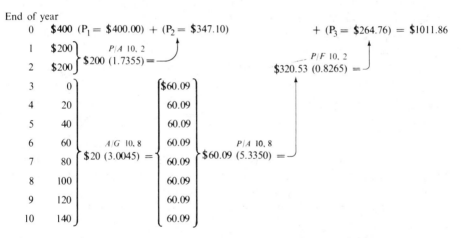

FIGURE 5.4. CALCULATION OF EQUIVALENCE USING THE GRADIENT FACTOR.

The equivalence at the beginning of year 1 of the cash flow shown in Figure 5.4 is calculated as follows:

$$P_1 \qquad\qquad\qquad\qquad = \$400.00$$

$$P_2 = \$200(\overset{P/A\,10,\,2}{1.7355}) \qquad\qquad = \$347.10$$

$$P_3 = \$20(\overset{A/G\,10,\,8}{3.0045})(\overset{P/A\,10,\,8}{5.3349})(\overset{P/F\,10,\,2}{0.8265}) = \$264.76$$

$$P = P_1 + P_2 + P_3 \qquad\qquad = \$1011.86$$

In the previous two examples the equivalences of the cash flows were found at the beginning of year 1. In many cases the equivalence amount that is desired is a single payment at some point in time other than the present or it may be an equal payment series over some time span. The calculations required to find various cash flow equivalents are quite similar to those shown in Figures 5.3 and 5.4. For instance, to find 10 equal annual payments that are equivalent to the cash flow in Figure 5.4 it is only necessary to convert the single payment equivalent of $1011.86 at the beginning of year 1 into the desired equivalent by

$$A = P(\overset{A/P\, i,\, n}{})$$
$$= \$1011.86(\overset{A/P\, 10,\, 10}{0.1628}) = \$164.73.$$

Because there are a number of different calculations that will lead to the same solution for this type of problem, an effort should be made to minimize the amount of computation required. For example in this problem each cash payment could have been converted to its equivalent at the beginning of year 1 by using the appropriate single-payment present-worth factor. This approach would have required 9 factors while only 6 factors were required in the calculation shown above. The ability to reduce the amount of calculation for a particular cash flow will develop from experience.

5.4. CALCULATION OF THE INTEREST RATE ON BONDS

A bond is a financial instrument setting forth the conditions under which money is borrowed. It consists, usually, of a pledge of a borrower of funds to pay a stated amount or percent of interest on the par or face value at stated intervals and to repay the par value at a stated time. Bonds are commonly written with par values in multiples of $100 or $1000. A typical $1000 bond may embrace a promise to pay its holder $60, for example, one year after purchase and each succeeding year until the principal amount or par value of $1000 is repaid on a designated date. Such a bond would be referred to as a 6% bond with interest payable annually. Bonds may also provide for interest payments to be made semiannually or quarterly. Since pledges to pay as they are embodied in bonds have value, bonds are bought and sold. The market price of a bond may range above or below its par or face value depending on prevailing market conditions.

Suppose that an individual is considering the purchase at $950 of a $1000, 6% bond with interest payable annually and the face value due at the end of 7 years. What will be the equivalent rate of interest earned on such an arrangement if all payments stipulated on the bond are met? Table 5.2 shows the anticipated disbursements and receipts.

Table 5.2. BOND RECEIPTS AND DISBURSEMENTS.

End of Year	Disbursements	Receipts
0	$950	—
1	—	$60
2	—	60
3	—	60
4	—	60
5	—	60
6	—	60
7	—	1060

The question may be answered by finding the interest rate that makes the expenditure of $950 at the present equivalent to the present worth of the receipts. This is

$$\$950 = \$60(\overset{P/A\,i,\,7}{}) + \$1,000(\overset{P/F\,i,\,7}{}).$$

Solve for i by trial and error. The present worth of receipts at 6%

$$= \$60(\overset{P/A\,6,\,7}{5.5824}) + \$1,000(\overset{P/F\,6,\,7}{0.6651})$$
$$= \$1,000.$$

The present worth of receipts at 7%

$$= \$60(\overset{P/A\,7,\,7}{5.3893}) + \$1,000(\overset{P/F\,7,\,7}{0.6228})$$
$$= \$946.16.$$

The value of i that makes the present worth of the receipts equal to $950 lies between 6% and 7%. By interpolation

$$i = 6\% + 1\%\left[\frac{\$1000 - \$950}{\$1000 - \$946.16}\right]$$
$$= 6\% + 0.93\% = 6.93\%.$$

PROBLEMS

1. An individual is purchasing a $3,000 automobile which is to be paid for in 24 monthly installments of $145. What nominal interest rate is being paid for this financing arrangement?

2. A no-load mutual stock fund has grown at a rate of 20% compounded annually since its beginning. If it is anticipated that it will continue to grow at this rate, how much must be invested every year so that $40,000 will be accumulated at the end of 8 years?

3. A series of equal quarterly payments of $550 extends over a period of 8 years. What is the amount at the present that is equivalent to this series at 7% interest compounded annually; compounded quarterly; compounded continuously?

4. A continuous flow of funds of $1,600 per year is deposited into a sinking fund. What amount will be accumulated at the end of 5 years if the interest rate is 12% compounded annually; compounded monthly; compounded continuously?

5. A series of equal quarterly payments of $750 for 25 years is equivalent to what present amount at an interest rate of 8% compounded quarterly; compounded continuously?

6. (a) An interest rate of 10% compounded annually is desired on an investment of $8,600. How many years will be required to recover the capital with the desired interest if $1,750 is received each year?

   (b) A building is priced at $37,000. If a down payment of $14,000 is made and a payment of $1,500 every six months thereafter is required, how many years will be necessary to pay for the building? Interest is charged at the rate of 10% compounded semiannually.

7. What single amount at the end of the fifth year is equivalent to a uniform annual series of $1,000 per year for 12 years? The interest rate is 8% compounded annually.

8. An individual's salary is now $10,000 per year and he anticipates retiring in 30 more years. If his salary is increased by $600 each year and he deposits 10% of his yearly salary into a fund that earns 7% interest compounded annually, what will be the amount accumulated at the time of his retirement?

9. A series of 10 annual payments of $900 is equivalent to three equal payments at the end of years 12, 15, and 20 at 9% interest compounded annually. What is the amount of these three payments?

10. A series of payments—$8,000, first year; $7,500, second year; $7,000, third year; $6,500, fourth year; and $6,000, fifth year—is equivalent to what present amount at 10% interest compounded annually; compounded continuously? Solve this problem using the gradient factors, and then solve it using only the single payment present-worth factors.

11. As usually quoted, the prepaid premium of insurance policies covering loss by fire and storm for a 3-year period is 2.5 times the premium for one year of coverage. What rate of interest does a purchaser receive on the additional present investment if he purchases a 3-year policy now rather than three 1-year policies at the beginning of each of the years?

12. A young couple have decided to make advance plans for financing their 3-year-old son's college education. Money can be deposited at 7% compounded annually. What annual deposit on each birthday from the 4th to the 17th inclusive must be made to provide $3,000 on each birthday from the 18th to the 21st inclusive?

13. A petroleum engineer estimates that the present production of 300,000 barrels of oil per year from a group of 10 wells will decrease at the rate of 15,000 barrels per year for the next 19 years. Oil is estimated to be worth $3 per barrel for the next 13 years and $3.50 per barrel thereafter. If the interest rate is 12% compounded annually, what is the equivalent present amount of the prospective future receipts from the wells?

14. An engineering firm is seeking a loan of $170,000 to finance production of a newly patented product line. Due to a good reception of the product at its introductory showing, the bank has agreed to loan the firm an amount equal to 90% of the present worth of firm orders received for delivery during the next 5 years. The orders are as follows:

Year 1	20,000
Year 2	16,000
Year 3	12,000
Year 4	8,000
Year 5	4,000

If the product will sell for $4 each, will the present worth of the orders received justify the loan required? Interest is 10% compounded annually.

15. A city that was planning an addition to its water supply and distribution system contracted to supply water to a large industrial user for 10 years under the following conditions: The first five years of service were to be paid for in advance, and the last five years of service were to be paid for at a rate of $15,000 a year payable at the beginning of each year.

Two years after the system is in operation the city finds itself in need of funds and desires that the company pay off the entire contract so that the city can avoid a bond issue.

(a) If the city uses 5% interest compounded annually in calculating a fair receipt on the contract, what amount can they expect?

(b) If the company uses 10% interest compounded annually, how much is the difference between what the company would consider a fair value for the contract and what the city considers to be a fair value to pay for the contract?

16. A man has borrowed $10,000 which he will repay in 60 equal monthly installments. After his twenty-fifth payment he desires to pay the remainder of the loan in a single payment. At 24% interest compounded monthly what is the amount of the payment?

17. A manufacturing company purchased electrical services to be paid for $50,000 now and $10,000 per year beginning with the sixth year. After two years service the company, having surplus profits, requested to pay for another five years service in advance. If the electrical company elected to accept payment in advance, what would each company set as a fair settlement to be paid if (a) the electrical company considered 12% compounded annually as a fair return, and (b) the manufacturing company considered 10% a fair return.

18. A man has the following outstanding debts:

(a) $5,000 borrowed 4 years ago with the agreement to repay the loan in 60 equal monthly payments. (There are 12 payments outstanding). Interest on the loan is 9% compounded monthly.

(b) Twenty-four monthly payments of $200 owed on a loan on which interest is charged at the rate of 1% per month on the unpaid balance.

(c) A bill of $1,000 due in 2 years.

A loan company has offered to pay his debts if he will pay them $143.14 per month for the next 5 years. What monthly rate of interest is he paying if he accepts the loan company's offer? What is the nominal rate he is paying? What is the annual effective interest rate he is paying?

19. A chapter of a social fraternity is being organized and is in need of housing facilities. A local real estate man agrees to lease the fraternity a suitable house and pay all maintenance costs for $22,000 a year. The fraternity can purchase a building site for $12,000 and construct a house for $140,000. Annual maintenance, taxes, and insurance will cost the fraternity $3,000 if they own their house. With interest at 9%, how many years will it require before the new house would pay for itself, assum-

ing that the building site will continue to be valued at $12,000 and that all other costs would be the same regardless of whether the house is leased or built? Assume the house salvage value is zero.

20. A manufacturer pays a patent royalty of $0.80 per unit of a product he manufactures, payable at the end of each year. The patent will be in force for an additional 5 years. For this year, he manufactures 8,000 units of the product per year, but it is estimated that output will be 10,000, 12,000, 14,000, and 16,000 in the four succeeding years. He is considering asking the patent holder to terminate the present royalty contract in exchange for a single payment at present or asking the patent holder to terminate the present contract in exchange for equal annual payments to be made at the beginning of each of the five years. If 6% interest is used, what is (a) the present single payment and (b) the beginning-of-the-year payments that are equivalent to the royalty payments in prospect under the present agreement?

21. A man is planning to retire in 40 years. He wishes to deposit a regular amount every three months until he retires so that beginning one year following his retirement he will receive annual payments of $8,000 for the next 20 years. How much must he deposit if the interest rate is 8% compounded quarterly?

22. A company is considering the purchase of an air compressor. The compressor has a first cost of $2,500 and the following end-of-year maintenance costs:

Year	1	2	3	4	5	6	7	8
Maintenance Costs	$400	$400	$400	$500	$600	$700	$800	$900

What is the present equivalent value of this series of costs if interest is 12%?

23. A city power plant wishes to install a feed-water heater in their steam generation system. It is estimated that the increase in efficiency will pay for the heater one year after it is installed, and a contractor has promised he can install the heater in 5 months. If the venture is undertaken and it is found that the heater does pay for itself in a year by saving $700 per month, what amount was paid to the contractor at the last of each month of construction? Assume that the saving of $700 occurs at the last of each month, and that the money paid to the contractor could have been invested elsewhere at 6% compounded monthly.

24. Mr. A possesses a mine property estimated to contain 120,000 tons of coal. The mine is now leased to a coal company, which pays Mr. A $0.80 royalty per ton of coal removed. Coal is removed at the rate of 20,000 tons per year. The rate is expected to continue until the mine is exhausted, at which time the mine property is estimated to be worth $15,000. Mr. A now employs a checker whose duty is to measure the coal removed and to bill the coal company for the royalty on the coal removed. The checker receives $5,100 per year.

(a) If interest is at 8% compounded annually and taxes are neglected, for how much can Mr. A afford to sell the property?

(b) If interest is at 6% compounded annually, how much can the coal company afford to pay for the property?

(c) What is the most important factor causing the difference in the results obtained in (a) and (b)?

25. Oil reservoir engineers estimate the annual production of an oil well for the next 12 years to be as follows:

Year End	Annual Production in Barrels
1	23,800
2	15,100
3	10,600
4	6,700
5	4,300
6	2,600
7	1,400
8	790
9	430
10	260
11	140
12	70

Assuming that the oil will sell for $3.10 per barrel during the 12-year period, what is the present equivalent of estimated future production, if interest is compounded annually at 8%?

26. What single payment at the end of year 5 is equivalent to a uniform flow of payments of $2,000 per year beginning at the start of year 3 and ending at the end of year 15. Interest is 7% compounded continuously.

27. What uniform flow of payments for 5 years is equivalent to a series of equal end-of-year payments of $600 for 8 years at 7% compounded continuously?

28. An increasing annual uniform gradient series begins at the end of the second year and ends after the fifteenth year. What is the value of the gradient g that makes the gradient series equivalent to a uniform flow of payments of $800 per month for 7 years at 10% compounded continuously?

29. A man desires to make an investment in bonds, provided he can realize 8% on his investment. How much can he afford to pay for a $1,000 bond that pays 6% interest annually and will mature 12 years hence?

30. A bond is offered for sale for $1,040. Its face value is $1,000 and the interest is 8% payable annually. What rate of interest will be received if the bond matures 9 years hence?

31. A $1,000, 7% bond is offered for sale for $900. If interest is payable annually and the bond will mature in 7 years, what interest rate will be received?

32. How much can be paid for a $1,000, 7% bond with interest paid semiannually, if the bond matures 8 years hence? Assume the purchaser will be satisfied with 6% interest compounded semiannually, since the bonds were issued by a very stable and solvent company.

33. A $1,000 bond will mature in 10 years. The annual rate of interest is 8% payable semiannually. If the bond can be purchased for $940 what annual interest compounded semiannually will be received?

III

ECONOMIC ANALYSIS
OF ALTERNATIVES

6

Bases for Comparison
of Alternatives

All the decision criteria that are to be considered in this book incorporate some index, measure of equivalence, or *basis for comparison* that summarizes the significant differences between investment alternatives. It is important that the difference between a decision criterion and a basis for comparison be distinguished. A basis for comparison is an index containing particular information about a series of receipts and disbursements representing an investment opportunity. The most common bases for comparison are the present-worth amount, the annual equivalent amount, the capitalized amount, the future-worth amount, and the rate of return. Other bases for comparison that are considered in this chapter are the payout period, and the prospective value.

The reduction of alternatives to a common base is necessary so that apparent differences become real differences, with the time value of money considered. When expressed in terms of a common base, the real differences become directly comparable and may be used in decision making.

6.1. DESCRIBING INVESTMENT OPPORTUNITIES

An investment opportunity is usually described by the actual cash receipts and disbursements that are anticipated if the investment is undertaken. The representation of the amounts and timing of these cash receipts and disbursements is referred to as the investment's *cash flow*.

Consider the following investment opportunity. An individual is planning to place $1,000 in a savings account that pays him 5% interest at the end of each year. The saver anticipates that he will withdraw his initial investment at the end of 4 years. The cash flow that represents the saver's investment opportunity is shown in Table 6.1. Notice that a negative amount represents a cash cost or disbursement for the saver while a positive amount indicates a cash income or receipt.

Table 6.1. CASH FLOWS AS VIEWED BY TWO PARTIES

End of Year	Cash Flow from the Saver's Viewpoint	Cash Flow from the Bank's Viewpoint
0	$-1,000	$1,000
1	50	-50
2	50	-50
3	50	-50
4	1,050	-1,050

Table 6.1 also illustrates the cash flow from the bank's point of view. The bank receives $1,000 from the saver at the beginning of the first year and it must make payments of $50 at the end of each year including the last when the bank returns the principal to the saver. Thus, it is quite important that the proper point of view is considered when cash flows are being developed for decision-making purposes.

When an investment opportunity has both cash receipts and disbursements occuring simultaneously, a net cash flow is usually calculated for the investment opportunity. The net cash flow is the arithmetic sum of the receipts (+) and the disbursements (−) that occur at the same point in time. The utilization of net cash flows in decision making implies that the net dollars received or disbursed have the same effect on an investment decision as does the separate consideration of an investment's total receipts and disbursements.

To facilitate describing investment cash flows the following notation will be adopted. Let F_{jt} = net cash flow for investment proposal j at time t. If $F_{jt} < 0$ then F_{jt} represents a net cash cost or disbursement. If $F_{jt} > 0$ then F_{jt} represents a net cash income or receipt.

For example consider a cash flow with the cash receipts and disbursements shown in Table 6.2. The values of the F_{jt}'s are shown in the right hand column.

Table 6.2. AN ILLUSTRATION OF A NET CASH FLOW

End of Year	Receipts	Disbursements	F_{jt}
0	$ 0	-$5,000	-$5,000
1	4,000	- 2,000	2,000
2	5,000	- 1,000	4,000
3	0	- 1,000	- 1,000
4	7,000	0	7,000

6.2. PRESENT-WORTH AMOUNT

In determining a basis for comparison of investment alternatives one likely candidate would be an index that reflects the differences between alternatives by considering the time value of money. In Chapter 5 the idea of equivalence was discussed and it was seen that for a known cash flow and given interest rate an equivalent amount for the cash flow can be calculated at any point in time. Thus it is possible to calculate a single equivalent amount at any point in time that is equivalent in *value* to a particular cash flow. Since this single value summarizes the value of a cash flow it seems appropriate to use such a value as a basis for comparison.

The present-worth amount is an amount at the present ($t = 0$) that is equivalent to an investment's cash flow for a particular interest rate i. Thus, the present-worth of investment proposal j at interest rate i with a life of n years can be expressed as

$$PW_j(i) = F_{j0}(\overset{P/F\ i,0}{}) + F_{j1}(\overset{P/F\ i,1}{}) + F_{j2}(\overset{P/F\ i,2}{}) + \ldots + F_{jn}(\overset{P/F\ i,n}{})$$

$$PW_j(i) = \sum_{t=0}^{n} F_{jt}(\overset{P/F\ i,t}{}).$$

But since $(\overset{P/F\ i,t}{}) = (1 + i)^{-t}$

$$PW_j(i) = \sum_{t=0}^{n} F_{jt}(1 + i)^{-t}. \qquad (-1 < i < \infty)[1].$$

The present-worth amount has a number of features that makes it suitable as a basis for comparison. First, it considers the time value of money according to the value of i selected for the calculation. Second, it concentrates the equivalent value of any cash flow in a single index at a particular point in time ($t = 0$). Third, the value of the present-worth amount is always unique no matter what may be the investment's cash flow pattern. That is, any sequence of receipts and disbursements will give a unique present-worth amount for a particular value of i.

In addition, the present-worth amount is the equivalent amount by which the equivalent receipts of a cash flow exceed or fail to equal the equivalent disbursements of that cash flow. Figure 6.1 illustrates the range of present-worth values for a particular cash flow by plotting the present-worth amount *vs.* the interest rate for the cash flow described in Table 6.3.

By examining the $PW(i)$ function in Figure 6.1 considerable information useful for decision-making purposes can be ascertained about the investment opportunity. For the range of interest rates ($0 \le i < 22\%$) it is observed that

---

[1] For the range ($-1 < i < 0$) no interest is received and the principal invested is not completely repaid. Discussions of interest rate for most problems can be confined to the range ($0 \le i < \infty$).

Table 6.3. CALCULATION OF THE PRESENT-WORTH FOR A CASH FLOW USING DIFFERENT INTEREST RATES

End of Year	Cash Flow	i	$PW(i) = -\$1{,}000 + \$400\left\{\sum\limits_{t=1}^{4}\left[\dfrac{1}{(1+i)^t}\right]\right\}$
0	−$1,000	0%	$600
1	400	10%	268
2	400	20%	35
3	400	22%	0
4	400	30%	−133
		40%	−260
		50%	−358
		∞	−1,000

$PW(i)$ is positive indicating that the equivalent receipts at the present exceed the equivalent disbursements. On the assumption that the cash flow estimates of Table 6.3 eventually prove to be correct, the significance of the $PW(i)$ function is that for a particular value of i, say 10%, the investment can be said to produce a 10% return on the unrecovered investment plus an equivalent receipt of $270 at the beginning of the first year.

If the plot had been extended for $i > 50\%$ it would be observed that the curve would be asymptotic to $PW(i) = -\$1{,}000$, the value of F_{jo}. The reason for this result becomes clear when the definition of $PW(i)$ is re-examined. As the interest rate is increased every cash flow in the future is discounted to the present by a factor of the general form $1/(1+i)^t$. As i approaches infinity it is evident that the discounting factors will approach zero for all points in time except for $t = 0$. Thus $F_{jo} = -\$1{,}000$ is the only cash flow that is not reduced to zero when $i = \infty$.

For cash flows that have initial disbursements followed by a series of positive receipts the $PW(i)$ function possesses characteristics similar to those of Figure 6.1. That is, the function will be decreasing as i increases from zero and it will intersect the horizontal axis only once. Since most of the cash flows that occur in actual decision situations have this pattern of early disbursements followed by future receipts, most of the discussions concerning the selection of investment alternatives is confined to proposals with $PW(i)$ functions similar to the one in Figure 6.1. However it must be recognized that there are many real world investments that have cash flows which produce $PW(i)$ functions dissimilar to Figure 6.1. One such function is shown in Figure 6.2.

Capitalized Equivalent. A special case of the present-worth basis of comparison is the so called *capitalized equivalent* ($CE(i)$). In engineering economic analysis this term represents a basis of comparison that consists of finding a single amount at the present which at a given rate of interest will be equiva-

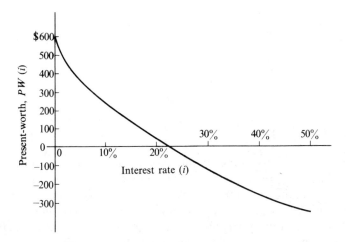

FIGURE 6.1. PRESENT-WORTH AS A FUNCTION OF INTEREST RATE FOR CASH FLOW IN TABLE 6.3.

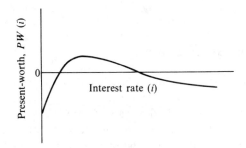

FIGURE 6.2.

lent to the net difference of receipts and disbursements if a given cash flow pattern is repeated in perpetuity.

This concept should not be confused with the accountant's concept of "capitalizing" an expenditure in the books of account. For the accountant an expenditure is capitalized if it is recorded as an asset or prepaid expense rather than being recorded as an expense at the time it is incurred.

To calculate the capitalized equivalent for an investment or a series of investments that are expected to produce cash flows from the present to infinity, the most common method is to first convert the actual cash flow into an equivalent cash flow of equal annual payments A that extends to infinity. Then the equal annual payments are discounted to the present by using the equal payment series present-worth factor.

$$CE(i) = PW(i) \text{ where the cash flow extends forever}$$

$$CE(i) = A(\overset{P/A\,i,\,\infty}{}) = A\left[\frac{(1+i)^\infty - 1}{i(1+i)^\infty}\right]$$

$$= A\left[\frac{1 - \dfrac{1}{(1+i)^\infty}}{i}\right] = \frac{A}{i}.$$

Thus

$$CE(i) = \frac{A}{i}.$$

An intuitive understanding of this last relationship is obtained by considering what present-worth amount invested at i will enable an investor to periodically withdraw an amount A forever. If the investor withdraws more than amount A each period, he will be withdrawing a portion of the initial principal. If this initial principal is being consumed with each withdrawal, it will eventually be reduced to zero and additional withdrawals will be impossible. However when the amount being withdrawn each period equals the interest earned on the principal for that period, the principal remains intact. Thus, the series of withdrawals can be continued forever.

6.3. ANNUAL EQUIVALENT AMOUNT

The annual equivalent amount is another basis for comparison that has characteristics similar to the present-worth amount. This similarity is evident when it is realized that any cash flow can be converted into a series of equal annual payments by first calculating the present-worth amount for the original series and then multiplying the present-worth amount by the interest factor ($\overset{A/P\,i,\,n}{}$). Thus the annual equivalent amount for interest rate i and n years can be defined as

$$AE(i) = PW(i)(\overset{A/P\,i,\,n}{})$$

$$= \left[\sum_{t=0}^{n} F_{jt}(1+i)^{-t}\right]\left[\frac{i(1+i)^n}{(1+i)^n - 1}\right].$$

There are two important features of this relationship that need to be understood. First, if the values of i and n are finite the relationship reduces to $AE(i) = PW(i)$ times a constant. Therefore when different cash flows are evaluated for a particular value of i and a particular value of n, the comparison of their annual equivalent amounts will yield the same relative results as those obtained from making the comparison on the basis of the present-worth amount. That is, the ratio of the annual equivalent amounts for two different cash flows will equal the ratio of the present-worth amounts for the respective cash flows.

Secondly, as long as i and n are finite the values of $AE(i)$ and $PW(i)$ will be zero for the same value of i. Graphically this means that the intersections of the horizontal axis $AE(i) = 0$ by the $AE(i)$ function will occur at the same value of i for which the $PW(i)$ function intersects the horizontal axis $PW(i) = 0$. Thus, the present-worth amount and the equivalent annual amount can be said to be *consistent* bases for comparison. Any particular decision criterion that uses either of these bases for direct comparison will yield the same selection of alternatives for fixed values of i and n.

The $AE(i)$ basis of comparison will sometimes be preferred to $PW(i)$ because of computational advantages that may arise when it is assumed that a particular cash flow pattern will repeat itself. The cash flow in Table 6.4 illustrates a repeating cash flow pattern that results from renewing an investment every two years. The cash flows inclosed by lines indicate the receipts and disbursements associated with each investment proposal. For example the $1,000 disbursement at the end of year 2 is actually the disbursement at the *beginning* of year 3 for another proposal that will return $400 and $900 at the end of year 3 and year 4, respectively.

When converting each investment into an annual equivalent series over the 2 year period of the investment's life, the same calculations are required for each investment. It is only necessary to calculate the annual equivalent amount over the life of one investment proposal. When the investment is to be repeated the same annual equivalent amount will be repeated over the time span for which the investment's cash flow is repeated. For example the calculation of the annual equivalent amount is $61.93 per year for an investment requiring a $1,000 disbursement followed by receipts of $400 and $900 in one and two years, respectively. If this investment is repeated seven times as shown in Table 6.4, the annual equivalent over the fourteen year time span is still $61.93 per year.

The calculation of the present-worth amount for the seven investments of Table 6.4 would require the discounting of all payments in years 1 through 14 back to the present. In this instance the annual equivalent amount basis for comparison would be computationally more efficient than finding the present-worth of each receipt and disbursement.

Of course if computational efficiency and the present-worth amount are both desired for decision-making it is a simple matter to first determine the annual equivalent as described above and then convert that quantity to a present-worth amount.

The Annual Equivalent Cost of an Asset. An asset such as a machine is a unit of capital. Such a unit of capital loses value over a period of time in which it is used in carrying on the productive activities of an enterprise. This loss of value of an asset represents actual piecemeal consumption or expenditure of capital.

Table 6.4. THE USE OF ANNUAL EQUIVALENT FOR REPEATED CASH FLOWS

End of Year	Disbursements	Receipts	Annual Equivalent for Each Investment Proposal
0	−$1,000		
1		$400	$\left[-\$1,000 + \$400(\overset{P/F\,10,1}{0.9091}) + \$900(\overset{P/F\,10,2}{0.8265})\right](\overset{A/P\,10,2}{0.5762}) = \begin{cases}\$61.93 \\ \$61.93\end{cases}$
2	− 1,000	900	
3		400	$\left[-\$1,000 + \$400(\overset{P/F\,10,1}{0.9091}) + \$900(\overset{P/F\,10,2}{0.8265})\right](\overset{A/P\,10,2}{0.5762}) = \begin{cases}\$61.93 \\ \$61.93\end{cases}$
4	− 1,000	900	
·	·	·	· · ·
·	·	·	· · ·
·	·	·	· · ·
12	− 1,000	900	
13		400	$\left[- 1,000 + \$400(\overset{P/F\,10,1}{0.9091}) + \$900(\overset{P/F\,10,2}{0.8265})\right](\overset{A/P\,10,2}{0.5762}) = \begin{cases}\$61.93 \\ \$61.93\end{cases}$
14	− 1,000	900	

Two monetary transactions are associated with the procurement and eventual retirement of a capital asset; its first cost and salvage value. From these amounts it is possible to derive a simple formula for the equivalent annual cost of the asset for use in economy studies. Let

P = first cost of the asset
F = estimated salvage value
n = estimated service life in years.

Then, the annual equivalent cost of the asset may be expressed as the annual equivalent first cost less the annual equivalent salvage value, or

$$P(\overset{A/P\ i,\ n}{}) - F(\overset{A/F\ i,\ n}{}).$$

But since

$$(\overset{A/F\ i,\ n}{}) = (\overset{A/P\ i,\ n}{}) - i$$

by substitution

$$P(\overset{A/P\ i,\ n}{}) - F\left[(\overset{A/P\ i,\ n}{}) - i\right]$$

and

$$(P - F)(\overset{A/P\ i,\ n}{}) + Fi.$$

As an example of the use of this important formula, consider the following situation. An asset with a first cost of $5,000 has an estimated service life of 5 years and an estimated salvage value of $1,000. For an interest rate of 6% the annual equivalent cost is

$$(\$5,000 - \$1,000)(\overset{A/P\ 6,\ 5}{0.2374}) + \$1,000(0.06) = \$1,010.$$

It should be recognized that the cost of an asset is made up of the cost resulting from the loss in value plus the cost of interest on unrecovered capital. The asset in this example suffered a loss in value of $800 per year for 5 years. In addition, the annual equivalent cost of interest on the unrecovered capital was $210 per year for 5 years.

Capital assets are purchased in the belief that they will earn more than they cost. One part of the prospective earnings is considered to be *capital recovery*. Capital invested in an asset is recovered in the form of income derived from the services rendered by the asset and from its sale at the end of its useful life. If the asset provided services valued at $800 per year during its life, and if $1,000 was received from its sale, a total of $5,000 would be recovered.

A second part of the prospective earnings will be considered to be *return*. Since capital invested in an asset is ordinarily recovered piecemeal, it is necessary to consider the interest on the unrecovered balance as a cost of ownership. Thus, an investment in an asset is expected to result in income sufficient not only to recover the amount of the original investment, but also to provide for a return on the diminishing investment remaining in the asset at any time during its life. This gives rise to the phrase *capital recovery plus return* which will be used in subsequent chapters.

6.4. FUTURE-WORTH AMOUNT

The future-worth basis for comparison is an equivalent amount of a cash flow calculated at a future time for some interest rate. The future-worth amount for proposal j at some future time n years from the present is

$$FW_j(i) = F_{j0}(\overset{F/P\,i,\,n}{\qquad}) + F_{j1}(\overset{F/P\,i,\,n-1}{\qquad}) + \ldots F_{j,n-1}(\overset{F/P\,i,\,1}{\qquad}) + F_{j,n}(\overset{F/P\,i,\,0}{\qquad})$$

$$FW_j(i) = \sum_{t=0}^{n} F_{jt}(\overset{F/P\,i,\,n-t}{\qquad}).$$

But since

$$(\overset{F/P\,i,\,n-t}{\qquad}) = (1 + i)^{n-t}$$

$$FW_j(i) = \sum_{t=0}^{n} F_{jt}(1 + i)^{n-t}.$$

Another method of calculating the future-worth amount is to first determine the present-worth amount of the cash flow and then to convert to its future equivalent n years hence. Thus, the future-worth amount for a proposal j can be expressed as

$$FW_j(i) = PW_j(i)(\overset{F/P\,i,\,n}{\qquad}).$$

From this relationship it is clear that for given finite values of i and n the future-worth amount is merely the present-worth amount times a constant. As a consequence the relative differences between alternatives on the basis of present-worth will be the same as the relative differences between alternatives compared on the basis of future-worth as long as i and n are fixed. Therefore, an alternative which has a present-worth three times as large as another alternative's present-worth will also have a future-worth three times as large as the future-worth of that alternative.

The future-worth amount, annual equivalent amount and the present-worth amount are consistent bases of comparison. As long as i and n are fixed and proposals A and B are being compared the following relationships will hold.

$$\frac{PW_A(i)}{PW_B(i)} = \frac{AE_A(i)}{AE_B(i)} = \frac{FW_A(i)}{FW_B(i)}.$$

Because the present-worth amount, the annual equivalent amount, and the future-worth amount are all measures of equivalence differing only in the times at which they are stated, it is not surprising that they are consistent bases for comparison of investment alternatives. Therefore it should be expected that any decision criterion that directly compares present-worth amounts could just as well employ future-worth amounts or annual equivalent amounts as bases for comparison without affecting the selection outcome.

6.5. RATE OF RETURN

The internal rate of return or rate of return, as it is frequently called, is a widely accepted index of profitability. It is defined as the interest rate that reduces the present-worth amount of a series of receipts and disbursements to zero. That is, the rate of return for investment proposal j, is the interest rate i_j^* that satisfies the equation

$$0 = PW_j(i_j^*) = \sum_{t=0}^{n} F_{jt}(1 + i_j^*)^{-t}$$

where proposal j has a life of n periods. The rate of return must lie in the interval $(-1 < i^* < \infty)$ in order to be economically relevant. For most practical problems it will be sufficient to consider the range of the rate of return in the interval $(0 < i^* < \infty)$.

From the previous discussion of the relationships between the annual equivalent amount, the future-worth amount and the present-worth amount it follows that the rate of return (i_j^*) for a cash flow will also satisfy the expressions $0 = AE_j(i_j^*)$ and $0 = FW_j(i_j^*)$.

The Meaning of Rate of Return. In economic terms the rate of return represents the percentage or rate of interest earned on the *unrecovered* balance of an investment. The unrecovered balance of an investment can be viewed as the portion of the initial investment that remains to be recovered after interest payments and receipts have been added and deducted, respectively, up to the point in time being considered.

By examining the calculations displayed in Table 6.5 the fundamental meaning of rate of return should become evident. Each of the two cash flows in Table 6.5 can be viewed as an arrangement where someone has borrowed $1,000 with an agreement to pay 10% on the unpaid or unrecovered balance and to reduce the unpaid balance to zero at the end of the loan's time limit. These cash flows could also represent the purchase of productive assets where these assets will yield a rate of profit of 10% on the amount of dollars that are unrecovered or "tied-up" in the assets during their lifetime.

If $U_t =$ the unrecovered balance at the beginning of period t, the unrecovered balance for any time period can be found from the recursive equation

$$U_{t+1} = U_t(1 + i) + F_t$$

Table 6.5. TWO CASH FLOWS DEMONSTRATING THE FUNDAMENTAL MEANING OF RATE OF RETURN

End of Year	Cash Flow at End of Year t	Unrecovered Balance at Beginning of Year t	Interest Earned on the Unrecovered Balance During Year t	Unrecovered Balance at the Beginning of Year $t+1$
		Proposal $A(i_A^* = 10\%)$		
t	$F_{A,t}$	U_t	$U_t(0.10)$	$U_t(1 + 0.10) + F_{A,t} = U_{t+1}$
0	$-\$1,000$	—	—	$-\$1,000$
1	400	$-\$1,000$	$-\$100$	-700
2	370	-700	-70	-400
3	240	-400	-40	-200
4	220	-200	-20	0
		Proposal $B(i_B^* = 10\%)$		
t	$F_{B,t}$	U_t	$U_t(0.10)$	$U_t(1 + 0.10) + F_{B,t} = U_{t+1}$
0	$-\$1,000$	—	—	$-\$1,000$
1	100	$-\$1,000$	$-\$100$	$-1,000$
2	100	$-1,000$	-100	$-1,000$
3	100	$-1,000$	-100	$-1,000$
4	1100	$-1,000$	-100	0

where F_t = the payment received at the end of period t
$\quad\quad i$ = the interest rate earned on the unrecovered balance during period t,
and U_1 = the initial amount of loan or unrecovered balance.

The unrecovered balances related to each cash flow appear as negative values in Table 6.5 indicating that they are amounts owed by the borrower or the amounts unrecovered by the lender.

A common misinterpretation of what the rate of return (i^*) measures is observed when the rate of return of a project is viewed as being the rate of interest earned on the initial outlay required by that project. Proposal A in Table 6.5 has a rate of return of 10% but it is clear that the cash flow of the proposal does not yield a 10% return on the $1,000 initial outlay for the full four year life of the proposal. In fact, Proposal A earns $100, $70, $40 and $20 in years 1, 2, 3, and 4, respectively, on an investment that decreases from an initial commitment of $1,000 to $0 at the end of the 4th year.

Proposal B represents a special type of cash flow for which the rate of return (i^*) does indicate the return earned on the initial investment. That is, the rate of return is 10% and the $100 amounts earned each year are 10% of the initial investment. The reason for this result is the fact that for this unique cash flow the unrecovered balance throughout the proposal's life always equals the amount of the proposal's initial outlay.

Thus the fundamental concept of rate of return emerges; it is the rate of interest earned on the unrecovered balance of an investment so that the unrecovered balance is zero at the end of the investment's life.

Computing the Rate of Return. The computation of rate of return generally requires a trial and error solution. For example, to calculate the rate of return for the cash flow shown below it is necessary to find the value i^* that sets the present-worth amount to zero.

End of Year t	Cash Flow F_t
0	$-\$1,000$
1	-800
2	500
3	500
4	500
5	1200

That is, find the value of i that satisfies

$$0 = PW(i)$$

$$= -\$1,000 - \$800(\overset{P/F\,i,\,1}{\ }) + \$500(\overset{P/A\,i,\,4}{\ })(\overset{P/F\,i,\,1}{\ }) + \$700(\overset{P/F\,i,\,5}{\ }).$$

Rather than trying to solve for i^* directly from this equation, a trial and error solution should be attempted. Try $i = 0\%$

$$PW(0) = -\$1,000 - \$800(1) + \$500(4)(1) + \$700(1)$$

$$PW(0) = \$900.$$

With the present-worth amount greater than zero at $i = 0$ the next step is to examine the cash flow in order to see how the next rate selected will affect the present-worth amount. Since all the positive cash flows are further in the future than the negative cash flows, an increase in the interest rate will reduce the present-worth of the receipts more than the present-worth of the outlays. Thus, the total present-worth amount will be decreased toward zero. Try $i = 12\%$

$$PW(12) = -\$1,000 - \$800(\overset{P/F\,12,\,1}{\ }) + 500(\overset{P/A\,12,\,4}{\ })(\overset{P/F\,12,\,1}{\ })$$
$$+ 700(\overset{P/F\,12,\,5}{\ })$$

$$PW(12) = \$32.$$

Since $PW(12)$ is still greater than zero try a larger interest rate. With $i = 15\%$

$$PW(15) = -\$1,000 - \$800(\overset{P/F\,15,\,1}{0.8616}) + \$500(\overset{P/A\,15,\,4}{0.8696})(\overset{P/F\,15,\,1}{0.8696})$$
$$+ \$700(\overset{P/F\,15,\,5}{0.4972})$$

$$PW(15) = -\$116.$$

Thus, it is determined that the rate of return will lie between 12% and 15%. By interpolation

$$i^* = 12\% + 3\%\left[\frac{32 - 0}{32 - (-116)}\right] = 12\% + 3\%\left[\frac{32}{148}\right] = 12.6\%.$$

Since the solution for the rate of return of a cash flow with a life of n periods is the solution of an nth degree polynomial, there exist various mathematical methods that systematically converge on the roots or values of i that satisfy the equation. Thus the trial and error approach is not the only method of solving for rate of return.

To reduce the computational effort required to determine the rate of return computer solutions of the present-worth equations are often used. As a result the computational effort associated with the rate of return is a significant consideration only where manual calculations are required.

One of the obvious differences between rate of return and the other bases previously discussed is that no interest rate need be known in order to determine the rate of return. The present-worth amount, annual equivalent amount, and the future-worth amount are all functions of an interest rate and in order to calculate a particular value of these bases a particular value of i must be known. For investment situations where knowledge about the future and future interest rates is highly uncertain, rate of return can be a workable way to compare the economic desirability of alternative investments.

By definition the rate of return is related to the present-worth amount as shown in Figure 6.3. The value of i where the $PW(i)$ function intersects the horizontal axis is i^* the value of i that sets the present-worth amount equal to zero. Note that for the $PW(i)$ function in Figure 6.3 the present-worth is positive for all values of i less than i^*. This relationship also exists for the annual equivalent and future-worth amounts.

Cash Flows with no Rate of Return. It should be recognized that there are certain cash flows for which *no* rate of return exists in the interval $(-1 <$

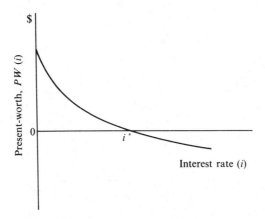

FIGURE 6.3. RATE OF RETURN AND ITS
RELATIONSHIP TO PRESENT WORTH.

$i < \infty$). The most common example of this situation is where the cash flow consists of either all receipts or all disbursements with the initial receipt or disbursement occuring at the beginning of year 1.

In practice investment proposals are frequently described by cost cash flows when the alternatives are assumed to provide the same service, benefit, or revenue. Since it is impossible to calculate a meaningful rate of return for such a cash flow pattern means other than direct calculation of rate of return must be utilized for decision making with cost cash flows. These methods are discussed in Chapter 7.

Cash Flows with a Single Rate of Return. Because of the desirability of having a present-worth function with a single rate of return, and the form of the function shown in Figure 6.3, it is important to have a method for predicting whether a particular cash flow can produce such a present-worth function. One useful generalization that can be made is that any proposal cash flow with an initial disbursement or a series of disbursements starting at the present *followed* by a series of positive receipts will always have a present-worth function similar to Figure 6.3 if the absolute sum of the receipts is *greater* than the absolute sum of the disbursements. Table 6.6 presents the cash flows of two proposals (*A* and *B*) that satisfy these requirements and two cash flows (*C* and *D*) that do not.

Table. 6.6. FOUR CASH FLOW PATTERNS

End of Year	Proposal *A*	Proposal *B*	Proposal *C*	Proposal *D*
0	−$1,000	−$1,000	−$2,000	−$1,000
1	500	− 500	0	4,700
2	400	− 500	10,000	− 7,200
3	300	− 500	0	3,600
4	200	1,500	0	0
5	100	2,000	−10,000	0

For Proposal *A* the sum of the receipts ($1,500) is greater than the sum of the disbursements ($1,000) and for Proposal *B* the sum of the receipts ($3,500) exceeds the sum of the disbursements ($2,500). Thus these two cash flows have a single rate of return with a present-worth plot of the form shown in Figure 6.3. Most practical proposals have estimated cash flows that have patterns similar to Proposal *A* and Proposal *B* since most investments require initial commitment of funds which is followed by the income series resulting from the productivity of the project.

Proposals *C* and *D* represent a more unusual class of cash flow patterns. These cash flows consist of disbursements, receipts, more disbursements, receipts, etc. Such cash flows do not follow the pattern of the class of cash flows that include Proposals *A* and *B* so there is no assurance that their present-worth plots will resemble the present-worth function of Figure 6.3.

Cash Flows with Multiple Rates of Return. Figure 6.2 exhibited a situation where the cash flow has more than one value of i for which the present-.worth amount will equal zero. This result should be expected since the present-worth equation is an nth degree polynomial of the form

$$PW(i) = 0 = F_0 + F_1 x + F_2 x^2 + F_3 x^3 + \ldots + F_n x^n$$

where $x = \dfrac{1}{(1 + i)}$.

For this polynomial there may be n different roots or values of x which satisfies this equation. In order to have a rate of return that has economic relevance the value of i must lie in the interval $(-1 < i < \infty)$. Thus x must be in the interval $(0 < x < \infty)$. The number of positive real roots of x that satisfy the above equation equals the number of meaningful rates of return that will be produced by a cash flow.

For decision-making purposes cash flows which have a unique rate of return and behave similarly to the example shown in Figure 6.3 are much simpler to handle than cash flows with multiple rates of return. When multiple rates of return occur, questions arise such as "Which rate of return is the correct one?" and "Are the decision rules most frequently used for investment selection applicable when multiple rates of return occur?" The answers to these and other questions that relate to decision criteria are deferred until the next chapter.

The question that should be answered at this time is, "What effect does the cash flow pattern have on the number of positive real roots of the above equation?" By understanding something about these relationships an intelligent determination can be made concerning the conditions under which the rate of return is an appropriate basis for comparison.

A rule that can be helpful in identifying the possibility of multiple rates of return is Descartes' rule of signs for an nth degree polynomial. This rule states that the number of real positive roots of an nth degree polynomial with real coefficients is never greater than the number of changes of sign in the sequence of its coefficients

$$F_0, F_1, F_2, F_3, \ldots, F_{n-1}, F_n$$

and if less, always by an even number. For example, the sequence of signs of the cash flows for Proposal A and Proposal B shown in Table 6.6 changes only once while the sequence of signs for Proposal C and D changes two and three times, respectively. The sequence of signs for Proposal C has one change from the initial negative value to positive at the end of year 2. (A zero cash flow can be considered signless for the purpose of applying the rule of signs.)

The sequence of signs remains unchanged until the end of year 5 when the sequence changes from positive back to negative.

For both Proposal *A* and Proposal *B* the rule of signs indicates that there is no more than one rate of return in the interval $(-1 < i < \infty)$. This result is verified since both cash flows are known to have a unique rate of return as their present-worth functions are of the form shown in Figure 6.3. However, it should be understood that although the rule of signs may indicate a maximum of one rate of return there is no assurance that the present-worth function will be of the form shown in Figure 6.3.

The rule of signs indicates for Proposal *C* that the maximum possible number of positive, real roots is two. Figure 6.4 depicts the present-worth

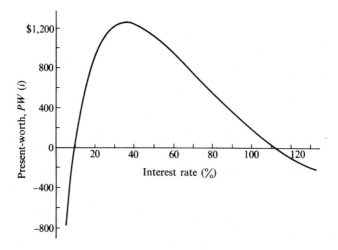

FIGURE 6.4. PRESENT-WORTH VS. *i* FOR CASH FLOW *C* IN TABLE 6.6.

function for Proposal *C* and it is seen that in fact this cash flow does have two distinct interest rates for which the present-worth is zero. The two rates of return for Proposal *C* are 9.8% and 111.5%.

It must be understood that the rule of signs can give only an indication as to the possibility of multiple rates of return because the rule only predicts the *maximum* number of *possible* rates of return that may occur. Thus numerous cash flows exist that have multiple changes in their sequence of signs but still possess a present-worth function similar to Figure 6.3.

Therefore the decision maker, who develops a decision procedure based on the assumption that the alternatives being considered will have a present-worth function similar to Figure 6.3, must assure himself that the cash flows do indeed produce such a function. The first test to make is to observe if the cash flow has a series of disbursements that is followed by a series of receipts and the arithmetic sum of the series is positive. If the cash flow has such a pattern then the present-worth function is confirmed to be of the normal form. If the cash flow does not meet this requirement or if the rule of signs indicates that there is the possibility of multiple rates of return, the decision maker should plot $PW(i)$ vs. i for those investment proposals. Using the plot of the present-worth function one can properly interpret the desirability of the proposed investment by observing the values of i for which the present-worth amount is greater than zero. For Proposal D in Table 6.6 the present-worth plot shown in Figure 6.5 reveals that this proposal is desirable for

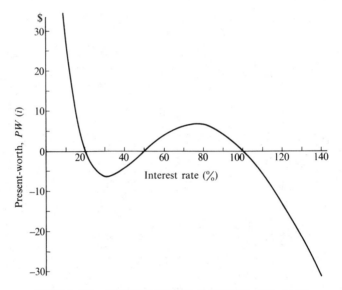

FIGURE 6.5. PRESENT-WORTH FUNCTION FOR CASH
FLOW D, TABLE 6.6.

values of i less than 20% and for values of i between 50% and 100%. To attempt to interpret multiple rates of return using normal procedures will in most cases be meaningless without the supplementary use of the present-worth function.

It is a simple matter to construct a cash flow pattern that produces selected multiple rates of return. Suppose one desires to find a cash flow pattern that has rates of return of 20%, 50%, and 100%. All that is required is to multiply the factors that yield these rates of return. The result of such a multiplication is an equation that represents a future-worth calculation. This equation must be equal to zero since rates of return are defined as those interest rates that set the present-worth or future-worth of a cash flow equal to zero. An example of such a calculation is shown below.

$$FW(i^*) = 0 = [(1 + i) - 1.2][(1 + i) - 1.5][(1 + i) - 2.0]$$
$$0 = (1 + i)^3 - 4.7(1 + i)^2 + 7.2(1 + i) - 3.6$$

Multiplying by $-\$1,000$ gives $0 = -\$1,000(1 + i)^3 + 4,700(1 + i)^2$
$$-\$7,200(1 + i) + \$3,600.$$

The coefficients in this last equation are the individual cash receipts and disbursements of Proposal D described in Table 6.6. The present-worth plot in Figure 6.5 confirms that the rates of return of this cash flow are 20%, 50%, and 100%.

6.6. PAYOUT PERIOD

Expressions like "This investment will pay itself in less than three years" are common in industry and they are indicative of the tendency to evaluate assets in terms of a *payback* or *payout* period. The payout period is most commonly defined as the length of time required to recover the first cost of an investment from the net cash flow produced by that investment for an interest rate equal to zero. That is, if F_o = first cost of the investment and if F_t = the net cash flow in period t then the payout period is defined as the value of n that satisfies the equation

$$0 = F_o + \sum_{t=1}^{n} F_t.$$

It should be noted that in actual practice there are other methods that are variations of the above approach that are also referred to as payout period. However, the pertinent limitations and merits of most payout period methods as a basis for comparison can be made clear by examining the payout period as defined.

For example Table 6.7 presents the cash flows for three investment proposals for each of which the payout period is 3 years. A look at the proposals presented in Table 6.7 reveals the fact that the payout period as a measure of investment desirability does have some serious shortcomings. Certainly no one would feel that under normal circumstances these three proposals

Table 6.7. THREE PROPOSALS WITH A PAYOUT PERIOD OF THREE YEARS

End of Year	Proposal A	Proposal B	Proposal C
0	$-\$1,000$	$-\$1,000$	$-\$700$
1	500	200	-300
2	300	300	500
3	200	500	500
4	200	1000	0
5	200	2000	0
6	200	4000	0
Sum of Cash Flow	600	7000	0

have equal economic desirability although they have equal payback periods.

In general the most serious deficiencies of the payback period are that it fails to consider:

1. The time value of money;
2. The consequences of the investment following the payback period including the magnitude and timing of the cash flows and the expected life of the investment.

Because of the limitations just mentioned the payout period tends to favor shorter lived investments. Experience has generally indicated that this bias is unjustifiable and in many cases economically unsound.

Nevertheless it must be said that the payout period does give some measure of the rate at which an investment will liquidate its initial outlay. For those situations where there is a high degree of uncertainty concerning the future and a firm is interested in its cash position and borrowing commitments, the payout period can supply useful information about investments that are under consideration. As a result this measure of investment desirability is frequently used to supplement the bases for comparison that have been discussed earlier.

6.7. PROSPECTIVE VALUE

A recently developed basis for comparison requires the knowledge of two interest rates before it can be calculated.[2] Called the prospective value (PV) this measure of investment desirability considers the fact that in practice the rate representing a minimum desirable return from investments is nearly always greater than the rate at which money can be invested so that it is available on demand. For a proposal j with life n the prospective value is expressed as

$$PV_j(\bar{m}, i_\delta) = F_{jo}(1 + i_\delta)(1 + \bar{m})^{n-1} + \sum_{t=1}^{n} F_{jt}(1 + \bar{m})^{n-t}$$

where $\bar{m}$ is an average marginal rate at which it is anticipated the net receipts will be invested in the future and i_δ is a rate at which highly liquid funds can be invested i.e., a bank savings account.

The basic assumption that is unique to this approach is the supposition that if proposal j is not accepted the money available for investment, F_{jo}, will be invested at an interest rate i_δ for one period and then withdrawn and invested at the rate $\bar{m}$ for the remainder of the $n - 1$ periods. Thus the future-worth of this opportunity foregone by the acceptance of proposal j is repre-

---

[2]Oakford, R. V. and Thuesen, G. J., "The Maximum Prospective Value Criterion," *The Engineering Economist*, Vol. 13, No. 3, 1968.

sented by the term $F_{jo}(1 + i_\delta)(1 + \bar{m})^{n-1}$. The other terms $\sum_{t=1}^{n} F_{jt}(1 + \bar{m})^{n-t}$ represent the future-worth of the money to be realized from the investment of F_{jo} dollars if proposal j is undertaken and the net receipts from that proposal are invested at rate $\bar{m}$ until the end of its life.

The sum of these two parts represent the net future-worth of accepting proposal j. That is, if proposal j is undertaken the future-worth to be derived is $\sum_{t=1}^{n} F_{jt}(1 + \bar{m})^{n-t}$. This amount must be reduced by the future-worth of the opportunity foregone when proposal j is accepted. The future-worth of the foregone opportunity is $F_{jo}(1 + i_\delta)(1 + \bar{m})^{n-1}$. Since F_{jo} is normally a disbursement and therefore negative, this reduction is accomplished simply by summing the two amounts.

By following this line of reasoning it is seen that if $PV_j(\bar{m}, i_\delta) > 0$ the opportunity associated with the investment of money in proposal j is worth more than the opportunity that is to be foregone. Similarly if $PV_j(\bar{m}, i_\delta) < 0$ it is more profitable to invest in highly liquid funds for one period and then invest those proceeds in the future at an anticipated rate of interest $\bar{m}$.

In order to simplify this basis for comparison it is generally easier to compute this measure discounted to the present rather than examing it as a future-worth amount. Thus we can write

$$P_j(\bar{m}, i_\delta) = PV_j(\bar{m}, i_\delta)(\overset{P/F \; \bar{m}, \, n}{}) = PV_j(\bar{m}, i_\delta)[1/(1 + \bar{m})^n]$$

$$P_j(\bar{m}, i_\delta) = [F_{jo}(1 + i_\delta)(1 + \bar{m})^{n-1} + \sum_{t=1}^{n} F_{jt}(1 + \bar{m})^{n-t}][1/(1 + \bar{m})^n]$$

$$P_j(\bar{m}, i_\delta) = F_{jo}\frac{(1 + i_\delta)}{(1 + \bar{m})} + \sum_{t=1}^{n} F_{jt}(1 + \bar{m})^{-t}.$$

It is seen from this expression that the present-worth amount which is defined as

$$PW_j(i) = \sum_{t=0}^{n} F_{jt}(1 + i)^{-t}$$

is a special case of $P_j(\bar{m}, i_\delta)$ when $\bar{m}$ and $i_\delta = i$. When $\bar{m} = i$ and $\bar{m} > i_\delta$ the only difference between the present-worth amount and $P_j(\bar{m}, i_\delta)$ is in the first term. For $P_j(\bar{m}, i_\delta)$ this term is $F_{jo}[(1 + i_\delta)/(1 + \bar{m})]$ and for the present-worth amount it is F_{jo}.

Nevertheless this difference is significant since $P_j(\bar{m}, i_\delta)$ can be greater than zero even if the rate of return i_j^* is less than the rate $\bar{m}$. When $P_j(\bar{m}, i_\delta) > 0$ the acceptance of proposal j produces more revenue than the opportunity that would be undertaken if proposal j was not accepted. For the present-worth amount, $PW_j(i)$, can never be greater than zero as long as the proposals' rates of return are less than i, the interest rate used in the present-worth calculation. (This statement is based on the assumption that the

$PW_j(i)$ function for the proposals being considered is of the form described in Figure 6.3.) Of course for proposals with $i^* > \bar{m}$ or i, $P_j(\bar{m}, i_\delta)$ and $PW_j(i)$ will always be greater than zero.

Under what conditions is a proposal with a rate of return in the interval $(i_\delta < i^* < \bar{m})$ deemed economically desirable? That is, what causes $P_j(\bar{m}, i_\delta)$ to be greater than zero for a proposal with $i_\delta < i^* < \bar{m}$? It is the speed with which the proposed investment promises to generate its returns; i.e., the faster a proposal liquidates itself the greater the chance of $P_j(\bar{m}, i_\delta)$ being greater than zero.

A simple example of this would be a proposal with $i^* = 8\%$ where the investment proposal is a cash flow with a single disbursement now and a single receipt one period from now. If $i_\delta = 5\%$ and $\bar{m} = 15\%$ it would be a sound decision to accept the proposal earning 8% rather than investing the money in a bank account at 5%. The 5% investment or the 8% investment are equally liquid since the money is available for investment in the very next period.

When the lives of the proposals are longer than one period the decision is not so obvious. For example Proposals A, B and C shown in Table 6.8 have the same initial cost, the same rate of return (10%), but different lives. The values of $P_j(\bar{m}, i_\delta)$ for these proposals are also shown in Table 6.8. It is assumed that $i_\delta = 5\%$ and $\bar{m} = 15\%$.

Table 6.8. THREE PROPOSALS WITH VARYING LIQUIDITY

End of Year	Proposal A	Proposal B	Proposal C
0	$-\$1,000$	$-\$1,000$	$-\$1,000$
1	576	402	315
2	576	402	315
3		402	315
4			315
$P_j(0.15, 0.05)$	\$23	\$5	$-\$14$

PROBLEMS

1. An investor is considering a business opportunity which requires the receipts and disbursements shown below. Calculate the net cash flow representing this investment and then find the present worth of the net cash flow for an interest rate of 10%.

End of Year	Disbursements (thousands)	Receipts (thousands)
0	$100	$ 0
1	10	5
2	10	25
3	5	50
4	0	50
5	0	40

2. Al, Bill, and Charlie started a business, and the following financial transactions occurred:

End of Year	Transactions
0	A, B, and C each invested $10,000
1	B paid A $2,000; business paid C $3,000
2	C paid A $1,000; B paid A $1,000; business paid $2,000 each to B and C
3	Business paid $2,000 each to A, B, and C; B paid $8,000 each to A and C for their interest in business
4	B received $36,000 from the business for his interest in the business

Show the *net cash flow* each period for each individual and the business.

3. Calculate and graph the present-worth amount as a function of the interest rate for the following cash flow.

End of Year	0	1	2	3	4	5
Cash Flow	−$10,000	$3,000	$3,000	$3,000	$3,000	$3,000

(a) Assume the interest rate is compounded annually.
(b) Assume the interest rate is compounded continuously.
(c) Repeat Problem 3 for the annual equivalent amount.
(d) Repeat Problem 3 for the future-worth amount.

4. For the following cash flow calculate and graph the present worth as a function of the interest rate.

End of Year	0	1	2	3	4	5	6	7
Cash Flow	−$40,000	$6,000	$7,000	$8,000	$9,000	$10,000	$11,000	$12,000

(a) Assume the interest rate is compounded annually.
(b) Assume the interest rate is compounded continuously.
(c) Repeat Problem 4 for the annual equivalent amount.
(d) Repeat Problem 4 for the future-worth amount.

5. Given the cash flow shown below find the present-worth amount as a function of interest rate and graph the results.

End of Year	0	1	2	3
Cash Flow	−$1,000	$800	$600	$1,100

(a) Assume the interest rate is compounded annually.
(b) Assume the interest rate is compounded continuously.
(c) Repeat Problem 5 for the annual equivalent amount.
(d) Repeat Problem 5 for the future-worth amount.

6. Graph the present-worth amount as a function of interest rate for the following cash flows.

(a)

End of Year	0	1	2	3	4
Cash Flow	−$6,000	$1,500	$1,500	$1,500	$1,500

(b)

End of Year	0	1	2	3	4
Cash Flow	$500	$1,000	$1,500	−$2,000	−$2,500

7. A man is considering giving an endowment to a university in order to provide payments of $7,000, $6,000, $5,000, and $4,000, respectively, at the end of the first, second, third, and fourth quarters during a year. If the interest rate is 8% compounded quarterly, what is the capitalized equivalent that must be deposited now so that the quarterly payments can be repeated forever?

8. A tunnel to transport water through a mountain range requires periodic maintenance. If the maintenance costs are as shown below for each 6-year maintenance cycle what is the capitalized equivalent of these expenses? The interest rate is 7%.

End of Year	1	2	3	4	5	6
Maintenance Costs	$20,000	$20,000	$20,000	$30,000	$30,000	$30,000

9. Consider the cash flows given below and assume that $i = 10\%$.

	End of Year				
	0	1	2	3	4
Cash Flow					
A	$-100	$40	$40	$40	$40
B	-100	20	20	60	60

Calculate the present-worth amounts, annual equivalents, and future-worth amounts of these two cash flows. Next calculate PW_A/PW_B, AE_A/AE_B, and FW_A/FW_B. Then compare these ratios.

What important implication can be drawn from this comparison?

10. Investment proposals A and B have the net cash flows shown below.
    (a) Compare the present worth of A with the present worth of B for $i = .05$. Which has the higher value?
    (b) Now let $i = .15$ and compare the two. Which has the higher value?
    (c) On the same axis, graph the present-worth amounts for each of these proposals as a function of the interest rate.

End of Year	0	1	2	3
A	-$1,000	$600	$500	$ 300
B	- 1,000	300	200	1,000

11. If a cash flow is composed of a series of disbursements in its early life followed by a series of receipts that total to an amount greater than the disbursements, what can be said about the relationship between rate of return and the present-worth amount?

12. Find the rate of return for the following cash flows by using the interest factors.

Year	0	1	2	3	4
(a)	-$10,000	$5,000	$5,000	$5,000	$5,000
(b)	-$ 200	$ 100	$ 200	$ 300	$ 400
(c)	-$ 500	$ 500	$1,000	$1,000	$1,000
(d)	-$ 100	$ 25	$ 25	$ 25	$ 20
(e)	-$ 1,000	$ 800	$ 400	$ 700	$ 600

13. Graph the present-worth of the following cash flow as a function of interest rate. What is the rate of return for this cash flow?

End of Period	Net Cash Flow
0	−1000
1	+2230
2	−1242

14. Graph the present-worth amount as a function of the interest rate for each cash flow shown below. The range of interest-rate values for which these graphs should be drawn extends from 0% to 50%. From the graphs determine the rates of return for each cash flow.

	End of Year				
	0	1	2	3	4
Cash Flow					
(a)	− $4,000	$ 0	$ 0	$40,000	− $40,000
(b)	− 1,000	2,500	− 1,540		
(c)	−10,000	50,000	−93,000	77,500	− 24,024
(d)	− 1,000	3,900	− 5,030	2,145	
(e)	− 1,000	3,600	− 4,320	1,728	
(f)	− 1,000	1,000	− 100	1,000	

15. Apply Descartes' rule of signs to the cash flows in Problem 14 and find the maximum number of real, positive roots that can be found for each cash flow.

16. Design a cash flow having a present-worth function that intersects the horizontal axis only once although Descartes' rule of signs indicates that the cash flow will have a maximum number of positive, real roots that exceeds one.

17. If the prospective value, $PV_j(\bar{m}, i_s)$, for investment proposal j is (a) positive, (b) negative, what is indicated about the economic desirability of proposal j. The desirability of proposal j is measured with respect to what investment strategy?

18. Calculate for the following cash flows the present-worth amount and the prospective value, $PV(\bar{m}, i_s)$ for $i = \bar{m} = 15\%$ and $i_s = 6\%$. Use the interest factors to reduce the amount of computation required.

	End of Year				
	0	1	2	3	4
Cash Flow					
(a)	−$ 1,000	$ 360	$ 360	$ 360	$ 360
(b)	− 8,000	1,300	2,300	3,300	4,300
(c)	− 11,000	2,000	2,000	2,000	5,000

19. As the difference between $\bar{m}$ and i_s increases, what is the effect on the prospective value, $PV(\bar{m}, i_s)$? Does such an increase make a proposal appear more or less desirable?

20. A proposal with an initial cost of $5,000 is expected to earn a rate of return of 15%. If a company is calculating the economic worth of this proposal using the prospective value for $\bar{m} = 20\%$ and $i_s = 5\%$, what is the longest time over which the proposal can receive equal annual payments and still appear economically desirable? What is the answer to the preceding question if the rate of return on the proposal is expected to be 8%?

7

Decision Making
Among Alternatives

Once a set of investment alternatives has been submitted for consideration the question that arises is: What investment opportunity should be selected? To answer this question it is necessary to study various decision criteria that describe how investment decisions should be made.

A decision criterion is a rule or procedure that describes how to select investment opportunities so that particular objectives can be achieved. The degree to which these objectives are realized depends on the efficacy of the decision criterion. Therefore, it is important that the strengths and weaknesses of the most commonly used decision criteria be fully understood. To develop this understanding, this chapter examines decision criteria with respect to the fundamentals of decision making.

7.1. TYPES OF INVESTMENT PROPOSALS

An *investment proposal* in this book is considered to be a single project or undertaking which is being considered as an investment possibility. It is important to distinguish an investment proposal from an *investment alternative* which is defined to be a decision option. According to the above definitions every investment proposal can be considered to be an investment alternative. However an investment alternative can consist of a group or set of investment proposals. It can also represent the option of "doing nothing." Thus, if a decision maker is considering two proposals, *A* and *B*, it is possible for him to have four decision options or alternatives, as shown on the following page.

Alternatives	X_A[1]	X_B
Do nothing (Reject A, Reject B)	0	0
Accept only A	1	0
Accept only B	0	1
Accept A and B	1	1

Independent Proposals. When the acceptance of a proposal from a set of proposals has no effect on the acceptance of any of the other proposals contained in the set, the proposal is said to be *independent*. Even though few proposals in a firm are truly independent, for practical purposes it is quite reasonable to assume that certain proposals are independent. For instance, the decision to air-condition a company's production facility would normally be considered to be independent of the decision to undertake an advertising campaign since each investment has a different function.

It is almost always possible to recognize some relationship between proposals that are functionally different. For example it might be expected that the proposed investment to air-condition the plant will lead to a lower product cost through increased productivity. This decrease in cost may have an effect on the selling price of the product which in turn may affect consumer demand. The advertising campaign is also intended to influence consumer demand. Even though such relationships may exist between investment proposals, they are usually very difficult to trace and in many cases their effect is negligible. Therefore unless such relationships are rather direct they are usually ignored. In addition, if the amounts of money available for investment in the production or marketing operations are not immediately transferable it is reasonable to assume that any financial dependency between the proposals can also be disregarded.

Usually if proposals are functionally different and there are no other obvious dependencies between them it is reasonable to consider the proposals as independent. For example, proposals concerning the purchase of a numerically controlled milling machine, a security system, office furniture, and fork lift trucks would be considered independent under most circumstances.

Dependent Proposals. For many decision problems a group of investment proposals will be related to one another in such a way that the acceptance of one of the proposals will influence the acceptance of the others. Such interdependencies between proposals occur for a variety of reasons.

First, if the proposals contained in a set of proposals are related so that the acceptance of one proposal from the set precludes the acceptance of any of the other proposals in the set, the proposals are said to be *mutually*

---

[1]For each investment proposal there is a binary variable X_j that will have the value 0 or 1 indicating that proposal j is rejected (0) or accepted (1). Each row of binary numbers represents an investment alternative. This convention is used throughout this chapter.

exclusive. Usually a situation where proposals are mutually exclusive occurs when a decision maker is attempting to fulfill a need and there are a number of proposals each of which will satisfy that need.

For example a road building contractor may require additional earth moving capability. As shown in Table 7.1 there may be a number of types of equipment, each of which could perform the function desired. Although these proposals may have different first costs and different operating characteristics, they are still considered to be mutually exclusive for decision-making purposes since the selection of one precludes the selection of the others.

Table 7.1. A SET OF MUTUALLY EXCLUSIVE PROPOSALS

Proposal	Type of Equipment
A1	Roadgrader—Model B, Manufacturer 1
A2	Roadgrader—Model C, Manufacturer 1
A3	Roadgrader—Model A, Manufacturer 2
A4	Caterpillar Tractor—Model D, Manufacturer 3
A5	Caterpillar Tractor—Model D, Manufacturer 4

Another type of relationship between proposals arises from the fact that once some initial project is undertaken there are a number of other auxiliary investments that become feasible as a result of the initial investment. Such auxiliary proposals are called *contingent* proposals because their acceptance is conditional on the acceptance of another proposal. Thus the purchase of a computer magnetic tape drive unit is contingent on the purchase of the computer central processing unit. The construction of the third floor of a building is contingent on the construction of the first and second floors. A contingent relationship is a one-way dependency between proposals. That is, the acceptance of a contingent proposal is dependent on the acceptance of some prerequisite proposal but the acceptance of the prerequisite proposal is independent of the contingent proposals.

Where there are limitations on the amount of money available for investment and the initial cost of all the proposals exceeds the money available for investment, financial interdependencies are introduced between proposals. These interdependencies are usually complex and they will occur whether the proposals are independent, mutually exclusive or contingent. Thus whenever a budget constraint is imposed on some decision problem it is important to realize that interdependencies which are not obvious are being introduced. For instance assume the three proposals shown in Table 7.2 are independent. If the decision maker has only $5,500 to spend it is seen that the acceptance of Proposal C eliminates Proposals A and B from possible acceptance. In addition, the acceptance of Proposal A or Proposal B now precludes the acceptance of Proposal C. Depending on the budget size, the number of

Table 7.2. THREE INDEPENDENT PROPOSALS

End of Year	Cash Flow		
	Proposal A	Proposal B	Proposal C
0	−$1,000	−$3,000	−$5,000
1	600	1,500	2,000
2	600	1,500	2,000
3	600	1,500	2,000

proposals, and their first costs these interdependencies can be rather complex. For the example just cited the possible arrangements of proposals that are feasible with regard to the budget constraint are shown in Table 7.3.

Table 7.3. ACCEPTABLE ARRANGEMENTS OF INDEPENDENT PROPOSALS FOR A BUDGET CONSTRAINT OF $5,500

Possible Arrangements of Proposals	Proposals			Total Money Required for each Arrangement	Budget Remaining
	X_A	X_B	X_C		
1	0	0	0	0	$5500
2	1	0	0	$1,000	4500
3	0	1	0	3,000	2500
4	0	0	1	5,000	500
5	1	1	0	4,000	1500
6	1	0	1	6,000	− 500*
7	0	1	1	8,000	−2500*
8	1	1	1	9,000	−3500*

*Infeasible arrangements

7.2. MUTUALLY EXCLUSIVE ALTERNATIVES AND DECISION MAKING

To facilitate the discussion of decision criteria in the next section the selection of investment proposals will be viewed as a problem of selecting a single decision alternative from a set of alternatives. Because the acceptance of one of these alternatives will preclude the acceptance of any of the other alternatives being considered, the alternatives are considered to be *mutually exclusive*.

Comparing Mutually Exclusive Alternatives. When comparing mutually exclusive alternatives it is the future *difference* between the alternatives that is relevant for determining the economic desirability of one compared to the other. It is this fundamental concept that is the basis for the discussion concerning decision criteria in this chapter.

The reason why the difference between alternatives is so fundamental in the comparison of alternatives is demonstrated by the comparison of alternatives $A1$ and $A2$ shown in Table 7.4. To compare the two alternatives described in Table 7.4 it is sufficient to examine the cash flow that represents the difference between $A1$ and $A2$ because the advantage or disadvantage of Alternative $A2$ over Alternative $A1$ is completely described by the cashflow representing $A2$ subtracted from $A1$.

Table 7.4. DIFFERENCES BETWEEN MUTUALLY EXCLUSIVE ALTERNATIVES

End of Year	Alternative $A1$	Alternative $A2$	Cash Flow Difference $(A2 - A1)$
0	$-\$1,000$	$-\$1,500$	$-\$500$
1	800	700	$-$ 100
2	800	1,300	500
3	800	1,300	500

The cash flow representing Alternative $A2$ can be viewed as being the sum of two separate and distinct cash flows. One of these cash flows is identical to the cash flow of Alternative $A1$. The other is the cash flow representing the difference between Alternative $A1$ and Alternative $A2$. If the cash flow representing the differences between the alternatives is economically desirable, then alternative $A2$ consisting of a cash flow that is the sum of a cash flow identical to Alternative $A1$ and a desirable cash flow is clearly economically superior to Alternative $A1$. On the other hand if the differences between alternatives is considered to be undesirable, Alternative $A2$ is inferior to Alternative $A1$.

Forming Mutually Exclusive Alternatives. In Section 7.1 it was pointed out that investment proposals can be independent, mutually exclusive, or contingent and that additional interdependencies between proposals can result if there is a limited amount of money to invest. To devise special rules to incorporate each of these different relationships in a decision criterion would produce a complicated, difficult to apply criterion.

In order to provide a simple method of handling the various types of proposals and to provide some insight into mathematical programming formulations of the decision problem a general approach is undertaken. This approach requires that all investment *proposals* be arranged so that the selection decision involves the consideration of the cash flows of *mutually exclusive alternatives* only.

All that is required is the enumeration of all the feasible combinations of the proposals under consideration. Each combination of proposals represents a mutually exclusive alternative since each combination is unique and the acceptance of one combination of proposals precludes the acceptance of any of the other combinations. The cash flow of each alternative is determined simply by adding, period by period, the cash flows of each proposal contained in the alternative being considered.

For example consider the three independent investment proposals described in Table 7.2. Table 7.5 shows the cash flow that would be realized for each mutually exlusive alternative produced from those proposals. As previously explained each row of binary numbers represent a mutually exclusive alternative where the value of X_j indicates whether proposal j is rejected (0) or accepted (1).

Table 7.5. MUTUALLY EXCLUSIVE ALTERNATIVES FOR THREE INDEPENDENT PROPOSALS

Mutually Exclusive Alternatives	Proposals X_A X_B X_C			Cash Flow End of Year			
	X_A	X_B	X_C	0	1	2	3
1	0	0	0	$ 0	$ 0	$ 0	$ 0
2	1	0	0	−1,000	600	600	600
3	0	1	0	−3,000	1,500	1,500	1,500
4	0	0	1	−5,000	2,000	2,000	2,000
5	1	1	0	−4,000	2,100	2,100	2,100
6	1	0	1	−6,000	2,600	2,600	2,600
7	0	1	1	−8,000	3,500	3,500	3,500
8	1	1	1	−9,000	4,100	4,100	4,100

Now suppose that the decision maker is considering two independent sets of mutually exclusive proposals. That is, proposals $A1$ and $A2$ are mutually exclusive while proposals $B1$ and $B2$ are mutually exclusive. However, the selection of any proposal from the set of proposals $A1$ and $A2$ is independent of the selection of any proposal from the set of proposals $B1$ and $B2$. For example the decision problem may be to select one lathe where two lathes are under consideration and to select one fork lift truck where

Table 7.6. MUTUALLY EXCLUSIVE ALTERNATIVES FOR TWO INDEPENDENT SETS OF MUTUALLY EXCLUSIVE PROPOSALS

Mutually Exclusive Alternatives	Proposals			
	X_{A1}	X_{A2}	X_{B1}	X_{B2}
1	0	0	0	0
2	1	0	0	0
3	0	1	0	0
4	0	0	1	0
5	0	0	0	1
6	1	0	1	0
7	1	0	0	1
8	0	1	1	0
9	0	1	0	1

two fork lift trucks are being considered. The decision options available to the decision maker can be represented as the set of mutually exclusive alternatives shown in Table 7.6.

If proposals are contingent it is again possible to arrange the proposals into a single set of mutually exclusive alternatives by enumerating the combination of proposals that are feasible considering the relationships that exist between the proposals. Suppose that three proposals are being considered where Proposal C is contingent on the acceptance of both proposals A and B, and Proposal B is contingent on the acceptance of Proposal A. Table 7.7 indicates the mutually exclusive alternatives that can be developed from a group of proposals with the described contingency relationships.

Table 7.7. MUTUALLY EXCLUSIVE ALTERNATIVES FOR CONTINGENT PROPOSALS

Mutually Exclusive Alternatives	Proposals		
	X_A	X_B	X_C
1	0	0	0
2	1	0	0
3	1	1	0
4	1	1	1

Now suppose there is a decision problem where there are independent, mutually exclusive and contingent proposals and in addition there is a limitation on the amount of money available for investment. As before, all the possible combinations of proposals will be listed and the cash flows for these mutually exclusive alternatives will be determined. The only additional procedure required to account for the constraint on the budget amount is the elimination of each mutually exclusive alternative that requires more money that the total amount available. An example of this procedure is demonstrated in Table 7.3 where a total budget of $5,500 was assumed for the three independent proposals in Table 7.2. By examing the first cost of each mutually exclusive alternative it is seen that alternatives 6, 7, and 8 are infeasible and therefore should be eliminated from consideration.

For problems that concern a small number of proposals the general technique just presented for arranging various types of proposals into mutually exclusive alternatives can be computationally practical. However for larger numbers of proposals the number of mutually exclusive alternatives becomes quite large and therefore this approach becomes computationally cumbersome. For instance the maximum number of mutually exclusive alternatives, N, that can be obtained where

$S =$ the maximum number of sets of proposals that are independent and $M_j =$ the maximum number of proposals within each set j where each proposal in that set is mutually exclusive is

$$N = \prod_{j=1}^{s} (M_j + 1) = (M_1 + 1)(M_2 + 1)(M_3 + 1) \ldots (M_s + 1)$$

Suppose a decision maker had the following proposals to consider:

$$
\begin{array}{cccccc}
A1 & A2 & A3 & A4 & A5 & A6 \\
B1 & B2 & B3 & & & \\
C1 & C2 & C3 & C4 & & \\
D1 & D2 & & & &
\end{array}
$$

where the proposals in each row are mutually exclusive and the set of proposals on each row is independent from any other set or row of proposals. That is, proposals $A1$ and $A2$ are mutually exclusive but proposals $A1$ and $B1$ are independent. The maximum number of mutually exclusive alternatives that can be obtained from the group of proposals in this example is found to be 420.

$$S = 4 \text{ (the number of rows)}$$
$$M_1 = 6, \ M_2 = 3, \ M_3 = 4, \ M_4 = 2$$

so that

$$
\begin{aligned}
N &= (M_1 + 1)(M_2 + 1)(M_3 + 1)(M_4 + 1) \\
&= (7)(4)(5)(3) = 420.
\end{aligned}
$$

Thus for this relatively small problem the number of mutually exclusive alternatives is sizable.

The important point that must be realized is that the approach just discussed makes possible the consideration of a variety of proposal relationships in a single form; the mutually exclusive alternative. Therefore any decision criteria that is designed to make decisions about mutually exclusive alternatives can also handle proposals that are independent, mutually exclusive, or contingent merely by arranging the proposals into mutually exclusive alternatives. In addition the imposition of a budget constraint can also be quite easily incorporated into such a decision criteria. Thus a conceptually simple and consistent approach has been presented encompassing a wide variety of investment situations.

7.3. DECISION CRITERIA FOR MUTUALLY EXCLUSIVE ALTERNATIVES

All the decision criteria discussed in this section have as their objective the maximization of profit or present-worth given that all investment alternatives must yield a return that exceeds some *minimum attractive rate of return* (MARR). This cut-off rate is usually the result of a policy decision made by the management of the firm.

The minimum attractive rate of return can be viewed as a rate at which the firm can always invest since it has a large number of opportunities that yield such a return. Thus whenever any money is committed to an invest-ment proposal an opportunity to invest that money at the MARR has been foregone. For this reason the minimum attractive rate of return is some-times considered to be an "opportunity" cost.

If this view of unlimited investment opportunities yielding a return at the MARR is extended into the future, it can be assumed that the proceeds produced by current investments can be invested at the minimum attractive rate of return. Under these circumstances the minimum attractive rate of return has been called a "reinvestment" rate since the future income received from current investments is thought to be invested or "reinvested" at this rate.

The minimum attractive rate of return should not be confused with the cost of capital; a composite rate that represents the cost of providing money from external sources through the sale of stock, the sale of bonds and by direct borrowing. Normally the minimum attractive rate of return is sub-stantially higher than the cost of capital. Where a firm's cost of capital may be 9% its minimum attractive rate of return may be 15%. This difference occurs because few firms are willing to invest in projects that are expected to earn slightly more than the cost of capital due to the risk elements in most projects and because of uncertainty about the future.

Present-Worth on Total Investment. This criterion is one of the most frequently used criteria for selecting an investment alternative from a set of mutually exclusive alternatives. Since the stated objective of the selection of alternatives is to choose the alternative with the maximum present-worth amount, the rules for this criterion are rather simple. All that is required is to calculate the present-worth amount for the cash flow representing each alternative. Then select the alternative that has the maximum present-worth amount provided this amount is positive. The present-worth amount must be positive to assure that the alternative yields a return that is greater than the minimum attractive rate of return.

To see the computational simplicity of this criterion, it is applied to the mutually exclusive alternatives described in Table 7.8. Using a MARR =

Table 7.8. CASH FLOWS REPRESENTING FOUR MUTUALLY EXCLUSIVE ALTERNATIVES

End of Year	Alternatives			
	Do Nothing	A1	A2	A3
0	$0	−$5,000	−$8,000	−$10,000
1–10	0	1,400	1,900	2,500

15% the calculations of the present-worth amounts gives

$$PW_o(15) \qquad\qquad\qquad\qquad\qquad\qquad = \$ \quad 0.00$$

$$PW_{A1}(15) = -\$\ 5{,}000 + \$1{,}400(\overset{P/A\ 15,\ 10}{5.0188}) = \$2{,}026.32$$

$$PW_{A2}(15) = -\$\ 8{,}000 + \$1{,}900(\overset{P/A\ 15,\ 10}{5.0188}) = \$1{,}535.72$$

$$PW_{A3}(15) = -\$10{,}000 + \$2{,}500(\overset{P/A\ 15,\ 10}{5.0188}) = \$2{,}547.00.$$

It is seen that the maximum value of the present-worth amounts for these four alternatives is \$2,547.00, the present-worth of Alternative $A3$. Although for this example the alternative selected happened to have the largest first cost, it is certainly possible for alternatives with the smaller first costs to have present-worths greater than those alternatives with the larger first costs. For example if Alternative $A3$ is excluded from consideration it is seen that Alternative $A1$ has a larger present-worth than Alternative $A2$ even though it requires less initial outlay.

When the receipts from a number of alternatives are assumed to be equal, it is common to describe the cash flows of the alternatives by showing only their costs. If the costs are shown as positive numbers, then the decision rule for this criterion is to select the alternative that minimizes the present-worth amount of the costs. The Do Nothing alternative is not considered when cost only cash flows are being used.

In Chapter 6 it is shown that the present-worth amount, the annual equivalent amount, and the future-worth amount are consistent bases for comparing alternatives. That is, if the present-worth amount of Alternative A is greater than the present-worth amount of Alternative B, then

$$AE_A(i) > AE_B(i)$$

and

$$FW_A(i) > FW_B(i).$$

Therefore if either the annual equivalent amount or the future-worth amount is substituted for the present-worth amount as the basis for comparison in this criterion the same solution will result. By applying the annual equivalent on total investment criterion or the future-worth on total investment criterion to the alternatives in Table 7.8 the selection of Alternative $A3$ is again selected as expected.

$$AE_o(15) \qquad\qquad\qquad\qquad\qquad\qquad = \$ \quad 0$$

$$AE_{A1}(15) = -\$5{,}000(\overset{A/P\ 15,\ 10}{0.1993}) + \$1{,}400 = \$403.50$$

$$AE_{A2}(15) = -\$8{,}000(\overset{A/P\ 15,\ 10}{0.1993}) + \$1{,}900 = \$305.60$$

$$AE_{A3}(15) = -\$10,000(\overset{A/P\ 15,\ 10}{0.1993}) + \$2,500 = \$507.00.$$

or

$$FW_o(15) \hspace{5cm} = \$ \quad 0.00$$

$$FW_{A1}(15) = -\$5,000(\overset{F/P\ 15,\ 10}{4.046}) + \$1,400(\overset{F/A\ 15,\ 10}{20.304}) = \$8,195.60$$

$$FW_{A2}(15) = -\$8,000(\overset{F/P\ 15,\ 10}{4.046}) + \$1,900(\overset{F/A\ 15,\ 10}{20.304}) = \$6,209.60$$

$$FW_{A3}(15) = -\$10,000(\overset{F/P\ 15,\ 10}{4.046}) + \$2,500(\overset{F/A\ 15,\ 10}{20.304}) = \$10,300.00.$$

An examination of the calculations of the future-worth amounts indicates that the receipts from the investment are actually invested at the minimum attractive rate of return from the time they are received to the end of the life of the alternative. Thus it is said that future-worth calculations *explicitly* consider the investment or "reinvestment" of the future receipts generated by investment alternatives. Because the three decision criteria just discussed are consistent and lead to the same selection of alternatives, it follows that use of the present-worth amount and the annual equivalent amount *implicitly* assumes the investment or "reinvestment" of alternatives' receipts at the minimum attractive rate of return.

At first glance it appears that the criterion, present-worth on total investment, violates the basic decision rule that requires the consideration of the differences between alternatives. (See Section 7.2.) The fact that the differences between the alternatives are reflected in the comparison of the present-worths on total investment becomes clear in the following example. Suppose there are two mutually exclusive alternatives $A1$ and $A2$ as shown below.

End of Year	Alternative A1	Alternative A2	A2 − A1
0	−$1,000	−$1,500	−$500
1–5	500	900	400

Alternative $A2$ can be visualized as consisting of a cash flow identical to Alternative $A1$ plus the cash flow representing the difference between alternatives $A2$ and $A1$. The cash flow portion of $A2$ that is identical to $A1$ will have the same present-worth as Alternative $A1$. Therefore the only difference between the present-worths for the total investment $A1$ and the total investment $A2$ is represented by the present-worth for the incremental cash flow $(A2 - A1)$. The realization that the present-worth on total investment criterion does in fact examine the differences in alternatives will be further substantiated in the discussion of the decision criterion, present-worth on incremental investment.

Present-Worth on Incremental Investment. When making decisions between mutually exclusive alternatives it is the differences between alternatives that are relevant for decision-making purposes. The present-worth on incremental investment criterion provides an example of this rule since it requires that the incremental differences between alternative cash flows actually be calculated.

When comparing one alternative to another the first task is to determine the cash flow representing the difference between the two cash flows. Then the decision whether to select a particular alternative rests on the determination of the economic desirability of the additional increment of investment required by one alternative over the other. The incremental investment is considered to be desirable if it yields a return that exceeds the minimum attractive rate of return. In other words, if the present-worth amount for the incremental investment is greater than zero, the increment is considered desirable and the alternative requiring this additional investment is deemed best. As long as the present-worth function for the cash flow representing the difference between alternatives, is of the form shown in Figure 6.3, any positive present-worth assures that such a cash flow will yield a return greater than the MARR.

If the cash flows representing the differences between alternatives do not have present-worth plots similar to the one shown in Figure 6.3, the rules of this criterion most probably will not be applicable. This deficiency is not generally serious since for most decision problems the cash flows are such that their differences consist of a disbursement followed by a series of receipts. It has been pointed out previously that such a cash flow pattern will produce a present-worth function of the form shown in Figure 6.3.

To apply this decision criterion to a set of mutually exclusive alternatives such as shown in Table 7.8 the following steps must be utilized.

1. List the alternatives in ascending order of their first cost or initial disbursements.

	Alternatives			
End of Year	Do Nothing	A1	A2	A3
0	$0	−$5,000	−$8,000	−$10,000
1–10	0	1,400	1,900	2,500

2. Select as the initial "current best" alternative the one which requires the smallest first cost. In most cases the initial "current best" alternative will be the alternative, Do Nothing, as it is in this example. All too frequently investment alternatives are compared without including the possibility of not undertaking the project at all. The exclusion of the Do Nothing alternative can lead to the investment of a scarce resource, money, in unproductive activities. That is, activities that yield a return that is less than the MARR.

3. Compare the initial "current best" alternative and the first "challenging" alternative. The challenger is always the next highest alternative in order of first cost that has not been previously involved in a comparison. The comparison is accomplished by examining the differences between the two cash flows. If the present-worth of the incremental cash flow evaluated at the MARR is greater than zero the challenger becomes the new "current best" alternative. If the present-worth is less than or equal to zero the "current best" alternative remains unchanged and the challenger in the comparison is eliminated from consideration. The new challenger is the next alternative in order of first cost that has not been a challenger previously. Then the next comparison is made between the alternative that is the "current best" and the alternative that is currently the challenger.

4. Repeat the comparisons of the challengers to the "current best" alternative as described in Step 3. These comparisons are continued until every alternative other than the initial "current best" alternative has been a challenger. The alternative that maximizes present-worth and provides a rate of return that exceeds the MARR is the last "current best" alternative.

Steps 3 and 4 lead to the following calculations for the alternatives being considered in Table 7.8. Assume that the MARR is equal to 15%.

The first comparison to be made in this example is between Alternative $A1$ (the first challenger) and the Do Nothing alternative (the initial "current best" alternative). The subscript notation in $PW_{A1-0}(15)$ indicates the present-worth amount is for the cash flow representing the difference between Alternative $A1$ and Do Nothing. The Do Nothing alternative is signified by zero.

$$PW_{A1-0}(15) = -\$5,000 + \$1,400(\overset{P/A\ 15,\ 10}{5.0188}) = \$2,026.32$$

Note that when comparing an alternative to the Do Nothing alternative the cash flow representing the incremental investment is the same as the cash flow on the total investment. Thus for such a situation the calculations of the present-worth on total investment criterion are exactly the same as for this criterion.

Because the present-worth amount of the differences between the cash flows is greater than zero ($2026.32), Alternative $A1$ becomes the new "current best" alternative as dictated by Step 3. The second challenger becomes Alternative $A2$. Alternative $A2$ is then compared to Alternative $A1$ on an incrementel basis as follows:

$$PW_{A2-A1}(15) = -\$3,000 + \$500(\overset{P/A\ 15,\ 10}{5.0188}) = -\$490.60.$$

Since this value is negative Alternative $A2$ is dropped from further consideration and Alternative $A1$ remains the "current best" alternative. The third challenger is Alternative $A3$. Comparing the "current best" with the next challenger yields

$$PW_{A3-A1}(15) = -\$5,000 + \$1,100(\overset{P/A\ 15,\ 10}{5.0188}) = \$520.68.$$

The present-worth on the additional investment required by Alternative $A3$ over Alternative $A1$ is positive and therefore that increment is economically desirable. Thus Alternative $A3$ becomes the "current best" alternative and the list of alternatives has been exhausted so that there is no new challenger possible. According to Step 4 when all challengers have been considered, the "current best" alternative is the alternative that maximizes present-worth and provides a return greater than the MARR. Therefore Alternative $A3$ is the optimum selection from the set of alternatives shown in Table 7.8.

The present-worth on incremental investment criterion is also appropriate for comparing alternatives that are described by cash flows that exclude the alternative's earnings. Table 7.9 presents the cash flows for a set of alternatives where it is assumed that the services provided by the alternatives are identical. Since these cash flows do not reflect the income produced by these alternatives, the Do Nothing alternative is not a meaningful alternative, i.e., no one would ever accept a cash flow that resulted in only disbursements if the opportunity to not invest was available. Thus it is assumed that it is mandatory to select one of the alternatives listed in Table 7.9.

Table 7.9. NET CASH FLOWS FOR FOUR ALTERNATIVES PROVIDING THE SAME SERVICE

End of	Alternatives			
Year	$B1$	$B2$	$B3$	$B4$
0	−$10,000	−$12,000	−$12,000	−$15,000
1	−2,500	−1,500	−1,200	−400
2	−2,500	−1,500	−1,200	−400
3	1,000*	1,500	1,500	3,000

*Positive values can arise as the result of the salvage value received when the asset is sold at the end of its life.

Applying the present-worth on incremental investment criterion to these alternatives produces the following calculations for the MARR = 10%.

$$PW_{B2-B1}(10) = -\$2,000 + \$1,000(\overset{P/A\ 10,\ 2}{1.7355}) + \$500(\overset{P/F\ 10,\ 3}{0.7513}) = \$111.15$$

Since this value is positive Alternative $B2$ becomes the "current best" alternative and Alternative $B1$ is eliminated from further consideration. The next comparison pits Alternative $B3$ against Alternative $B2$ as

$$PW_{B3-B2}(10) = \$0 + \$300(\overset{P/A\ 10,\ 2}{1.7355}) + \$0 = \$520.65.$$

Thus Alternative $B3$ is accepted and Alternative $B2$ is dropped from consideration. Now compare Alternative $B4$ to Alternative $B3$ yielding

$$PW_{B4-B3}(10) = -\$3,000 + \$800(\overset{P/A\,10,\,2}{1.7355}) + \$1,500(\overset{P/F\,10,\,3}{0.7513}) = -\$484.65.$$

The present-worth of the increment $(B4 - B3)$ is negative and therefore Althernative $B4$ is unacceptable and it is eliminated from further consideration. Since there are no more alternatives to become challengers the decision process is completed and Alternative $B3$, the last "current best" alternative, is the optimum selection.

The above calculations do consider each increment of investment but compared to the present-worth on total investment criterion the computations are more time consuming. It has been shown by example that both criteria lead to the same solution. In fact it is quite easy to prove generally that both methods will yield the same selection of alternatives. All that is required is to show that the present-worth of any Alternative B minus the present-worth of any Alternative A is equal to the present-worth of the difference between Alternative A and B. That is, demonstrate that

$$PW_B(i) - PW_A(i) = PW_{B-A}(i).$$

By definition

$$PW_j(i) = \sum_{t=0}^{n} F_{jt}(1 + i)^{-t}$$

so that

$$
\begin{aligned}
PW_B(i) - PW_A(i) &= \sum_{t=0}^{n} F_{Bt}(1 + i)^{-t} - \sum_{t=0}^{n} F_{At}(1 + i)^{-t} \\
&= F_{B0} - F_{A0} + F_{B1}(1 + i)^{-1} - F_{A1}(1 + i)^{-1} + \ldots \\
&\quad + F_{Bn}(1 + i)^{-n} - F_{An}(1 + i)^{-n} \\
&= F_{B-A,0} + F_{B-A,1}(1 + i)^{-1} + \ldots + F_{B-A,n}(1 + i)^{-n} \\
&= \sum_{t=0}^{n} F_{B-A,t}(1 + i)^{-t} \\
&= PW_{B-A}(i).
\end{aligned}
$$

Thus the relationship between the decision rules of the two present-worth criteria should be evident. If the objective is to maximize present-worth and $PW_B(i) > PW_A(i)$, the present-worth on total investment criterion says to select Alternative B. If $PW_B(i) > PW_A(i)$ then $PW_{B-A}(i)$ must be positive and the decision rule for the present-worth on incremental investment criterion is to accept B rather than A if $PW_{B-A}(i) > 0$.

The calculation of the present-worths on total investment for the alternatives described in Table 7.9 demonstrates that the incremental approach does select the alternative that minimizes total cost.

$$PW_{B1}(10) = -\$10,000 - \$2,500(\overset{P/A\ 10,\,2}{1.7355}) + \$1,000(\overset{P/F\ 10,\,3}{0.7513}) = -\$13,587.45$$

$$PW_{B2}(10) = -\$12,000 - \$1,500(\overset{P/A\ 10,\,2}{1.7355}) + \$1,500(\overset{P/F\ 10,\,3}{0.7513}) = -\$13,476.30$$

$$PW_{B3}(10) = -\$12,000 - \$1,200(\overset{P/A\ 10,\,2}{1.7355}) + \$1,500(\overset{P/F\ 10,\,3}{0.7513}) = -\$12,955.65$$

$$PW_{B4}(10) = -\$15,000 - \$400(\overset{P/A\ 10,\,2}{1.7355}) + \$3,000(\overset{P/F\ 10,\,3}{0.7513}) = -\$13,440.30$$

As has been pointed out for the present-worth on total investment criterion, the substitution of the annual equivalent amount or the future-worth amount for the present-worth amount as the basis for comparison for incremental decision making will lead to consistent solutions. The relationships shown below confirm this fact and they can be proved in a manner similar to that used for the present-worth amount.

$$AE_B(i) - AE_A(i) = AE_{B-A}(i)$$

and

$$FW_B(i) - FW_A(i) = FW_{B-A}(i)$$

Using the incremental approach and the annual equivalent amount requires the following calculations for the set of alternatives presented in Table 7.8.

$$AE_{A1-0}(15) = -\$5,000(\overset{A/P\ 15,\,10}{0.19925}) + \$1,400 = \$403.75$$

$$AE_{A2-A1}(15) = -\$3,000(\overset{A/P\ 15,\,10}{0.19925}) + \$500 = -\$97.75$$

$$AE_{A3-A1}(15) = -\$5,000(\overset{A/P\ 15,\,10}{0.19925}) + \$1,100 = \$103.75$$

By following the decision rules for an incremental analysis the optimum solution is to select Alternative $A3$. This is the same decision given by the total investment criteria in the preceeding section.

Rate of Return on Incremental Investment. This particular decision criterion is based on the same type of incremental analysis used by the criterion, present-worth on incremental investment. The only difference in the decision rules between these two criteria is the decision rule in Step 3 that determines whether an increment of investment is economically desirable. For the rate of return on incremental investment criterion the increment of investment is considered desirable if the rate of return resulting from the increment is greater than the minimum attractive rate of return ($i_{B-A}^* > $ MARR). Figure 7.1 shows the present-worth function of the increment $B - A$ and as long as it has the general form shown in Figure 7.1 the decision rules for both criteria are consistent.

To apply rate of return on an incremental basis it is first necessary to rank

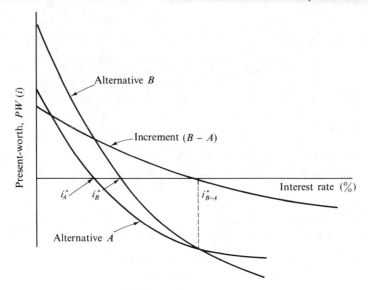

FIGURE 7.1. THE PRESENT-WORTH FUNCTION OF AN
INCREMENT OF INVESTMENT.

the alternatives in order by increasing first cost and then to select the initial
"current best" alternative. Using the set of alternatives in Table 7.8, Steps 3
and 4 of the incremental analysis procedure require the following calculations.
Find the value i^* of i so that the equation representing the present-worth
of the incremental cash flow is set equal to zero. Again the MARR is assumed
to be 15%. For increment $A1 - 0$

$$0 = -\$5,000 + \$1,400(\overset{P/A\,i,\,10}{})$$

$$i^*_{A1-0} = 25.0\%.$$

The rate of return on the increment is greater than the MARR so Alternative $A1$ becomes the initial "current best" alternative and the Do Nothing
alternative is dropped from further consideration. Next compare Alternative
$A2$ to Alternative $A1$. For the increment $(A2 - A1)$

$$0 = -\$3,000 + \$500(\overset{P/A\,i,\,10}{})$$

$$i^*_{A2-A1} = 10.5\%.$$

The rate of return of this increment is less than the MARR so Alternative
$A1$ remains the "current best" and Alternative $A2$ is rejected. Then compare
Alternative $A3$ to Alternative $A1$, the "current best" alternative. For increment $A3 - A1$

$$0 = -\$5,000 + \$1,100(\overset{P/A\ i,\ 10}{})$$

$$i^*_{A3-A1} = 17.6\%.$$

Alternative $A3$ becomes the "current best" alternative and Alternative $A1$ is removed from consideration. Since all the alternatives have been compared Alternative $A3$, the last "current best" alternative, is the optimum solution. This is the same solution given by the present-worth on total investment criterion and the present-worth on incremental investment criterion.

In general the three criteria discussed to this point will yield identical solutions for most types of investment decision problems. If present-worth plots other than the general form shown in Figure 7.1 occur for the cash flows being considered these three criteria could lead to inconsistent results unless a more detailed analysis is undertaken. For most problems encountered in practice the three criteria that have been discussed will select the alternative that maximizes total present-worth and which provides a return greater than the MARR.

One very important point to recognize from this example is that selecting the alternative with the highest rate of return on its total cash flow may *not* lead to the alternative that will maximize the total present-worth at the MARR. The rates of return for the *total* cash flows of the alternatives in Table 7.8 are

$$i^*_0 = 15\%, \quad i^*_{A1} = 25\%, \quad i^*_{A2} = 19.9\%, \quad i^*_{A3} = 21.9\%.$$

If Alternative $A1$ is selected because it has the maximum rate of return, the total present-worth will *not* be maximized for a MARR $= 15\%$. It has already been shown that Alternative $A3$ will maximize the present-worth for that minimum attractive rate of return.

The reason for the inconsistency introduced by examining the rates of return on total investment can be demonstrated in a number of ways. First the relationship between rate of return on total cash flows and rate of return on incremental cash flows is not the same as the relationship between present-worth amounts. That is, for two alternatives A and B

$$i^*_B - i^*_A \text{ does not necessarily equal } i^*_{B-A}.$$

Therefore even though using rate of return on an incremental basis does produce the optimum solution there is no relationship between the incremental approach and the total cash flow approach to assure an optimum solution from the latter approach.

Secondly, a review of the decision making procedures for the incremental approach shows that the acceptance of the increment of investment ($A3 - A1$) yields a rate of return of 17.6%. This rate is higher than the MARR of 15% but it is lower than the rate of return of 25% that would be earned if

Alternative $A1$ is accepted. Since it is always desirable to continue investing additional funds as long as they earn more than the MARR, the increment $(A3 - A1)$ is acceptable. Therefore Alternative $A3$ is more desirable than Alternative $A1$ since a return of 25% is earned on the portion of its cash flow that is identical to Alternative $A1$ and a 17.6% return is earned on the portion of its cash flow represented by the increment $(A3 - A1)$. It is expected that the combination of these two cash flows would yield a return less than 25% and greater than 17.6%. In fact, Alternative $A3$ has a rate of return of 21.9% for its total cash flow.

Third, Figure 7.2 indicates a situation where the rate of return on total investment produces a solution contrary to the objective of maximizing present-worth for a particular MARR. For any MARR that has a value less than the interest rate at which the present-worths of Alternative A and B are equal, (i_{B-A}^*), Alternative B maximizes the present-worth amount. However the rates of return for alternatives A and B indicate that i_A^* is greater

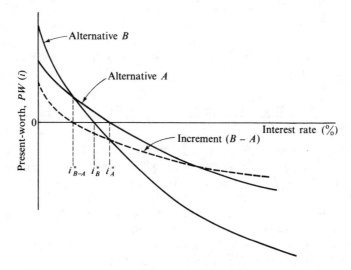

FIGURE 7.2. RATES OF RETURN ON TOTAL INVESTMENT
AND INCREMENTAL INVESTMENT.

than i_B^*; a contradiction to our objective to maximize present-worth for the MARR.

The application of rate of return on incremental investment where the income cash flows are assumed to be equal requires no changes in the criterion's decision rules. For example the set of alternatives described in Table 7.9 are analyzed in the following way. After placing the alternatives in order of ascending first cost each increment of investment is compared to the "current best" alternative. In this case the initial "current best" alternative is Alternative $B1$ and the MARR is 10%

$$0 = PW_{B2-B1}(i) = -\$2,000 + \$1,000(\overset{P/A\,i,\,2}{\quad}) + \$500(\overset{P/F\,i,\,3}{\quad})$$

$i^*_{B2-B1} = 13.5\%$ (B_2 becomes "current best" alternative)

$$0 = PW_{B3-B2}(i) = \$0 + \$300(\overset{P/A\,i,\,2}{\quad}) + \$0$$

$i^*_{B3-B2} = \infty$ (B_3 becomes "current best" alternative)

$$0 = PW_{B4-B3}(i) = -\$3,000 + \$800(\overset{P/A\,i,\,2}{\quad}) + 1,500(\overset{P/F\,i,\,3}{\quad})$$

$i^*_{B4-B3} = 1.5\%$ (B_3 remains "current best" solution).

Alternative $B3$ is the optimum selection as has been previously shown.

7.4. APPLYING DECISION CRITERIA WHEN MONEY IS LIMITED

Up to now the decision criteria that have been discussed have been applied to sets of mutually exclusive alternatives and it has been assumed that sufficient money is available to undertake all of the proposals. Now it will be assumed that the total money available for investment is a fixed amount thus restricting the number of feasible alternatives.

To illustrate the technique required to incorporate a budget constraint in the decision process, the present-worth on total investment criterion is applied to the proposed investments described in Table 7.10. Each set of

Table 7.10. CASH FLOWS FOR FIVE INVESTMENT PROPOSALS

Proposal	First Cost	Net Income (Years 1–10)	Salvage Value (Year 10)	$PW_j(8)$
$A1$	$-\$10,000$	$\$2,000$	$\$1,000$	$\$3,883$
$A2$	$-12,000$	$2,100$	$2,000$	$3,018$
$B1$	$-20,000$	$3,100$	$5,000$	$3,160$
$B2$	$-30,000$	$5,000$	$8,000$	$7,255$
$C1$	$-35,000$	$4,500$	$10,000$	-173

proposals with the same letter designation ($A1$, $A2$) is considered to be independent of the other sets having different letter designations ($B1$, $B2$). Therefore no more than one proposal with the same letter can be accepted while proposals having different letters can be accepted together. The problem is to select the proposal or proposals that maximize total present-worth if the amount of money available for investment is $35,000 and the minimum attractive rate of return equals 8%.

No changes at all are required in order to include the budget constraint in the decision process. All that is required is that the proposals be re-arranged into mutually exclusive alternatives as described in Section 7.2. To reduce the number of mutually exclusive combinations that must be consider-ed it is usually worthwhile to first calculate the present-worth for the cash flows of each proposal. If any proposal has a present-worth that is not positive it can be eliminated from consideration immediately. Clearly any proposal with a negative present-worth when combined with other proposals to make a mutually exclusive alternative will reduce the total present-worth of that alternative.

The present-worth amounts at 8 % for each proposal are shown in Table 7.10 in the far right-hand column. All the present-worths are positive except for Proposal $C1$ which can be immediately dismissed from further considera-tion. The remaining proposals are arranged into mutually exclusive alter-natives and these alternatives along with their cash flows and present-worth amounts are shown in Table 7.11.

Next those alternatives that have a first cost that exceeds the budget amount of $35,000 must be dropped from consideration. Thus the alternative to accept $A1$ and $B2$ and the alternative to accept $A2$ and $B2$ are eliminated since they are infeasible.

The last step is to select the remaining alternative that maximizes the present-worth. It is seen that Alternative 5 representing the acceptance of Proposal $B2$ meets this objective with a present-worth equal to $7,255. Thus the optimum solution to this problem is to select Proposal $B2$ and to reject all the other proposals. It is assumed that the $5,000 that is still available for investment after Alternative $B2$ is undertaken will be invested at the minimum attractive rate of return.

Usually the present-worth on total investment criterion is the most efficient criterion to use when solving this type of problem by enumerating all the mutually exclusive alternatives. It is important to realize that it would be just as easy and just as correct to use annual equivalent or future-worth on the basis of total investment.

In addition the incremental approaches described for present-worth, annual equivalent, future-worth and rate of return would also produce the same solution. The only difference from the approach used in these examples would be that the mutually exclusive alternative formed from the proposals under consideration must be placed in order of ascending first cost. In Table 7.11 this would mean switching Alternatives 7 and 8. Then additional cal-culations would be required to find the incremental cash flows representing the differences between the mutually exclusive alternatives. The decision rules would be exactly the same as those used in the incremental analysis of the examples in Section 7.4.

Table 7.11. CASH FLOWS FOR THE MUTUALLY EXCLUSIVE ALTERNATIVES FROM PROPOSAL IN TABLE 7.10

Alterna-tives	Proposals				Proposals Accepted	First Cost	Net Income Years 1–10	Salvage Value Year 10	$PW_j(8)$
	X_{A1}	X_{A2}	X_{B1}	X_{B2}					
1	0	0	0	0	None	$ 0	$ 0	$ 0	$ 0
2	1	0	0	0	A1	−10,000	2,000	1,000	3,883
3	0	1	0	0	A2	−12,000	2,100	2,000	3,018
4	0	0	1	0	B1	−20,000	3,100	5,000	3,160
5	0	0	0	1	B2	−30,000	5,000	8,000	7,255**
6	1	0	1	0	A1, B1	−30,000	5,100	6,000	7,043
7	1	0	0	1	A1, B2	−40,000	7,000	9,000	11,138*
8	0	1	1	0	A2, B1	−32,000	5,200	7,000	6,178
9	0	1	0	1	A2, B2	−42,000	7,100	10,000	10,273*

*Infeasible because the alternative requires more money than is available for investment ($35,000).

**Alternative that maximizes present-worth for a limited budget.

Linear Integer Programming Formulation of the Budget Constraint Problem. It is obvious that where there are large numbers of proposals under consideration the number of mutually exclusive alternatives that can be formed is quite large. (See Section 7.3.) For instance the number of mutually exclusive alternatives that can be formed from 100 independent proposals is approximately 1.268×10^{30}. Therefore to solve many problems of practical significance it is necessary to be able to utilize mathematical techniques that can consider all possible alternatives without having to make calculations for each alternative.

One such technique is linear integer programming. This technique requires that the problem under consideration be formulated according to a particular format. Once the problem has been properly formulated there are a number of solution procedures or algorithms that are available to solve problems of such a structure. These algorithms usually converge to the optimum solution in a highly efficient manner. Many of these algorithms are available as computer programs and the problem solutions generally require only a few minutes of computer time. Because an analysis of the mathematics of these algorithms is beyond the scope of this book, the emphasis of this discussion is on the formulation of the investment decision problem as a linear integer programming problem.

The general format of the linear integer programming problem is written as follows:

$$\text{Maximize } Z = c_1 x_1 + c_2 x_2 + \ldots + c_n x_n$$

subject to

$$a_{11} x_1 + a_{12} x_2 + \ldots + a_{1n} x_n \leq b_1$$
$$a_{21} x_1 + a_{22} x_2 + \ldots + a_{2n} x_n \leq b_2$$
$$\vdots \qquad\quad \vdots \qquad\qquad \vdots$$
$$a_{m1} x_1 + a_{m2} x_2 + \ldots + a_{mn} x_n \leq b_m$$
$$x_i \text{ for } i = 1, 2, \ldots n \text{ must be an integer} \geq 0$$

The c's, a's and b's are constants and the x's are the decision variables representing the values to be determined. The solution to this type of problem is given by the values of the x's such that Z will be maximized and all the constraints (the equations under "subject to") are satisfied.

The problem of making decisions between alternatives is easily converted to the linear integer programming format. The decision variables X_j rather than being any integer greater than or equal to zero are confined to the integers 0, (reject Proposal j) or 1, (accept Proposal j). The value of c_j is the present-worth of Proposal j at the minimum attractive rate of return, $PW_j(\text{MARR})$. The constraints for this type of problem are the result of two different types of relationships. One type of constraint reflects the limitation of the amount of money available for investment. That is, the total first cost of the proposals undertaken must not exceed the amount of the budget. The second type of constraint reflects the relationships between proposals such as whether the proposals are mutually exclusive, independent or contingent. Thus, for the decision problem related to the proposals in Table 7.10 the integer programming formulation is as follows:

Maximize $Z = \$3,883X_{A1} + \$3,018X_{A2} + \$3,160X_{B1} + \$7,255X_{B2} - \$173X_{C1}$

subject to

(budget constraint)

$$\$10,000X_{A1} + \$12,000X_{A2} + \$20,000X_{B1} + \$30,000X_{B2}$$
$$+ \$35,000X_{C1} \leq \$35,000$$

(mutually exclusive) constraint

$$X_{A1} + \qquad X_{A2} \qquad\qquad\qquad \leq 1$$

(mutually exclusive) constraint

$$X_{B1} + \qquad X_{B2} \qquad \leq 1$$

(zero—one constraint) $X_j = 0$ or 1 for all proposals.

Solving this problem with existing techniques yields the solution

$$X_{A1} = 0,\ X_{A2} = 0,\ X_{B1} = 0,\ X_{B2} = 1,\ X_{C1} = 0$$

Thus Proposal $B2$ is accepted and all the other proposals are rejected as previously shown in Table 7.11. The value of Z for this solution is $7,255. See Appendix A for a more detailed discussion of linear programming and the investment problem.

7.5. OTHER DECISION CRITERIA

There exist decision criteria other than those previously discussed for selecting alternatives from a set of alternatives. Two of these criteria are discussed in this section to illustrate some significant differences that exist between the various criteria that are used in economic analyses. The two criteria that are examined are the rank on rate of return criterion, and the maximum prospective value criterion.

Rank on Rate of Return. The decision rules of this criterion are accurately described by its name; rank on rate of return. The rate of return is calculated for each proposal and then the proposals are ranked in descending order of rate of return. The proposal with the highest rate of return is ranked first, the proposal with the second highest second, etc. The decision rule is to move down the ranked proposals accepting each proposal until there are no more proposals with a rate of return greater than the MARR. Although this criterion is widely used it has some important deficiencies.

For example, ranking on the rate of return will guarantee to select the set of proposals that maximize the total present-worth amount only if all the proposals are independent and there is no limitation on the money available for investments. The fact that rank on rate of return is a reliable criterion only for the conditions just described is easily demonstrated by a comparison with the decision criterion, rate of return on incremental investment.

First it must be recognized that since there are no interdependencies between the proposals the decision about each proposal is independent of the decisions regarding any of the other proposals. Associated with each independent decision are two mutually exclusive alternatives; accept the proposal or reject the proposal (Do Nothing). Thus if rate of return is applied to such a set of proposals on an incremental basis, each proposal can be independently accepted or rejected on the basis of the rate of return on its increment of investment. This increment of investment represents the difference between accepting the proposal and doing nothing. As a result the rate of return on the increment of investment is the *same* as the rate of return on total investment.

Recall that the decision rule for rate of return on incremental investment is to accept each increment of investment as long as its rate of return exceeds the MARR. This decision rule is effectively the same as the rank on rate of return decision rule that says to accept every proposal with a total investment rate of return greater than the MARR. Therefore under the stated conditions it is seen that rank on rate of return is the same as rate of return on incremental investment. It has been previously shown that the latter criterion does guarantee an optimum solution as long as the cash flows has a normal present-worth function.

However if mutually exclusive relationships are introduced or a budget constraint limits the money available for investment this ranking scheme faulters. As soon as interdependencies exist between the proposals it becomes necessary to compare one alternative to another rather than comparing each proposal to the MARR as is done when all proposals are independent. It has already been demonstrated in the discussion of the rate of return on incremental investment criterion in Section 7.3 that comparing one alternative to another on the basis of rate of return on total investment may lead to non-optimum solutions. (See Figure 7.2)

In addition if the proposals are considered to be indivisible (no fractional portion of the project can be accepted) the ranking procedure of this criterion can lead to nonoptimum solutions. This is easily seen by considering the following three independent proposals with a budget constraint of $50,000 and a MARR $= 10\%$.

Proposal	First Cost	Rate of Return
A	$ 1,000	25%
B	20,000	24%
C	30,000	23%

If these proposals were selected in strict order of ranking then proposals A and B would be selected. However it should be clear that it is economically more desirable to have $30,000 earning a return of 23% than to have $1,000 yielding a 25% return while 10% is earned on the remaining $29,000. Therefore the optimum selection from this set of proposals consists of Proposal B and Proposal C. It is this sort of difficulty that results from any decision criterion that uses a ranking procedure. Where there are numerous independent proposals being considered and each of their first costs is a small proportion of the total budget this effect is mitigated.

Maximum Prospective Value Criterion. The objective of this criterion is to select from a set of proposed investments those proposals that maximize the total prospective value. In Section 6.8 the basis for comparison use by the maximum prospective value criterion (MPV) is defined as

$$P_j(i_\delta, \bar{m}) = F_{jo} \frac{(1 + i_\delta)}{(1 + \bar{m})} + \sum_{t=1}^{n} F_{jt}(1 + \bar{m})^{-t}.$$

The selection of the two rates of interest $(i_\delta, \bar{m})$ required for the calculation of the prospective value is quite important to the successful implementation of this decision criterion. The rate of interest i_δ represents the return that can be received on the money that is invested so that it is available for investment at the time the next decision is made. Usually this rate reflects the return that can be earned on bank accounts, short-term government securities, or other similar investments that are easy to convert to cash.

The rate of interest $\bar{m}$ represents the rate at which the cash flow *differences* between the two best alternative set of proposals are expected to be invested during the future. A detailed discussion of the rate selection for this criterion is not presented here but it is available elsewhere.[3] For most practical investment decisions $\bar{m}$ can be approximated by the minimum attractive rate of return. Following are three conditions for which the MARR is a good approximation to $\bar{m}$:

---

[3]Oakford, R. V., and Thuesen, G. J., "The Maximum Prospective Value Criterion," *The Engineering Economist*, Vol. 13, No. 3, 1968.

1. When many proposals are being considered simultaneously.
2. When each proposal requires only a small proportion of the money available for investment.
3. When the proposals return their investment on a regular periodic basis.

The effectiveness of the MPV criterion has been tested and compared against the present-worth on total investment criterion, the rank on rate of return criterion, and other criteria.[4] In general this criterion does a better job of maximizing the capital growth rate because the criterion considers the advantage of investing now or not investing in light of future opportunities that are anticipated. Also the two rate formulation of this criterion incorporates relationships that are frequently found in actual practice.

The selection of proposals from large numbers of proposals where all cash flows have only an initial disbursement followed by a series of receipts can be accomplished by the formulation and solution of a linear integer programming problem. When the budget amount is B and the X_j's are the decision variables the linear integer programming formulation gives

$$\text{Maximize } Z = \sum_j P_j(i_\delta, \bar{m})X_j$$

subject to

$$-\sum_j F_{jo}X_j \leq B.$$

$$X_j = 0 \text{ or } 1 \text{ for all } j.$$

Any relationships between proposals will require additional constraints i.e. mutually exclusive, contingent, etc.

7.6. COMPARISON OF ALTERNATIVES WITH UNEQUAL SERVICE LIVES

Up to now all the examples that have been utilized to demonstrate the application of the various decision criteria have consisted of alternatives that have equal service lives. Often it is necessary to compare alternatives for which the time span of service will not be equal. In such situations it is necessary to make certain assumptions about the service interval so that the techniques of decision making just discussed are applicable.

When comparing alternatives with unequal lives the principle that all alternatives under consideration must be compared over the same time

---

[4]Oakford, R. V. and Thuesen, G. J., "The Effectiveness of the Maximum Prospective Value Criterion," *The Proceedings of the 19th Annual Institute Conference and Convention American Institute of Industrial Engineers*, May 1968.

span is basic to sound decision making. The time span over which alternatives are considered must be equal so that the effect of undertaking one alternative can be considered to be identical to the effect of undertaking any of the other alternatives. Clearly the direct comparison of Alternative A with a five year life and Alternative B with an eleven year life fails to consider the possible investments that could be undertaken during the six years following Alternative A's termination.

There are two basic approaches that can be used so that alternatives with different lives can be compared over an equal time span. The first approach confines the consideration of the effects of the alternatives being evaluated to some study period that is usually the life of the shortest-lived alternative. To illustrate this approach suppose a decision must be made as to which alternative should be selected from the two alternatives described in Table 7.12. It is assumed that these two alternatives provide the same service for each year that they are in existence.

Table 7.12. TWO ALTERNATIVES WITH UNEQUAL LIVES

End of Year	Alternative $A1$	Alternative $A2$
0	$-\$15,000$	$-\$20,000$
1	$-7,000$	$-2,000$
2	$-7,000$	$-2,000$
3	$-7,000$	$-2,000$
4	$-7,000$	—
5	$-7,000$	—

The study period for this example is chosen to be three years, the life of Alternative $A2$. Using the annual equivalent on total investment for an interest rate of 7% yields

$$AE_{A1}(7) = \overset{A/P\,7,\,5}{-\$15,000(0.2439)} - \$7,000 = -\$10,659 \text{ per year.}$$

The $15,000 first cost of Alternative $A1$ is distributed over its entire life to find its equivalent cost per year.

For Alternative $A2$

$$AE_{A2}(7) = \overset{A/P\,7,\,3}{-\$20,000(0.3811)} - \$2,000 = -\$9,622 \text{ per year.}$$

The cost advantage of Alternative $A2$ over Alternative $A1$ is $1,037 per year for the first three years. For years 4 and 5 Alternative $A1$ costs $10,659 more than Alternative $A2$ which provides no service for those last two years. Since the study period has been selected as three years the cost advantage of Alternative $A2$ over Alternative $A1$ is stated as $1,037 per year for three years. The costs occurring after the study period are disregarded since the equivalent costs are being compared only for the study period indicated.

The costs occuring after the study period would be considered when Alternative $A2$'s successor is to be compared to continuing with Alternative $A1$. The decision about $A2$'s successor is assumed to be separable from the original decision when the study period approach is used. An implication of this approach is that for any alternatives with a life longer than the study period the unrecovered balance of their first cost at the end of the study period is the assumed salvage value for these alternatives. For the example just discussed the assumed salvage for Alternative $A1$ after three years would have to be

$$\overset{A/P\,7,\,5}{\$15,000(0.2439)}\overset{P/A\,7,\,2}{(1.8080)} = \$6,615.$$

The second approach to the problem of unequal lives is to estimate the future sequence of events that are anticipated for each alternative being considered so that the time span is the same for each alternative. Two methods that are frequently used to accomplish this end are

1. The explicit consideration of future alternatives over the same time span.
2. The assumption that an investment opportunity will be replaced by an identical alternative until a common multiple of lives is reached.

To illustrate the first method suppose that it is anticipated that after Alternative $A2$ in Table 7.12 is terminated the service it was providing is continued by incurring costs of $15,000 at the end of years 4 and 5. Now the service is provided over equal time spans of five years and the annual equivalent costs for Alternative $A2$ and the additional expenditures required in years 4 and 5 are

$$AE(7) = \left[-\$20,000 - \overset{P/A\,7,\,3}{\$2,000(2.6243)}\right]\overset{A/P\,7,\,5}{(0.2439)} - \overset{F/A\,7,\,2}{\$15,000(\,2.070\,)}\overset{A/F\,7,\,5}{(0.1739)}$$
$$= -\$11,558.$$

The annual equivalent cost for Alternative $A1$ has been computed for a life of five years to be $10,659 per year. Now Alternative $A1$ has an annual cost advantage of $11,558 less $10,659 over Alternative $A2$ and its replacement. This advantage is stated as $899 per year for five years.

The second method that can be used to equate alternatives with unequal lives is to assume that each opportunity will be replaced by itself until a common multiple of lives is reached. The annual equivalent should be used when such an assumption is made since it is computationally the most efficient approach. As pointed out in Table 6.4, it is necessary only to compute the annual equivalent amount over an alternative's life when determining the annual equivalent amount for the alternative and its repeated cash flows. Thus under the assumption of repeated replacements the annual equivalents for the two alternatives in Table 7.12 are

$$AE_{A1}(7) = \overset{A/P\,7,\,5}{-\$15,000(0.2439)} - \$7,000 = -\$10,659 \text{ per year}$$

and

$$AE_{A2}(7) = \overset{A/P\,7,\,3}{-\$20,000(0.3811)} - \$2,000 = -\$9,622 \text{ per year.}$$

The lowest common multiple of years for these two alternatives is fifteen. Therefore, when using this method of examining alternatives over equal time spans the cost advantage of Alternative $A2$ over Alternative $A1$ is stated as $1,037 per year for fifteen years. If in fact the alternatives are replaced with similar alternatives as assumed, this approach is sound. However, it is infrequent that a sequence of alternatives will repeat themselves since technological progress can lead to improved alternatives in the future. This method of comparing alternatives tends to overstate the differences between the alternatives when it assumes that the differences will occur over a time span that exceeds the service lives of the current alternatives.

To use present-worth calculations for the method just discussed requires additional computation. The annual equivalent can be calculated for the life of each alternative and it is then converted to a present-worth amount over the same time period.

$$PW_{A1}(7) = \overset{P/A\,7,\,15}{-10,659(\,9.1079\,)} = -\$97,081$$

$$PW_{A2}(7) = \overset{P/A\,7,\,15}{-\,9,622(\,9.1079\,)} = -\$87,636$$

An even more laborious way of making these present-worth calculations is to describe the repeated cash flows so that the receipts and disbursements of the alternative and its successor are known year by year over the number of years that is the common multiple. Then the present-worth of the cash flow representing each alternative course of action can be properly calculated. It should be clear that to calculate the present-worth for cash flows of unequal duration is incorrect. That is, for the example just presented the following calculations are incorrect for comparing alternatives $A1$ and $A2$ of Table 7.12.

$$PW_{A1}(7) = \overset{P/A\,7,\,5}{-\$15,000 - \$7,000(4.1002)} = -\$43,701$$

$$PW_{A2}(7) = \overset{P/A\,7,\,3}{-\$20,000 - \$2,000(2.6243)} = -\$25,248$$

Such a calculation and comparison implies that for years 4 and 5 Alternative $A2$ will provide a service or income equal to that of Alternative $A1$ at no cost. Thus, when present-worth comparisons are made it is essential that the alternatives be compared over the same time span.

PROBLEMS

1. The estimated annual incomes and costs of a prospective venture are as follows:

Year End	Income	Cost
0	$ ·0	$1,200
1	700	100
2	700	200
3	800	250

Determine if this is a desirable venture by the (a) present-worth comparison, (b) the annual equivalent comparison, and (c) the rate-of-return comparison for a minimum attractive rate of return of 15%; 25%.

2. A prospective venture is described by the following receipts and disbursements:

Year End	Receipts	Disbursements
0	$ 0	$6,000
1	1,600	600
2	2,000	500
3	2,800	400
4	3,200	200

For an interest rate of 12% determine the desirability of the venture on the basis of:
(a) The present-worth cost comparison.
(b) The annual equivalent comparison
(c) The rate-of-return comparison

3. An engineering graduate estimated that his education had cost the equivalent of $16,000, as of the date of graduation, considering his increased expenses and loss of earnings while in college. He estimated that his earnings during the first decade after leaving college would be no greater than if he had not gone to college. If, by virtue of his added preparation, $3,000, $5,000, and $7,000 additional per year is earned in succeeding decades, what is the rate of return realized on his $16,000 investment in education?

4. A temporary warehouse with a zero salvage value at any point in time can be built for $8,000. The annual value of the storage space less annual maintenance and operating costs is estimated to be $1,260. If the interest rate is 12% and the warehouse is used 8 years, will this be a desirable investment? For what life will this warehouse be a desirable investment?

5. A silver mine can be purchased for $190,000. On the basis of estimated production, an annual income of $28,000 is foreseen for a period of 15 years. After 15 years, the mine is estimated to be worthless. What annual rate of return is in prospect? If the minimum attractive rate of return is 15%, should the mine be purchased?

6. A special lathe was designed and built for $75,000. It was estimated that the lathe would result in a saving in production cost of $10,500 per year for 20 years. With a zero salvage value at the end of 20 years, what was the expected rate of return? Actually, the lathe became inadequate after 6 years of use and was sold for $20,000. What was the actual rate of return?

7. As usually quoted, the prepaid premium of insurance policies covering loss by fire and storm for a 3-year period is 2.5 times the premium for one year of coverage. What

rate of interest does a purchaser receive on the additional present investment if he purchases a 3 year policy now rather than three 1-year policies at the beginning of each year? If your interest rate is 9% which alternative would be most desirable?

8. A manufacturer pays a patent royalty of $0.70 per unit of a product he manufactures, payable at the end of each year. The patent will be in force for an additional 4 years. Previously he manufactured 8,000 units of the product per year but it is estimated that output will be 10,000, 12,000, 14,000, and 16,000 in the 4 succeeding years. He is considering (a) asking the patent holder to terminate the present royalty contract in exchange for a single payment at present or (b) asking the patent holder to terminate the present contract in exchange for equal annual payments to be made at the beginning of each of the four years. If 10% interest is used, what is (a) the present single payment and (b) the beginning-of-the-year payments that are equivalent to the royalty payments in prospect under the present agreement?

9. The heat loss through the exterior walls of a building costs $206 per year. Insulation that will reduce the heat loss cost by 93% can be installed for $116, and insulation that will reduce the heat loss cost by 89% can be installed for $90. Determine which insulation is most desirable if the building is to be used for 8 years and if the interest rate is 10%.

10. An industrial firm can purchase a special machine for $20,000. A down payment of $2,000 is required and the balance can be paid in 5 equal year-end installments plus 7% interest on the unpaid balance. As an alternative the machine can be purchased for $18,000 in cash. If the firm's minimum attractive rate of return is 10%, determine which alternative should be accepted. Use the present-worth on incremental investment approach.

11. A needed service can be purchased for $90 per unit. The same service can be provided by equipment which costs $100,000 and which will have a salvage value of $20,000 at the end of 10 years. Annual operating expense will be $7,000 per year plus $25 per unit.

(a) If these estimates are correct, what will be the incremental rate of return on the investment if 400 units are produced per year?

(b) What will be the incremental rate of return on the investment if 250 units are produced per year?

(c) If the firm providing this service has an interest rate of 12% what would be the alternative to select for the production levels in part (a) and part (b)?

12. Every year the stationery department of a large concern uses 1,200,000 sheets of paper with three holes drilled for binding and 250,000 sheets that have the corners rounded. At present the drilling and corner cutting is done by a commercial printing establishment at a cost of $0.35 and $0.30 per thousand sheets, respectively.

Two alternatives are being considered. Alternative A consists of the purchase of a paper drill for $450, and alternate B consists of the purchase of a combination paper drill and corner cutter for $650. Obviously the two alternatives do not provide equal service. The following data apply to the two machines:

	Drill	Combined Drill and Cutter
Life	12 years	12 years
Salvage value	$50.00	$55.00
Annual maintenance	5.00	6.00

	Drill	Combined Drill and Cutter
Annual space charge	11.00	11.00
Annual labor to drill........................	32.00	36.00
Annual labor to cut corners	—	24.00
Interest rate	8%	8%

(a) Alter one or the other of the alternates given above so that they may be compared on an equitable basis. Calculate the equivalent annual cost of each of the revised alternatives.

(b) What other alternative or alternatives should be considered?

13. A manufacturing plant and its equipment are insured for $700,000. The present annual insurance premium is $0.86 per $100 of coverage. A sprinkler system with an estimated life of 20 years and no salvage value at the end of that time can be installed for $18,000. Annual operation and maintenance cost is estimated at $360. Taxes are 0.8% of the initial cost of the plant and equipment. If the sprinkler is installed and maintained, the premium rate will be reduced to $0.38 per $100 of coverage.

(a) How much of an incremental rate of return is in prospect if the sprinkler system is installed?

(b) If the minimum attractive rate of return is 15%, which alternative should be selected?

14. It is estimated that the annual heat loss cost in a small power plant is $310. Two competing proposals have been formulated which will reduce the loss. Proposal A will reduce heat loss cost by 60% and will cost $170. Proposal B will reduce heat loss cost by 55% and will cost $130. If the interest rate is 8%, and if the plant will benefit from the reduction in heat loss for 10 years, which proposal should be accepted?

(a) Use present-worth on total investment.

(b) Use present-worth on incremental investment.

(c) Use annual equivalent on total investment.

(d) Use annual equivalent on incremental investment.

(e) Use rate of return on incremental investment.

15. An engineering student who will soon receive his B.S. degree is contemplating continuing his formal education by working toward an M.S. degree. The student estimates that his average earnings for the next six years with a B.S. degree will be $10,000 per year. If he can get an M.S. degree in one year his earnings should average $11,200 per year for the subsequent five years. His earnings while working on the M.S. degree will be negligible and his additional expenses will be $2,400.

The engineering student estimates that his average per year earnings in the three decades following the initial six-year period will be $12,000 $14,200, and $16,400 if he does not stay for an M.S. degree. If he receives an M.S. degree his earnings in the three decades can be stated as $12,000 + x, $14,200 + x, and $16,400 + x. For an interest rate of 10% find the value of x for which the extra investment in formal education will pay for itself.

16. A 100-horsepower motor is required to power a large capacity blower. Two motors have been proposed with the following engineering and cost data.

	Motor A	Motor B
Cost	$3,600	$3,000
Life	12 years	12 years

	Motor A	*Motor B*
Salvage value	0	0
Efficiency $\frac{1}{2}$ load	85%	83%
Efficiency $\frac{3}{4}$ load	92%	89%
Efficiency full load.......................	89%	88%
Hours use per year at $\frac{1}{2}$ load	800	800
Hours use per year at $\frac{3}{4}$ load	1,000	1,000
Hours use per year at full load	600	600

Power cost per kilowatt-hour is $0.03. Annual maintenance, taxes, and insurance will amount to 1.4% of the original cost. Interest is 8%.

(a) What is the equivalent annual cost for each motor?

(b) What will be the return on the additional amount invested in Motor A?

17. Three mutually exclusive proposals requiring different investments are being considered. The life of all three alternatives is estimated to be 20 years with no salvage value. The minimum rate of return that is considered acceptable is 4%. The cash flows representing these 3 proposals are shown below.

Proposal	*A1*	*A2*	*A3*
Investment proposal	$-\$80,000$	$-\$50,000$	$-\$100,000$
Net income per year	6,440	5,100	9,490
Return on total investment	5%	8%	7%

Find the investment that should be selected using (a) rate-of-return on incremental investment, (b) present-worth on incremental investment, and (c) present-worth on total investment.

18. In a hydroelectric development under consideration, the question to be decided is the height of the dam to be built. The function of the dam is to create a head of water. Because of the width of the proposed dam site at different elevations, heights of the dam under consideration are 173, 194, and 211 feet; costs for these heights are estimated at $1,860,000, $2,320,000, and $3,020,000, respectively. The capacity of the power plant is based on the minimum flow of the stream of 1,760 cubic feet per second. This flow will develop $[(h \times 1,760 \times 62.4) \div 550] \times 0.75$ horsepower where h equals the height of the dam in feet. A horsepower-year is valued at $31. The cost of the power plant, including building and equipment, is estimated at $180,000 for the building and $34 per hp. of capacity for the equipment.

To be conservative, the useful life of the dam and buildings is estimated at 40 years with no salvage value. Life of the power equipment is also estimated at 40 years with no salvage value. Annual maintenance, insurance, and taxes on the dam and buildings are estimated at 2.8% of first cost. Annual maintenance, insurance, and taxes on the equipment are estimated at 4.7% of first cost. Operation costs are estimated at $38,000 per year for each of the alternatives. Determine the rate of return for each height and the rate of return on the added investment for each added height. To which height should the dam be built if 10% is required on all investments?

19. A firm is considering the purchase of a new machine to increase the output of an existing production process. Of all the machines considered the management has narrowed the field to the machines represented by the cash flows shown on the following page.

Machine	Initial Investment	Annual Operating Cost
1	$ 50,000	$22,500
2	60,000	19,894
3	75,000	17,082
4	85,000	14,854
5	100,000	11,374

If each of these machines provides the same service for 8 years and the minimum attractive rate of return is 11%, which machine should be selected ? Solve by using the rate of return on incremental investment. Compare this result to the result obtained by applying the annual equivalent on total investment.

20. A wholesale distributor is considering the construction of a new warehouse to serve a geographic region that he has been unable to serve until now. There are 6 cities where the warehouse could be built. After extensive study the expected income and costs associated with locating the warehouse in a particular city have been determined.

City	Initial Cost	Net Annual Income
A	$1,000,000	$407,180
B	1,120,000	444,794
C	1,260,000	482,377
D	1,420,000	518,419
E	1,620,000	547,771
F	1,700,000	555,575

The life of the warehouse is estimated to be 12 years. If the minimum attractive rate of return is 15% where should the wholesaler locate his warehouse ?
(a) Solve this problem using an incremental approach.
(b) Solve this problem using a total investment approach.
(c) What city would be selected if the alternative that maximized rate of return on total investment had been used ? Does this conform to the results in part (a) or part (b) ?

21. The state highway department is considering 6 locations for a new interstate highway. Listed below are the estimated construction costs, maintenance costs, and the user costs associated with each location.

Location	Construction Costs Per Mile	Annual Maintenance Costs Per Mile	Annual Users Cost Per Mile
A1	$ 800,000	$5,221	$240,000
A2	900,000	4,920	233,206
A3	1,000,000	4,630	227,789
A4	1,120,000	4,255	213,507
A5	1,200,000	3,540	197,613
A6	1,300,000	3,412	189,918

The life of the highway is expected to be 25 years with no salvage value. If the initial interest rate is 8%, which highway location is most desirable ?
(a) Solve using incremental analysis.
(b) Solve using total investment analysis.

22. A shipping firm is considering the purchase of a materials handling system for unloading ships at the dock. The firm has reduced their choice to 5 different systems, all of

which are expected to provide the same unloading speed. The initial costs and the operating costs estimated for each system are described below.

System	Initial Cost	Annual Operating Expenses
A7	$650,000	$ 91,130
B3	780,000	51,302
D8	600,000	105,000
K2	750,000	68,417
E5	720,000	74,945

The life of each system is estimated to be 5 years and the firm's minimum attractive rate of return is 16%. If the firm must select one of the materials handling systems, which one is the most desirable?

(a) Solve using the total investment approach.

(b) Solve using an incremental approach.

23. Assume that the proposals in Problem 17 are independent proposals rather than mutually exclusive proposals. If the minimum attractive rate of return is 4% and the amount of money available to invest is $180,000, which proposal or proposals should be selected? Which proposal or proposals should be selected if the amount of money available is $100,000?

24. The production manager of a plant has received the sets of proposals listed below from the supervisors of 3 independent production activities. The proposals related to a particular production activity are identified by the same letter and they are mutually exclusive. If the proposals are expected to have a life of 8 years with no salvage value and the minimum attractive rate of return is 15%, what proposals should be selected if the amount of money available for investment is (a) unlimited, (b) $50,000, and (c) $20,000.

Proposal	Initital Investment	Net Annual Income
Activity A		
A1	$10,000	$3,004
A2	20,000	6,423
A3	30,000	7,970
Activity B		
B1	$ 5,000	$1,006
B2	10,000	5,203
B3	15,000	6,209
B4	20,000	7,077
Activity C		
C1	$15,000	$4,506
C2	30,000	8,415

25. A regional sales manager has received 11 proposals for future expenditures from the 4 sales districts in his region. The proposals listed below are expected to span 10 years and the sales manager uses a minimum attractive rate of return of 12% to determine the acceptability of investment proposals. The proposals from each district are designated by a different letter. The acceptance of a proposal from one district does not affect the acceptance of proposals from the other districts unless money is limited. The proposals related to a particular district are mutually exclusive so it is impossible to select more than one proposal from a particular district. What proposal

or proposals should the sales manager select if the money available for investment is (a) unlimited, (b) $700,000, (c) $450,000, and (d) $350,000?

Proposals	Initial Costs	Net Annual Revenues
Q1	$100,000	$19,925
Q2	120,000	24,695
Q3	130,000	26,688
Q4	140,000	29,488
R1	150,000	35,778
R2	180,000	41,755
S1	200,000	32,550
S2	240,000	57,245
S3	300,000	48,825
T1	400,000	95,408
T2	500,000	123,415

26. A company is considering a group of research proposals that are related to either Product A, Product B, or Product C. It has been decided that one proposal will be selected from each set of proposals related to a particular product. The research proposals that are concerned with Product A are identified by the letter A, and those concerned with Product B are identified by the letter B, etc. The company expects the research to extend over a 5-year period. In the past the company has considered a return on investment of at least 10% to be satisfactory. Since it is believed that all projects will be equally beneficial to the company only the costs related to each project are shown below. If the money available is (a) unlimited, (b) $115,000, and (c) $95,000, what proposals should be selected?

Proposal	Initial Cost	Annual Expenses
A1	$40,000	$ 8,100
A2	60,000	2,134
B1	20,000	8,200
B2	25,000	6,528
B3	30,000	5,036
C1	15,000	16,200
C2	20,000	15,013
C3	40,000	7,840
C4	50,000	5,530

27. A refinery can provide for water storage with a tank on a tower or a tank of equal capacity placed on a hill some distance from the refinery. The cost of installing the tank and tower is estimated at $82,000. The cost of installing the tank on the hill, including the extra length of service lines, is estimated at $60,000. The life of the two installations is estimated at 40 years, with negligible salvage value for either. The hill installation will require an additional investment of $6,000 in pumping equipment, whose life is estimated at 20 years with a salvage value of $500 at the end of that time. Annual cost of labor, electricity, repairs, and insurance incident to the pumping equipment is estimated at $500. The interest rate is 5%.
(a) Compare the present-worth cost of the two plans.
(b) Compare the two plans on the basis of equivalent annual cost.
(c) Compare the two plans on the basis of their capitalized costs.
28. It is estimated that a manufacturing concern's needs for storage space can be met

by providing 240,000 square feet of space at a cost of $7.20 per square foot now and providing an additional 60,000 square feet of space at a cost of $40,000 plus $7.20 per square foot of space 6 years hence. A second plan is to provide 300,000 square feet of space now at a cost of $6.90 per square foot. either installation will have zero salvage value when retired some time after 6 years, and if taxes, maintenance, and insurance costs $0.10 per square foot, and the interest rate is 12%, which plan should be adopted?

29. A logging concern has two proposals under consideration which will provide identical service. Plan A is to build a water slide from the logging site to the saw mill at a cost of $250,000. Plan B consists of building a $100,000 slide to a nearby river and allowing the logs to float to the mill. Associated machinery at a cost of $75,000 and salvage value of $25,000 after 10 years will have to be installed to get the logs from the river to the mill. Annual cost of labor, maintenance, electricity, and insurance of the machinery will be $7,000. The life of the slides is estimated to be 30 years with no salvage value. The interest rate is 6%.

  (a) Compare the two plans on the basis of equivalent annual cost.

  (b) Compare the two plans on the basis of 30 years of service.

30. The Plasco Corporation is considering 6 investment alternatives for investment. The 6 proposals under consideration by management:

Proposal	Required Initial Investment	Net Annual Revenue	Expected Life (years)
A1	$1,300,000	$ 300,000	40
B1	1,600,000	500,000	35
C1	2,400,000	820,000	50
C2	2,600,000	840,000	38
D1	3,600,000	1,200,000	30
D2	5,000,000	1,570,000	30

It is expected the net salvage value at the end of the life of each proposal will be zero and the pretax minimum attractive rate of return used by Plasco is 25%. Which proposal or proposals should be accepted if there is no limitation on the amount of money available? Which if the budget is limited to $8,000,000? List any assumptions that you make. Proposals with the same letters are mutually exclusive, e.g. (Cl, C2).

31. Write the linear integer programming formulation of the decision problem described in (a) Problem 23, (b) Problem 24, (c) Problem 25, and (d) Problem 26.

8

Evaluating Replacement Alternatives

Mass production has been found to be the most economical method of satisfying human wants. However, mass production necessitates the employment of large quantities of producer goods which become consumed, inadequate, obsolete, or in some way become candidates for replacement. Decisions concerning the replacement of an asset would be simple if the future could be correctly predicted. If such were the case, the choice between an existing asset and its challenger would be based upon the differences in future receipts and disbursements as indicated by analysis that is directed toward reducing the differences to an equivalent basis for comparison. Unfortunately, there is no general rule that will yield accurate information about the future. Each situation must, therefore, be evaluated in the light of experience, knowledge, and judgment available at the time a decision is to be made.

It is not surprising to find many engineers actively engaged in replacement analysis. By virtue of their training, experience, familiarity with equipment, and objectivity, engineers are particularly well qualified to make recommendations concerning the replacement of physical assets. Engineers not directly engaged in replacement analysis will be interested in the methods employed since physical assets are an essential element in the process of want satisfaction.

8.1. REPLACEMENT SHOULD BE BASED UPON ECONOMY

When the success of an economic venture is dependent upon profit, replacement should be based upon the economy of future operation. Although production facilities are, and should be, considered as a means to an end; that is, production at lowest cost, there is ample evidence that motives other than economy often enter into analysis concerned with the replacement of assets.

The idea that replacement should occur when it is most economical rather than when the asset is worn out is contrary to the fundamental concept of thrift possessed by many people. In addition, existing assets are often venerated as old friends. People tend to derive a measure of security from familiar old equipment and to be skeptical to change, even though they may profess a progressive outlook. Replacement of equipment requires a shift of enthusiasm. When a person initiates a proposal for new equipment, he must ordinarily generate considerable enthusiasm to overcome inertia standing in the way of its acceptance. Later, enthusiasm may have to be transferred to a replacement. This is difficult to do, particularly if one must confess to having been overenthusiastic about the equipment originally proposed.

Part of the reluctance to replace physically satisfactory but economically inferior units of equipment has roots in the fact that the import of a decision to replace is much greater than that of a decision to continue with the old. A decision to replace is a commitment for the life of the replacing equipment. But a decision to continue with the old is usually only a deferment of a decision to replace that may be reviewed at any time when the situation seems clearer. Also a decision to continue with old equipment that results in a loss will usually result in less censure than a decision to replace it with new equipment that results in an equal loss.

The economy of scrapping a functionally efficient unit of productive equipment lies in the conservation of effort, energy, material, and time resulting from its replacement. The unused remaining utility of an old unit is sacrificed in favor of savings in prospect with a replacement. Consider, by way of illustration, a shingle roof. Even a roof that has many leaks will have some utility as a protection against the weather and may have many sound shingles in it. The remaining utility could be made use of by continual repair. But the excess of labor and materials required to make a series of small repairs over the labor and materials required for a complete replacement may exceed the utility remaining in the roof. If so, labor and materials can be conserved by a decision to replace the roof.

When a new unit of equipment is purchased, a number of additional expenses beyond its purchase price may be incurred to put the unit into operation. Such expense items may embrace freight, cartage, constuction of foundations, special connection of wiring and piping, guard rails, and personal services required during a period of run-in or adjustment. Expenses for such items as those mentioned are first-cost items and for all practial purposes represent an investment in a unit of equipment under consideration. For this reason all first-cost items necessary to put a unit of equipment into operation should be considered as part of the total original investment in the unit.

When a unit of equipment is replaced, its removal may entail considerable expense. Some of the more frequently encountered items of removal expense are dismantling, removal of foundations, haulage, closing off water and electrical connections, and replacing floors or other structural elements. The sum of such costs should be deducted from the amount received for the old unit to arrive at its net salvage value. It is clear that this may make the net salvage value a negative quantity. When the net salvage value of an asset is less than zero it is mathematically correct to treat it as a negative quantity in depreciation calculations.

The replacement of an existing asset with new equipment is worthwhile only when it results in a true reduction in cost. For this reason consideration of the costs of installation and removal is necessary if the value of a replacement is to be accurately evaluated and compared with the value of the existing asset.

8.2. BASIC REASONS FOR REPLACEMENT

There are two basic reasons for considering the replacement of a physical asset; physical impairment and obsolescence. Physical impairment refers only to changes in the physical condition of the asset itself. Obsolescence is used here to describe the effects of changes in the environment external to an asset. Physical impairment and obsolescence may occur independently or they may occur jointly in regard to a particular asset.

Physical impairment may lead to a decline in the value of service rendered, increased operating cost, increased maintenance cost, or a combination. For example, physical impairment may reduce the capacity of a bulldozer to move earth and consequently reduce the value of the service it can render. Fuel consumption may rise, thus increasing its operating cost, or the physical impairment may necessitate increased expenditure for repairs.

Little useful data are available relative to how such costs occur in relationship to length of service of assets. A storage battery may render perfect service and require no maintenance up to the moment it fails. Water pipes, on the other hand, may begin to acquire deposits on installation, which reduce their capacity in some proportion to the time they have been in service. Many assets are composites of a number of elements of different service lives. Roofs of buildings usually must be replaced before side walls. The basic structure of bridges ordinarily outlasts several deck surfacings.

Obsolescence occurs as a result of the continuous improvement of the tools of production. Often, the rate of improvement is so great that it is an economy to replace a physical asset in good operating condition with an improved unit. In some cases, the activity for which a piece of equipment has been used declines to the point that it becomes advantageous to replace it with a smaller unit. In either case, replacement is due to obsolescence and necessitates disposing the remaining utility of the present asset in order to allow for the employment of the more efficient unit. Therefore, obsolescence is characterized by changes external to the asset and is used as a distinct reason in itself for replacement where warranted.

8.3. THE PRESENT ASSET AND ITS REPLACEMENT

Two assets must be evaluated at the time replacement is being considered; the present asset and its challenger. The economic future of the present asset can be represented by a cash flow of estimated receipts and disbursements. Since the economic future of a possible replacement can be represented in the same way, the methods of analysis described in Chapter 7 are appropriate for comparing the cash flows of the present asset and its challenger. However, because in replacement analysis the alternatives being compared include old existing assets and new possible replacements there are certain characteristics of this type of decision problem that require special attention.

Usually the duration and the magnitude of cash flows for old existing assets and new assets are quite different. New assets characteristically have high capital costs and low operating costs. The reverse is usually true for assets which are being considered for retirement. Thus, capital costs for an asset for replacement may be expected to be low and decreasing while operating costs are usually high and increasing.

In addition, the remaining life of an asset being considered for replacement is usually short and the future of the asset can be estimated with relative certainty. There is also the advantage that a decision not to replace it now may be reversed at any time in the future. Thus, a decision may be made on the basis of next year's cost of the old asset, and if it is not replaced, a new decision can be made on the basis of next year's cost a year later and so forth.

Because of the tax implications associated with the retirement and depreciation of an asset, the effects of taxes on various replacement alternatives can be significant. However the effect of taxes on a replacement decision does not affect the methods of comparing replacement alternatives since the cash flows describing the alternatives can just as easily be after-tax cash flows as before-tax cash flows. Chapters 12 and 13 describe how depreciation and taxes affect cash flows so that after-tax cash flows can be developed.

The method of treating data relative to an existing asset should be the same as that used in treating data relative to a possible replacement in an economy study. In both cases only the future of the assets should be considered and sunk costs should be disregarded. Thus the value of the existing old asset that should be used in a study of replacement is the value that it will have if it is retired.

8.4. EVALUATION OF REPLACEMENTS INVOLVING SUNK COSTS

The following example will be used to illustrate correct and incorrect methods of evaluating replacements where sunk costs are involved. Suppose that Machine *A* was purchased four years ago for $2,200. It was estimated to have a life of ten years and a salvage value of $200 at the end of its life. Its operating expense had been found to be $700 per year, and it appeared that the machine would serve satisfactorily for the balance of its estimated life. Presently a salesman is offering Machine *B* for $2,400. Its life is estimated at ten years and its salvage value at the end of its life is estimated to be $300. Operating costs are estimated at $400 per year.

The operation for which these machines are used will be carried on for many years in the future. Equipment investments are expected to justify a 15% minimum attractive rate of return in accordance with the policy of the company concerned. The salesman offers to take the old machine in on trade for $600. This appears low to the company, but the best offer received elsewhere is $450. All estimates relative to both machines above have been carefully reviewed and are considered sound.

In order to make a proper comparison of alternatives the analysis may be undertaken from the standpoint of a person who has a need for the service that Machine *A* or Machine *B* will provide but owns neither. In attempts to purchase a machine he finds that he can purchase Machine *A* for $600 and Machine *B* for $2,400. This analysis of which to buy will not be biased by the past since he was not part of the original transaction for Machine *A* and, therefore, will not be forced to admit a sunk cost. With this *outsider* viewpoint, the analysis is given below.

Comparison Based on Present Value. If Machine *A* is traded in for $600, a sunk cost of $2,200 − $600 or $1,600 is revealed. Also, if Machine *B* is purchased, Machine *A* will be "sold" for $600. The logical alternatives then are (1) to consider Machine *A* to have a value of $600 and to continue with it for six years and (2) to purchase Machine *B* for $2,400 and use it for ten years.

The equivalent annual cost to continue with Machine *A* for six years is calculated as follows:

$$\text{Annual capital recovery and return, } (\$600 - \$200)(\overset{A/P\,15,6}{0.2642}) + \$200(0.15) \quad \$135.68$$

Annual operating cost 700.00

$$\overline{} \$835.68$$

The equivalent annual cost to dispose of Machine *A*, purchase Machine *B* and use it for ten years is calculated as follows:

$$\text{Annual capital recovery and return, } (\$2,400 - \$300)(\overset{A/P\,15,\,10}{0.1993})$$

Annual capital recovery and return, ($2,400 − $300)(0.1993)
+ $300(0.15) $463.53
Annual operating cost 400.00
                                                                                    $863.53

If the alternative to continue with Machine A is adopted the annual saving prospect for the next six years is $863.53 − $835.68 = $27.85. For the next four years after that time the amount of savings will be dependent upon the characteristics of the machine that might have been purchased six years from the present to replace Machine A. If it is assumed that Machine A will be replaced after six years by a machine with the same costs as Machine B, the costs of the two alternatives will be the same after the first six years. More will be said in Sections 8.7 and 8.8 about the assumptions that can be made when the existing asset and its challenger have different service lives.

Calculation of Comparative Use Value. A second method of comparison, which is particularly good for demonstrating the correctness of the comparison above to skeptical people, is to calculate the value of the machine to be replaced which will result in an annual cost equal to the annual cost of operation with the replacement. In this calculation, let X equal the present value of Machine A for which annual cost with Machine A equals annual cost with Machine B. Then

$$(X - \$200)(\overset{A/P\,15,\,6}{0.2642}) + \$200(0.15) + \$700$$
$$= (\$2,400 - \$300)(\overset{A/P\,15,\,10}{0.1993}) + \$300(0.15) + \$400.$$

Solving for X results in

$$X = \$705.$$

Machine A has a comparative use worth in comparison with Machine B of $705. Thus it is obvious that Machine A should be retained if it can be disposed of for only $600. Compare this result with that obtained in the previous section. Note that $705 − $600 = $105 is equivalent to $27.73($\overset{P/A\,15,\,6}{3.784}$) = $105.

Fallacy of Adding Sunk Cost to a Replacement. In spite of the fact that sunk cost cannot be recovered, a face saving practice of charging the sunk cost of a machine to the cost of its contemplated replacement is often employed. This practice, human but unrealistic, will be illustrated by the following situation.

Three years ago A, who authorizes machine purchases in a manufacturing concern, was approached by B for authorization to purchase a machine. B pleaded his cause in glowing terms and with enthusiasm. He had many figures and arguments to prove that an investment in the machine he proposed would easily pay out. *A* was at first skeptical, but he also became enthusiastic about the purchase as the profits in prospect were calculated, and authorized the purchase. After three years *B* realized that the machine was not coming up to expectations and would have to be replaced, at a loss of $1,200.

B was well aware of the necessity of admitting this sunk cost when he went to *A* to get authorization for a replacement. He realized the difficulty of trying to establish confidence in his arguments for the replacement and at the same time admit an error in judgment that had resulted in a loss of $1,200. But he hit on the expedient of focusing attention on the proposed machine by emphasizing that the $1,200 loss could be added to the cost of the new machine and that the new machine had such possibilities for profit that it would pay out shortly, even though burdened with the loss of the previous machine. Such improper handling of sunk cost is merely deception designed to make it appear that an error in judgment has been corrected.

As a numerical example of the fallacy of adding sunk cost of an old machine to the cost of a replacement, consider the following replacement situation. Machine *C*, purchased for $3,400 a year ago, had an estimated life of six years and a salvage value of $400. Its operating cost is $3,200 per year. At the end of the first year a salesman offers Machine *D* for $4,600. This machine has an estimated life of five years, a salvage value of $600, and, owing to improvements it embodies, an operating cost, as shown by trial, of only $2,200. The salesman offers to allow $1,400 for Machine *C* on the purchase price of Machine *D*. The sunk cost is $2,000.

Annual cost with Machine *C*, on the basis of its present trade-in value and estimated salvage value five years hence is

$$\overset{A/P\,8,\,5}{}$$

Annual capital recovery and return ($1,400 − $400)(0.2505) + $400(0.08) $ 282
Annual operating cost   3,200
                                                                                                                          $3,482

Annual cost with Machine *D*, as incorrectly calculated when sunk cost of $2,000 of Machine *C* is added to the cost of Machine *D*, is

$$\overset{A/P\,8,\,5}{}$$

Annual capital recovery and return ($6,600 − $600)(0.2505) + $600(0.08) $1,551
Annual operating cost   2,200
                                                                                                                          $3,751

On the basis of this *incorrect* result, Machine *C* is continued for the next year on the erroneous belief that $3,703 less $3,482, or $269 is being saved annually. Annual cost with Machine *D* as correctly calculated is

$$\text{Annual capital recovery and return } (\$4,600 - \$600)\overset{A/P\,8,\,5}{(0.2505)} + \$600(0.08) \quad \$1,050$$
Annual operating cost 2,200
$$\overline{\$3,250}$$

On this correct basis, purchase of Machine D should result in an annual saving of $3,482 less $3,250, or $232.

Receipts and Disbursements as a Basis for Comparison. For a different viewpoint, the alternatives of continuing five years with Machine C and of replacing Machine C with Machine D and using the latter for the next five years can be compared on the basis of receipts and disbursements associated with them. The receipts and disbursements if Machine C is retained for the next five years are

Receipts and Disbursements	Year					
	0	1	2	3	4	5
Disbursements (considered to come at end of year)	....	$3,200	$3,200	$3,200	$3,200	$3,200
Receipts at end of year	....	....	....	....	....	$ 400

The present worth of the net cost of five years of service with Machine C is equal to

$$PW_C(8) = \$3,200(\overset{P/A\,8,\,5}{3.993}) - \$400\overset{P/F\,8,\,5}{(0.6806)} = \$12,506.$$

The receipts and disbursements if Machine D is purchased and used for 5 years are

Receipts and Disbursements	Year					
	0	1	2	3	4	5
Disbursements (considered to come at end of year)	$4,600	$2,200	$2,200	$2,200	$2,200	$2,200
Receipts at end of year	$1,400	....	....	....	....	$ 600

The present worth of the net cost of five years of service if Machine C is sold and Machine D is purchased and used for five years is equal to

$$PW_D(8) = \$4,600 - \$1,400 + \$2,200(\overset{P/A\,8,\,5}{3.993}) - \$600\overset{P/F\,8,\,5}{(0.6806)} = \$11,577.$$

The net present worth of the advantage of accepting the latter alternative is $12,506 less $11,577, or $929. This is equivalent to

$$AE_{D-C}(8) = \$929\overset{A/P\,8,\,5}{(0.2505)} = \$232$$

and is equal to the annual saving resulting from the previous correct comparison made above.

The method of using the present salvage value as the capital cost for an existing asset is generally preferred over the method just described. It is cumbersome to utilize the latter approach when a number of mutually exclusive alternatives are being considered in a replacement analysis. In addition the latter method may lead to errors where the life of the existing asset and its challenger are different. If in this situation the trade-in value of the existing asset and the initial cost of the challenger are associated with the cash flow of the challenger, the analyst must be careful not to conclude that both of these values are related to the same time period.

8.5. PATTERNS OF MAINTENANCE COSTS

The replacement problem may be further understood by considering the relationship of maintenance costs to capital costs for an asset. In order to simplify the discussion, maintenance costs will be classified as sporadic, constant, and constantly increasing.

Table 8.1. ECONOMIC HISTORY OF A MACHINE WITH SPORADIC MAINTENANCE COSTS

End of Year Number A	Maintenance Cost for End of Year Given B	Summation of Maintenance Costs, $\sum B$ C	Average Cost of Maintenance Through Year Given, $C \div A$ D	Average Capital Cost If Retired at Year End Given, $\$400 \div A$ E	Average Total Cost Through Year Given, $D + E$ F
1	$100	$ 100	$100	$400	$500
2	100	200	100	200	300
3	300	500	167	133	300
4	100	600	150	100	250
5	100	700	140	80	220
6	100	800	133	67	200
7	100	900	129	57	186
8	300	1,200	150	50	200
9	100	1,300	144	44	188
10	100	1,400	140	40	180

Sporadic Maintenance Cost. Assume that a machine is purchased for $400, and that its salvage value is zero at any age at which it may be retired. Assume that the interest rate is zero. Then the pertinent facts related to Machine A may be set down as in Table 8.1. This table brings out the fact that capital costs decrease in some inverse proportion to the length of life.

This is also true for interest rates other than zero and for any pattern of salvage value normally encountered.

The fact that maintenance costs are averaged in Column D, tends to smooth out the effect of sporadic large maintenance costs. In the example, the ratio of the cost of the asset and its maintenance cost is relatively high. In spite of this, the average total cost in Column F is generally downward. Unless there is a rising trend in sporadic maintenance cost, there will be no "minimum" cost in a given year that will not be bettered in a future year. But it is clear that if replacement is to be made, it is desirable to do so immediately prior to a large expenditure for maintenance.

Constant Maintenance Costs. Where maintenance costs are constant in succeeding years they will never justify replacement. Where no interest or salvage value is involved an equation for the average cost of a year of service can be written as follows:

$$C = \frac{P}{n} + M$$

where

C = average annual cost of capital recovery and maintenance;
P = initial cost of asset;
M = constant yearly cost of maintenance;
n = life of asset in years.

It is apparent that the value of C will never reach a minimum value.

For a case where interest and salvage value are involved, an expression for equivalent annual cost, C, may be written as follows:

$$C = (P - F)(\overset{A/P\ i,\ n}{}) + Fi + M.$$

A glance at a table of value for $(\overset{A/P\ i,\ n}{})$ shows that C will decrease with an increase in n if the salvage value remains constant through time. However if, from period to period, there are large decreases in F, this trend may be reversed.

Constantly Increasing Maintenance Costs. An understanding of the replacement problem may also be gained from considering situations in which maintenance costs increase constantly with the age of an asset. Assume that a machine has been purchased for $800, that its salvage value is zero at any age, that its maintenance cost is zero the first year and rises at a constant rate of $100 per year thereafter. If it is assumed that the interest rate is zero the facts concerning the machine may be represented by Table 8.2.

Because there is a rising trend in maintenance cost, there will be a minimum average total cost at some point in the life of the asset. This point occurred in the fourth year in the example presented.

Table 8.2. ECONOMIC HISTORY OF A MACHINE WITH CONSTANTLY INCREASING MAINTENANCE COSTS

End of Year Number A	Maintenance Cost for End of Year Given B	Summation of Maintenance Costs, $\sum B$ C	Average Cost of Maintenance Through Year Given, $C \div A$ D	Average Capital Cost If Retired at Year End Given, $\$800 \div A$ E	Average Total Cost Through Year Given, $D + E$ F
1	$ 0	$ 0	$ 0	$800	$800
2	100	100	50	400	450
3	200	300	100	267	367
4	300	600	150	200	350
5	400	1,000	200	160	360
6	500	1,500	250	133	383

8.6. THE ECONOMIC LIFE OF AN ASSET

The economic life of an asset is the time interval that minimizes the asset's total equivalent annual costs or maximizes its equivalent annual net income. The economic life is also referred to as the minimum cost life or the optimum replacement interval.

If the future could be predicted with certainty, it would be possible to accurately predict the economic life for an asset at the time of its purchase. The analysis would simply involve the calculation of the total equivalent annual cost at the end of each year in the life of the asset. Selection of the total equivalent annual cost that is a minimum would specify a minimum cost life for the asset.

As mathematically attractive as it may be to determine the economic life of assets, this end is primarily an ideal toward which to strive. Rarely is the economic life used to determine how frequently an asset should be replaced. There are several reasons for this. First the economic life is valid as a replacement interval only under the restrictive assumptions that all future replacements are the same as the replacement under consideration with regard to first cost, salvage value, operating expenses, and net income produced. Second, reasonably good data describing the costs of an asset are rarely available for an asset at the time of its purchase. A third reason is that the decision to retire an asset is not usually made at the time of its purchase. Decisions to retire assets almost always result from consideration of factors in existence shortly before the time of retirement.

Minimum Cost Life, Increasing Maintenance Cost. Where a rising trend in maintenance cost exists, it is possible to formulate an idealized model that will express the minimum cost life for an asset. Neglecting interest, the average annual cost for an asset with increasing maintenance cost may be expressed as follows:

$$C = \frac{P}{n} + Q + (n-1)\frac{m}{2}$$

where

C = average annual cost;
P = initial cost of asset;
Q = annual constant portion of operating cost of asset (is equal to first year operation cost, of which maintenance is a part);
m = the amount by which maintenance costs increase each year;
n = life of asset in years.

This expression, if differentiated with respect to n, set equal to zero and solved for n, results in the following:

$$\frac{dC}{dn} = -\frac{P}{n^2} + \frac{m}{2} = 0$$

$$n = \sqrt{\frac{2P}{m}}.$$

For the example presented in Table 8.2, P = $800, Q = 0, and m = $100. Therefore the minimum cost life is

$$n = \sqrt{\frac{2(\$800)}{\$100}} = 4 \text{ years}$$

as is shown in Table 8.2. The minimum cost shown in Table 8.2 may be verified as follows:

$$C = \frac{\$800}{4} + 3\left(\frac{100}{2}\right) = \$350.$$

The economic history of an asset whose first cost is $5,000, whose salvage value at any time is zero, and whose cost of maintenance is zero the first year and increases at a constant rate of $100 for an interest rate of 6%, is shown in Table 8.3. The minimum equivalent annual cost occurs when the cost of extending an asset's life one more year exceeds the equivalent annual cost to date. Since the salvage value is zero at any time the incremental cost of one additional year's operation is reflected in Column B. The incremental cost of providing service for year 12 ($1,100) exceeds the equivalent annual cost for 11 years ($1,076) and therefore the economic life of the asset is 11 years.

This table illustrates a method for determining the equivalent annual cost of maintenance and the equivalent annual cost of capital recovery and return for lives ranging from one to fourteen years. The sum of these costs is a minimum for a life of eleven years. The quantities in Columns G, H, and I have been plotted to reveal trends in Figure 8.1.

Table 8.3. EQUIVALENT ANNUAL COST OF MAINTENANCE PLUS CAPITAL RECOVERY WITH A RETURN OF AN ASSET FOR CONSTANTLY INCREASING MAINTENANCE

End of Year Number	Maintenance Cost at End of Year Designated	Present-Worth Factor for Year Designated, $(P/F i, n)$	Present-Worth as of Beginning of Year No. 1, of Maintenance for Year Designated, $B \times C$	Summation of Present Worths of Maintenance Through Year Designated, $\sum D$	Capital-Recovery Factor for Year Designated, $(A/P i, n)$	Equivalent Annual Cost of Maintenance Through Year Designated, $E \times F$	Equivalent Annual Cost of Capital Recovery and Return Through Year Designated, $F \times \$5,000$	Total Equivalent Annual Cost Through Year Designated, $G + H$
A	B	C	D	E	F	G	H	I
1	$ 0	0.9434	$ 0	$ 0	1.06000	$ 0	$5,300	$5,300
2	100	0.8900	89	89	0.54544	48	2,727	2,775
3	200	0.8396	167	256	0.37411	96	1,870	1,966
4	300	0.7921	237	493	0.28859	142	1,442	1,585
5	400	0.7473	298	791	0.23740	188	1,187	1,375
6	500	0.7050	352	1,143	0.20336	233	1,016	1,249
7	600	0.6651	399	1,542	0.17914	276	895	1,172
8	700	0.6274	439	1,984	0.16104	319	805	1,124
9	800	0.5919	473	2,457	0.14702	361	735	1,096
10	900	0.5584	502	2,960	0.13587	402	679	1,081
11	1,000	0.5268	526	3,487	0.12679	442	633	1,076
12	1,100	0.4970	546	4,033	0.11928	481	596	1,077
13	1,200	0.4688	560	4,588	0.11296	518	564	1,083
14	1,300	0.4423	574	5,162	0.10758	556	537	1,094

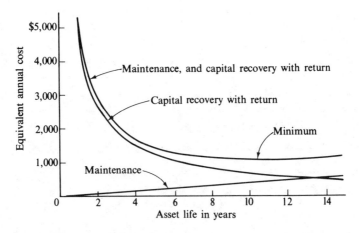

FIGURE 8.1.   MINIMUM COST LIFE OF AN ASSET.

Study of the total equivalent annual cost curve reveals that it is rather flat in the region of the minimum. It may, therefore, be concluded that a deviation of one or two years from the minimum cost life will result in relatively small increases in total equivalent annual cost.

The mathematical model for determining the economic life of an asset assumed an interest rate of zero. For the example of Table 8.3, with an interest rate of zero, the minimum cost life is

$$n = \sqrt{\frac{2P}{m}} = \sqrt{\frac{2(\$5,000)}{\$100}} = 10 \text{ years.}$$

Therefore, this equation may be used to approximate the economic life of an asset for cases involving interest.

8.7. REPLACEMENT WHERE FUTURE CHALLENGERS ARE UNKNOWN

Because most replacement decisions are concerned with the replacement of old assets by new ones, the economic alternatives being examined are seldom of equal duration. At first it may appear that the methods for handling alternatives with unequal lives described in Chapter 7 are appropriate for such an analysis. However, for those cases where it is difficult to quantify the cash flows that are expected from future challengers there may be substantial differences in the time span for which the alternative cash flows can be predicted. In addition an old existing asset may have little or no salvage value and as a result its capital costs may be neglible compared to those of the new challenger. Thus to determine a study period by selecting the time span of the shortest lived asset can lead to rather distorted results. Since the life of the asset being considered for replacement is frequently short, the use of relative short study periods discriminates against the challenger which normally has a longer life.

These conditions make the replacement analysis sensitive to the estimates of the lives used for the replace alternative and the retain alternative. Thus when comparing an existing asset and its possible replacement the lives that should be assumed are the lives that are most favorable to each asset. In other words the comparison should be made on the basis of each alternative's economic life.

If over an asset's life, its present salvage value always equals its future salvage value and its operating expenses are increasing, one year is the life most favorable to the asset. This fact is evident from the relationship describing the total costs of the asset for any year.

Total equivalent annual costs = Capital recovery with return +

Equivalent annual operating costs

The capital recovery portion of total costs will be constant for any asset for which $P = F$ no matter how long the asset is in service.

$$\text{Capital recovery with return} = (P - F)(\overset{A/P\,i,\,n}{}) + Fi$$

The equivalent annual operating costs for an asset will be ever increasing as long as each year's operating expense is greater than the proceeding year's expense. Thus for these two conditions the total equivalent annual costs will be minimized for the shortest time the asset might be reasonably retained. In this book this time will be considered to be the time between the present and the next time replacement would be reconsidered: one year.

It is common for the present salvage value and the future salvage value of old assets to be zero. In this case the economic life of the asset is one year and the cost of extending the use of the asset for one more year is just its anticipated operating costs for that year. Thus the existing asset's next year's costs can be directly compared to the minimum equivalent annual costs of the challenger. These costs would be determined in the manner demonstrated in Table 8.3. By comparing these costs it is possible to see the economic advantage of each alternative assuming each were kept in service for its particular economic life.

The implication of such a comparison is that the economic advantage reflected in the comparison of the minimum equivalent annual costs will be realized only if the alternative selected is retained for its economic life. Such an assumption may not be unrealistic if it is the existing asset that is retained. However uncertainties in the future make it more difficult to anticipate how long the challenger would actually be in service if it is decided to replace the existing asset. In addition there is uncertainty with respect to any future challenger that might be available one year from now. Technological improvements over the next year may produce an asset that is superior to the present challenger. Thus the economic differences must be considered in light of possible future events. Where it is thought that future challengers will be better than present challengers postponing replacement is favored. On the other hand if it appears that future challengers will be inferior to present ones the decision to replace is favored.

8.8. REPLACEMENT WITH EXPLICIT CONSIDERATION OF THE FUTURE

It is usually presumed that each event in history is dependent upon previous events. Thus in theory it is necessary, for accurate comparison of a pair of alternatives, to consider the entire future or a period from the present to a point in the future when the effect of both alternatives will be identical. It is rarely feasible to consider all links in the chain of events in the future.

It also is frequently impossible to be able to discern a point in the future at which the selection of one of a pair of alternatives in the present will have the same effect as the selection of the other.

In the paragraphs that follow, a general method for placing alternatives on a comparative basis involving the selection of more or less arbitrary study periods will be illustrated. With this method, comparison of alternatives is made on the basis of costs and income that occur during a selected period in the future. The effect of values occurring after the selected study period is eliminated by suitable calculations. Study periods for all alternatives in a given comparison should be equal.

As an illustration of the use of a selected study period, consider the following example. A certain operation is now being carried on with Machine A whose present salvage value is estimated to be $2,000. The future life of Machine A is estimated at five years, at the end of which its salvage value is estimated to be zero. Operating costs with Machine A are estimated at $1,200 per year. It is expected that Machine A will be replaced after five year by Machine B whose initial cost, life, final salvage value, and annual operating costs are estimated to be, respectively, $10,000, 15 years, zero, and $600. It should be realized that estimates relating to Machine B may turn out to be grossly in error.

The desirability of replacing Machine A with Machine C is being considered. Machine C's estimated initial cost, life, final salvage value, and annual operating costs are estimated to be, respectively, $8,000, 15 years, zero, and $900. The interest rate is taken to be 10%. Detailed investment and cost data for Machines A, B, and C are given in Table 8.4.

Analysis Based on a 15-year Study Period, Recognizing Unused Value. Because of the difficulty of making further estimates into the future, a study period of 15 years coinciding with the life of Machine C is selected. This will necessitate calculations that will bring both plans to equal status at the end of 15 years.

Under Plan I, the study period embraces five years of service with Machine A and ten years of service with Machine B, whose useful life extends five years beyond the study period. Thus, an equitable allocation of the costs associated with Machine B must be made for the period of its life coming within and after the study period. In assuming that annual costs associated with this unit of equipment are constant during its life, the present-worth cost of service during the study period may be calculated a follows.

The equivalent annual cost for Machine B during its life is equal to

$$AE_B(10) = \$10,000(\overset{A/P\ 10,\ 15}{0.1315}) + \$600 = \$1,915.$$

The present-worth cost of 15 years of service in the study period is equal to

Table 8.4. ANALYSIS BASED ON A SELECTED STUDY PERIOD

Year End Number	Plan I		Plan II	
	Machine Investment	Operating Costs	Machine Investment	Operating Costs
0	Machine A, $2,000		Machine C, $8,000	
1		$1,200		$900
2		1,200		900
3		1,200		900
4		1,200		900
5	Machine B, $10,000	1,200		900
6		600		900
7		600		900
8		600		900
9		600		900
10		600		900
11		600		900
12		600		900
13		600		900
14		600		900
15		600		900
16		600		
17		600		
18		600		
19		600		
20		600		

(The left margin spans years 0 through 15 with the bracketed label "15 years selected study period.")

$$PW_{\mathrm{I}}(10) = \$2,000 + \$1,200(\overset{P/A\,10,\,5}{3.791}) + \$1,915(\overset{P/A\,10,\,10}{6.144})(\overset{P/F\,10,\,5}{0.6209})$$
$$= \$13,855$$

On this basis the value remaining in Machine B at the end of the study period may be calculated as a matter of interest as follows:

$$\$10,000(\overset{A/P\,10,\,15}{0.1315})(\overset{P/A\,10,\,5}{3.791}) = \$4,935.$$

Under Plan II, the life of Machine C coincides with the study period. The present-worth cost of 15 years service in the study period is equal to

$$PW_{\mathrm{II}}(10) = \$8,000 + \$900(\overset{P/A\,10,\,15}{7.606}) = \$14,845.$$

On the basis of present-worth costs of $13,855 and $14,845 for a study period of 15 years, Plan I should be chosen.

Analysis Based on a 15-year Study Period, not Recognizing Unused Value. Values remaining in assets at the end of a selected study period are sometimes disregarded in order to simplify the calculations necessary for making a comparison. The effect of disregarding values remaining in an asset at the end of a study period is to assume that the asset will be retired at the end of the study period.

On the basis of this assumption the present-worth cost of 15 years of service for Plan I may be calculated as follows. The equivalent annual cost for Machine B during its life is equal to

$$AE_B(10) = \$10,000(\overset{A/P\ 10,\ 10}{0.1628}) + \$600 = \$2,228$$

and the present-worth cost of 15 years of service in the study period is equal to

$$PW_1(10) = \$2,000 + \$1,200(\overset{P/A\ 10,\ 5}{3.791}) + \$2,228(\overset{P/A\ 10,\ 10}{6.144})(\overset{P/F\ 10,\ 5}{0.6209})$$
$$= \$15,049$$

This result is significantly different from that calculated for 15 years of service for Plan I, where cognizance was taken of value remaining in Machine B at the end of the study period.

The practice of disregarding values remaining in an asset at the end of a study period introduces error equivalent to the actual value of the asset at that time. This practice is difficult to defend for it does not greatly reduce the burden of making comparisons, nor does it necessarily produce results in the direction of conservatism.

Analysis on the Basis of a-Five-Year Study Period. Lack of information often makes it necessary to use rather short study periods. For example, the characteristics of the successor to Machine A in Table 8.4 might be vague. In that case a study period of five years might be selected to coincide with the estimated retirement date of Machine A.

The equivalent annual cost of continuing with Machine A during the next five years is

$$AE_A(10) = \$2,000(\overset{A/P\ 10,\ 5}{0.2638}) + \$1,200 = \$1,727.60.$$

The equivalent annual cost of Machine C based on a life of 15 years is

$$AE_C(10) = \$8,000(\overset{A/P\ 10,\ 15}{0.1315}) + \$900 = \$1,952.00$$

The \$224.40 per year cost advantage of Machine A over Machine C can be interpreted in two ways. It can be said that the retention of Machine A will produce such savings for a five-year period if the future events after five years are ignored. That is, the commitment to Machine C for its remaining ten years is not reflected in the savings of \$224.40 per year for five years.

Another interpretation of this annual savings figure is to assume that each machine is replaced by an identical successor for the shortest period for which the machine lives are common multiples. Thus a saving of \$224.40 per year would be realized for fifteen years if Machine A is purchased and replaced by identical successors every five years as opposed to purchasing Machine C for fifteen years. Such an assumption concerning replacement by identical successors is implicit in making direct economic comparisons of annual equivalent amounts.

In general, the longer the study period, the more significant the results. But the longer the study period, the more likely that estimates are in error. Thus, the selection of a study period must be based on estimate and judgment.

Considering Sequences of Future Challengers. One approach to replacement analysis is to explicitly consider the consequences of retaining an existing asset and replacing it by a sequence of successors as compared to the consequences of accepting the present challenger and its sequence of successors. As a result the consequences of retaining the existing asset or accepting the challenger can be considered over a relatively long time span.

In order to quantify the cash flows that are to be expected from future challengers various assumptions have been made to describe the effects of technological change and physical impairment on these cash flows. As technological innovation and improvement continues it is expected that there will be a decrease in the cost of providing a future service similar to the service presently provided. Thus because of obsolescence the longer an asset is retained the greater the disparity between it and possible future replacements. These are not the only costs that increase relative to an asset's service life. Usually those costs associated with physical impairment also increase as the life of an asset is extended.

When considering the sequences of challengers that are assumed to be the successors of an existing asset or the present challenger it is necessary to determine how frequently future challengers should be replaced in light of the effects of obsolescence and physical impairment on their future cash flows. The replacement decision can then be made considering those alternatives that assure the most favorable replacement policy for the future successors.

Minimum Cost Life, Increasing Obsolescence Cost. If it is assumed that obsolescence takes place at a uniform rate, it may be treated mathematically in a way that gives an insight into its effect. For simplicity, assume zero salvage value and that an improved replacement becomes available at the beginning of each succeeding year. Assume that the installed costs of the original asset and each possible replacement are equal and are represented by P. Assume that the original asset is purchased in the year $19x1$, its annual operating cost is Q, and the annual operating cost of possible future replacements is as given in Table 8.5.

In Table 8.5, Q represents the actual operating cost of the original asset purchased at the beginning of $19x1$. The term b represents a net improvement in comparative performance of succeeding assets. The improvement represented by b consists of a reduction of operating costs and increases in

Table 8.5. EFFECT OF CONSTANTLY INCREASING OBSOLESCENCE ON ANNUAL OPERATING COSTS

Beginning of Year in Which Asset Is Purchased	Comparative Year-End Operating Costs of Assets Acquired in Year Designated						
	$19x1$	$19x2$	$19x3$	$19x4$	$19x5$	$19x6$	$19x7$
$19x1$	Q	Q	Q	Q	Q	Q	Q
$19x2$		$Q-b$	$Q-b$	$Q-b$	$Q-b$	$Q-b$	$Q-b$
$19x3$			$Q-2b$	$Q-2b$	$Q-2b$	$Q-2b$	$Q-2b$
$19x4$				$Q-3b$	$Q-3b$	$Q-3b$	$Q-3b$
$19x5$					$Q-4b$	$Q-4b$	$Q-4b$
$19x6$						$Q-5b$	$Q-5b$

value of services of a succeeding asset as compared to the original acquired in $19x1$. It should be recalled that an increase in income has the same effect as a decrease in cost and that neither Q nor b includes physical impairment.

From the standpoint of capital costs an asset purchased in $19x1$ should be used forever as a means of approaching a minimum capital recovery cost. But assets should be replaced yearly to take advantage of reduced operating costs. Thus, it appears that there is an optimum interval of replacement for the conditions given, and this optimum can be found as follows. Consider the two plans of replacement shown in Table 8.6. Under Plan A the assets are replaced annually, and under Plan B a single asset is used indefinitely. The costs of the two plans may be tabulated as given in Columns K and L.

Table 8.6. REPLACEMENT ALTERNATIVES INVOLVING OBSOLESCENCE

Year J	Plan A K	Plan B L	Plan A Minus Plan B M
$19x1$	$P + Q$	$P + Q$	$P - (P)$
$19x2$	$P + Q - b$	Q	$P - (+b)$
$19x3$	$P + Q - 2b$	Q	$P - (+2b)$
......	....	...	....
$19xn$	$P + Q - (n - 1)b$	Q	$P - [+(n - 1)b]$

The differences of Plan A and Plan B may be expressed as shown in Column M. Plan A minus Plan B will be a maximum for the interval of years for which the equivalent annual amount of the right-hand quantities in Column M will be a minimum. Except for P, the pattern of the right-hand quantities in Column M is identical with that in Column B in Table 8.3. The equivalent annual amount of all the right-hand quantities in Column M of the tabulation may be found by following the method used to obtain Column I in Table 8.3.

The optimum life span may be obtained from a counterpart of Column *I* by inspection. The life span so found is the interval at which assets should be replaced for the established conditions: namely, constant asset costs, constantly rising obsolescence, and no physical impairment. This fact can be proved by demonstrating that the differences in the equivalent annual costs of the right-hand quantities in Column *M* for any number of periods (n_1) and any other number of periods (n_2) are equal to the differences between the equivalent annual costs of replacing an asset every n_1 periods compared to replacing it every n_2 periods.

The left-hand quantities in Column *M* of the tabulation have no effect on determining the optimum life span as they represent a constant equivalent annual amount no matter what time span is under consideration. Care should be exercised to realize that costs designated by b, $2b$, and so forth, in this analysis are not real costs but merely comparative costs. Although they are useful in determining the interval of replacement, they will not become apparent as year to year disbursements of the asset whose optimum life they are used to determine.

Minimum Cost Life, Physical Impairment and Obsolescence Costs in Combination. For an illustration of the combined effect of physical impairment and obsolescence, consider a situation in which physical impairment increases at a rate m per year and obsolescence increases at a rate b per year. The first year's operating cost is Q. The first costs of assets are equal to P, and their salvage values at the end of any life are equal to zero. To analyze this situation, consider two plans: Plan *A* in which assets are replaced annually, and Plan *B* in which a single asset is used. The costs of the two plans are given in Columns *K* and *L* of Table 8.7.

Table 8.7. REPLACEMENT ALTERNATIVES INVOLVING PHYSICAL IMPAIRMENT AND OBSOLESCENCE

Year J	Plan A K	Plan B L	Plan A Minus Plan B M
19x1	$P + Q$	$P + Q$	$P - P$
19x2	$P + Q - b$	$Q + m$	$P - (b + m)$
19x3	$P + Q - 2b$	$Q + 2m$	$P - 2(b + m)$
......	....	....	....
19xn	$P + Q - (n - 1)b$	$Q + (n - 1)m$	$P - (n - 1)(b + m)$

The optimum life to keep the asset purchased under Plan *B* in 19x1 for the conditions outlined may be found by finding the life for which the equivalent annual amount of the right-hand quantities in Column *M* in Table 8.7 is a minimum by the method suggested for the case of constantly increasing maintenance. The life so found is the one which will result in minimum disbursements in the future for the conditions given, but it should be realized that quantities designated with b are not real costs of the asset in question but are comparative costs of the asset in question and a succession of future assets.

For situations of constantly increasing maintenance and obsolescence, with no salvage value and zero interest rate, the following equation applies:

$$C = \frac{P}{n} + Q + (n - 1)\frac{m + b}{2}$$

from which

$$n = \sqrt{\frac{2P}{m + b}}.$$

Let $b = rm$
Then

$$n = \sqrt{\frac{2P}{m(1 + r)}} = \frac{1}{\sqrt{1 + r}}\sqrt{\frac{2P}{m}}.$$

Values of $1/\sqrt{1 + r}$ for selected values of r are given in Table 8.8.

Table 8.8. EFFECT OF OBSOLESCENCE AND INCREASING MAINTENANCE ON OPTIMUM LIFE

Value of r	0.25	0.5	1	2	3
Value of $\dfrac{1}{\sqrt{1 + r}}$	0.89	0.82	0.71	0.59	0.50

The tabulation shows that the optimum life decreases as obsolescence, in proportion to maintenance, increases. For example, inclusion of an obsolescence rate equal to the maintenance rate corresponding to $r = 1$ reduces the optimum life by $1 - 0.71$, or 0.29%.

The use of the approach just discussed to determine the frequency of replacing future challengers is an integral part of the MAPI approach to replacement analysis.[1] The MAPI approach explicitly assumes various cash flow patterns for future challengers. Appendix B gives the basic assumptions and approaches used by the earlier MAPI formula. The material in Appendix B is presented to demonstrate one approach to explicit consideration of future events and it is not intended to be a complete discussion of existing MAPI methodology.

---

[1] A number of mathematical replacement models have been developed by the Machine and Allied Products Institute and they have been commonly referred to as the MAPI formulas.

8.9. CONSIDERATIONS LEADING TO REPLACEMENT

The main considerations leading to replacement may be classified as inadequacy, excessive maintenance, declining efficiency, and obsolescence. Any of the above may lead to replacement, but usually two or more are involved when replacement is considered. In the sections that follow, examples illustrating an approach to replacement analysis for each of these considerations will be presented.

Replacement Because of Inadequacy. A physical asset that is inadequate in capacity to perform its required services is a logical candidate for replacement. For example, a boring mill used almost exclusively to face and bore pulleys has a maximum capacity of machining pulleys 54 inches in diameter. At the time the mill was purchased, the largest pulley ordered was well below the capacity of the mill, but at the present time orders are being received for pulleys up to 72 inches in diameter and these orders seem to be on the increase.

Orders for pulleys between 54 and 72 inches are subcontracted to another concern. Not only is this costly but it occasions delays that are detrimental to the reputation of the company. The factor entering into consideration of replacement in this example is inadequacy. Although the present boring mill is up to date, efficient, and in excellent condition, consideration of its replacement is being forced by the need for a boring mill of greater capacity.

Where there is inadequacy, a usable piece of equipment, often in excellent condition, is on hand. Often, as in the case of the boring mill, the desired increased capacity can be met only by purchasing a new unit of equipment of the desired capacity.

In many cases, such as with pumps, motors, generators, and fans, the increased capacity desired can be met by purchasing a unit to supplement the present machine, should this alternative prove more desirable than purchasing a new unit of the desired capacity.

The method of comparing alternatives where inadequacy is the principal factor will be illustrated by the following example. One year after a 10 h.p. motor has been purchased to drive a belt coal conveyor, it is decided to double the length of the belt. The new belt requires 20 h.p. The needed power can be supplied either by adding a second 10 h.p. motor or by replacing the present motor with a 20 h.p. motor.

The present motor cost $420 installed and has a full load efficiency of 88%. An identical motor can now be purchased and installed for $440. A 20 h.p. motor having an efficiency of 90% can be purchased and installed for $780. The present 10 h.p. motor will be accepted as $270 on the purchase price of the 20 h.p. motor. Current costs $0.02 per kw-hr., and the conveyor system is expected to be in operation 2,000 hours per year.

Maintenance and operating costs other than for current of each 10 h.p. motor are estimated at \$35 per year and for the 20 h.p. motor at \$50 per year. Taxes and insurance are taken as 1% of the purchase price. Interest will be at the rate of 6%. The service lives of the new motors in the present application are taken as 10 years, with a salvage value of 20% of their original cost at that time. The present motor will be considered to have a total life of 11 years, an approximation that will introduce little practical error in the analysis. Most likely all motors will outlast the period of service they will have in the application under consideration.

Alternative A will involve the purchase of the 20 h.p. motor for \$780 and the disposal of the present motor for \$270. The annual cost for this alternative is computed as follows:

Capital recovery and return, ($780 − \$156)($\overset{A/P\,6,\,10}{0.1359}$) + \$156(0.06) .... \$ 94.16

Current cost, $\dfrac{20\text{ h.p.}}{0.90\text{ eff.}} \times \dfrac{0.746\text{ kw}}{\text{h.p.}} \times \dfrac{\$0.02}{\text{kw-hr.}} \times 2,000$ hr. 663.04

Maintenance and operating cost 50.00

Taxes and insurance, \$780 × 0.01............................... 7.80

Total equivalent annual cost \$815.00

Alternative B will involve the purchase of an additional 10 h.p. motor for \$440. The annual cost for this alternative is computed as follows:

Present 10 h.p. Motor:

Capital recovery and return, ($270 − \$84)($\overset{A/P\,6,\,10}{0.1359}$) + \$84(0.06) ...... \$ 30.32

Current cost $\dfrac{10}{0.88} \times 0.746 \times \$0.02 \times 2,000$ 339.10

Maintenance and operating cost 35.00

Taxes and insurance, \$420 × 0.01............................... 4.20

New 10 h.p. Motor:

Capital recovery and return, ($440 − \$88)($\overset{A/P\,6,\,10}{0.1359}$) + \$88(0.06) 53.12

Current cost, $\dfrac{10}{0.88} \times 0.746 \times \$0.02 \times 2,000$ 339.10

Maintenance and operating cost 35.00

Taxes and insurance, \$440 × 0.01............................... 4.40

Total equivalent annual cost \$840.24

On the basis of the analysis above, the advantage of replacing the 10 h.p. motor rather than supplementing it is equivalent to \$840 less \$815, or \$25 per year. Because of the incorrect decision to purchase a 10 h.p. motor a year ago, a sunk cost equal to \$420 less \$270 is incurred. This sunk cost has been revealed, rather than caused, by the present analysis. Since engineering economy analyses are concerned with the future, this sunk cost must not enter into the analysis.

The trade-in value of $270 was taken as the present value of the original 10 h.p. motor because if it is replaced $270 will be received for it. Thus, its value is a necessary element in the comparative analysis. The annual charge for taxes and insurance was based on the original cost, because taxes and insurance charges are usually based upon book value; but for simplicity no reduction was made in these items to correspond to the expected decline in book values.

The alternative to supplement the present motor will require an investment of $270 in the present motor plus $440 in a new 10 h.p. motor or a total investment of $710. The second alternative can be implemented by an investment of $780.

The fact that $270 can be realized from the sale of a capital asset and applied upon the purchase price of the 20 h.p. motor does not reduce the expenditure necessary to acquire the motor or the amount invested in it. Thus the analysis above reveals that an additional investment of $780 − $710 = $70 will result in a return of 6% on the additional investment plus $25 per annum.

Replacement Because of Excessive Maintenance. A machine rarely has all of its elements wear out at one time. Experience has proven that it is economical to repair many types of assets in order to maintain and extend their usefulness. Some repairs are of a current nature and minor in extent. Others are periodic and extensive.

An extensive periodic repair is not usually contemplated until it becomes necessary to extend the life of the unit of equipment in question. Usually, for example, an engine is not overhauled until its failure to provide acceptable service has occurred or is believed to be imminent. Thus the cost of an extensive periodic repair may be considered to be an expenditure to purchase additional service by extending the life of a unit of equipment. This view holds even when a program of preventative maintenance is in effect.

Before an expenditure for major repairs is made to extend the service life of a machine or structure, analysis should be made to determine if the needed service might be more conomically provided by other alternatives.

In this connection consider the following situation. The main roadway through an oil refinery, six-tenths of a mile long and twenty feet wide and made of concrete, is badly in need of repair to continue in service. The maintenance department of the refinery estimates that repairs which will extend the life of the roadway for three years can be made for $4,600. A contractor has offered to replace the present roadway with a type of pavement estimated to have a life of 20 years for $18,400.

Current maintenance cost on the repaired pavement is estimated to average $400 per year and that on the replacing pavement is estimated to average $140 per year. Other items are considered to be equal or negligible. The company's minimum attractive rate of return is considered to be 12%. The salvage value of the present pavement is considered to be nil if it is replaced. The annual cost comparison for the two alternatives follows:

Repair Pavement to Extend Its Life 3 Years

Capital recovery and return $4,600 ($\overset{A/P\ 12,\ 3}{0.4164}$) $1,915

Average annual repair cost 400

Total $2,315

Replace with Pavement with Estimated Life of 20 Years

Capital recovery and return $18,400 ($\overset{A/P\ 12,\ 20}{0.1339}$)..................... $2,463

Average annual repair cost 140

Total $2,603

Annual advantage of repairing over replacing .... $ 288

In some classes of equipment current repairs increase with age. Maintenance may be slight at first but increases at a progressive rate. Thus a point in time is ultimately reached where it is more economical to replace than to continue maintenance. To illustrate the economy of this situation, consider the following example.

A piping system in a chemical plant was installed at a cost of $32,000. This system deteriorated by corrosion until it was replaced at the end of six years. The salvage value of the system was nil. Maintenance records show that maintenance costs in the past have been as given in Column *B* of Table 8.9. It may be noted that annual cost of maintenance increases with lapse of time. This is typical of many classes of equipment and may be the primary reason for replacement. Total expenditures for repairs are given to the end of any year in Column *C*. The sum of the maintenance costs given in Column *C* and the original cost of the equipment is equal to the cost of providing the number of years' service designated in Column *A*.

Table 8.9. ANALYSIS OF MAINTENANCE COSTS

Year *A*	Cost of Maintenance for Year *B*	Sum of Maintenance Cost to End of Year, ΣB *C*	Cost of *n* Years of Service, $32,000 + C$ *D*	Average Annual Cost of Service to End of Year, $D \div A$ *E*
1	$ 1,260	$ 1,260	$33,260	$33,260
2	3,570	4,830	36,830	18,415
3	6,480	11,310	43,310	14,437
4	9,840	21,150	53,150	13,287
5	14,230	35,380	67,380	13,476
6	19,820	55,200	87,200	14,533

The piping system could have been scrapped at the end of any year. Column *E* gives the average annual cost of service that would have resulted from scrapping the system at the end of any year. Thus if the system had been scrapped at the end of the first year, the cost for a year of service would have been $33,260. If it had been scrapped at the end of the second year, the two years of service would have cost $36,830, as given in Column *D*, and the average annual cost would have been, $18,415.

The least average annual cost, $13,287, occurs for a four year life. If interest had been considered and equivalent annual costs had been used, the quantities in Column *E* would have been somewhat larger than those given. But the general pattern would have been much the same. Although the lowest annual cost in the example above occurs for a four year life, it does not necessarily follow that greatest economy would have resulted from scrapping the system after four years of service. The economy of replacement depends upon a number of additional factors such as the need for services of a piping system in the future, changes in levels of maintenance cost, and the characteristics and cost of a replacement. A decision to replace the present equipment should be based on an analysis of costs in prospect with the present equipment and with a possible replacement.

Replacement Because of Declining Efficiency. Equipment usually operates at peak efficiency initially and suffers a loss of efficiency with usage and age. A gasoline engine usually reaches its maximum efficiency after a short run-in period after which its efficiency declines as cylinder walls, pistons, piston rings, and carburetors wear and the ignition system deteriorates.

When loss of efficiency is due to the malfunctioning of only a few parts of a whole machine, it is often economical to replace them periodically and in this way maintain a high level of efficiency over a long life.

There are a number of facilities that decline in efficiency with use and age but which it is not feasible to repair. Pipes that carry hot water, for example, often fill with scale. As their internal diameter decreases, the amount of energy required to force a given quantity of water through them increases. Pipe lines often decline in efficiency as carriers of fluid or gas because of increasing loss by leakage due to external or internal corrosion with age. When it is not economical to restore efficiency by maintenance, the entire system should be replaced at intervals on the basis of economy. Consider the following example. The buckets on a conveyor are subject to wear that reduces the capacity of the conveyor in accordance with the data given in Table 8.10.

As the capacity of the buckets becomes smaller, it is necessary to run the conveyor for longer periods of time, thus increasing operating costs. When the buckets are in new condition, the desired annual quantity of material can be handled in 1,200 hours of operation. The hours of operation required for various efficiency levels are shown in Column *D*. At $6.40 per hour of

Table 8.10. ANALYSIS OF DECLINING EFFICIENCY

Year Number A	Efficiency at Beginning of Year B	Average Efficiency During Year C	Annual Hour of Operation $1200 \div C$ D	Annual Cost of Operation Exclusive of Replacement of Buckets, $D \times \$6.40$ E	Sum of Operation Costs to End of Year, $\sum E$ F	Average Annual Cost of Service to End of Year, $(\$960 + F) \div A$ G	Equivalent Annual Cost of Service to End of Year for 7% Interest H
1	1.00	0.97	1,237	$7,917	$ 7,917	$8,877	$8,973
2	0.94	0.91	1,319	8,442	16,359	8,659	8,701
3	0.88	0.86	1,395	8,928	25,287	8,749	8,773
4	0.84	0.82	1,463	9,363	34,650	8,903	8,904
5	0.80						

operation the annual cost of operation is given in Column E. The average annual cost in Column G is based on a bucket replacement cost of $960.

The example above is typical of many kinds of equipment whose efficiency declines progressively when it is not feasible to arrest the decline with maintenance. In the example the least cost of operation occurs when efficiency is permitted to decline to 88%, corresponding to a life of two years before replacement takes place.

Although least cost of operation occurs for a life of two years in the example above, this is not conclusive evidence that least cost of operation will result from a policy to replace buckets at two-year intervals unless the replacing buckets will duplicate the buckets being replaced in first and subsequent costs. But determination of a least-cost life for a unit of equipment and casual consideration of subsequent replacement is often sufficient and as far as it is practical to go in many situations.

Suppose an existing conveyor system is being considered for replacement by a conveyor with the operation characteristics shown in Table 8.10. If the salvage value of the old conveyor is nil and the estimated cost of operation for the next year is $8,900 with anticipated increases in future operating expenses, the old conveyor's economic life is one year. For an interest rate of 7% it is seen in Column H that the economic life of the new conveyor is two years. Thus the new conveyor should be selected since an equivalent saving of at least $199 ($8,900 − $8,701) per year will be realized if the new conveyor is kept for the optimum time. Even if the new conveyor is retained for a period longer than its economic life, say for 4 years, there still seems to be an advantage to replacing the old conveyor since its annual operating expenses are expected to increase.

Replacement Due to Obsolescence. As an illustration of the analysis involving replacement because of obsolescence, consider the following example. A manufacturer produces a hose coupling consisting of two parts. Each part is machined on a turret lathe purchased thirteen years ago for $3,700 including installation. A new turret lathe is proposed as a replacement for the old. Its installed cost will be $6,800.

The production times per 100 sets of parts with the new and old machine are as follows:

Part	Present Machine	New Machine
Connector............	2.92 hours	2.39 hours
Swivel	1.84 hours	1.45 hours
Total	4.76 hours	3.84 hours

The company's sales of the hose couplings average 40,000 units per year and are expected to continue at approximately this level. Machine operators are paid $4.21 per hour. The old and the proposed machine require equal floor space. The proposed machine will use power at a greater rate than the present one, but since it will be used fewer hours, the difference in cost is not considered worth figuring. This is also considered true of general overhead items. Interest is to be taken at 8%. The salesman for the new machine has found a small shop that will purchase the old machine for $900. The prospective buyer estimates the life of the new machine at 10 years and its salvage value at 10% of its installed cost of $6,800. The old turret lathe is estimated to be physically adequate for 10 more years and to have a salvage value of $250 at the end of that time.

The equivalent annual cost of operation if the present turret lathe is retained will be as follows:

$$\overset{A/P\,8,\,10}{}$$

Capital recovery and return, ($900 − $250)($\overset{A/P\,8,\,10}{0.1490}$) + $250(0.08) ...... $ 117
Direct labor, (4.76 ÷ 100)(40,000)($4.21) 8,016
$8,133

The equivalent annual cost of operation if the new turret lathe is purchased will be as follows:

Capital recovery and return, ($6,800 − $680)($\overset{A/P\,8,\,10}{0.1490}$) + $680(0.08) .... $ 966
Direct labor, (3.84 ÷ 100)(40,000)($4.21) 6,467
$7,433

The annual amount in favor of the new machine is $700. It should be noted that the new machine will be used (3.84 ÷ 100) × 40,000 = 1,536 hours per year. No cognizance is taken of the fact that it is available for use many more hours per year; the unused capacity is of no value until used. Since, however, the additional capacity is potentially of value and may prove a safeguard against inadequacy, it should be considered an irreducible in favor of the new machine.

Replacement Because of a Combination of Causes. In most situations, a combination of causes rather than a single cause leads to replacement consideration. As an item of equipment ages, its efficiency may be expected to decline and its need for maintenance to increase. More efficient units of equipment may become available. Moreover, it frequently happens that changes in activities result in a unit being either too large or too small for maximum economy.

Regardless of the cause or combination of causes that lead to consideration of replacement, analysis and decision must be based upon estimates of what will occur in the future. The past is irrelevant in the contemplated analysis.

PROBLEMS

1. A set of magnetic tape units for a computer cost $120,000 when the computer was purchased 2 years ago. New technology has made available an improved set of tape units that can increase the processing speed of the computer system by 15%. The manufacturer of the new tape units offers to allow 25% of the old unit's first cost as a trade-in value. The new tape units cost $350,000. It is anticipated that the present computer system will be completely replaced in 4 years by a new generation computer. The salvage values of the old and new tape units at that time are estimated to be $20,000 and $40,000, respectively. The computer will operate 8 hours a day for 20 days per month. If computer time saved is valued at $300 per hour and the interest rate is 12% compounded monthly, should the existing tape units be replaced? Maintenance costs are considered to be the same for both old and new units. If it is anticipated that the operating time is to be 10 hours per day, would this change the decision?

2. A soft drink bottler purchased a bottling machine 2 years ago for $16,800. At that time it was estimated to have a service life of 7 years with no salvage value. Annual operating cost of the machine amounted to $4,400. A new bottling machine is being considered which would cost $20,000 but would match the output of the old machine for an annual operating cost of $1,800. The new machine's service life is 5 years with no salvage value. An allowance of $4,000 would be made for the old machine on the purchase of the new machine. The interest rate is 6%.

    (a) List the receipts and disbursements for the next 5 years if the old machine is retained; if the new machine is purchased. Compare the present worth of receipts and disbursements.

    (b) Take the "outsider" viewpoint and calculate the equivalent annual cost for each of the 2 alternatives.

    (c) What is the use value of the old machine in comparison with the new machine?

    (d) Should the new machine be purchased? Why?

3. A municipality 3 years ago purchased a pump for its sewage treatment plant at $1,300. This pump had annual operating costs of $1,000 and these are expected to continue. This pump is expected to continue to operate satisfactorily for 5 additional years, at which time it may be expected to have negligible salvage value. The municipality has an opportunity to purchase a new pump for $2,000. The new pump is estimated to have a life of 5 years, negligible salvage value at the end of its life, and an annual operating cost of $400. If the new pump is purchased the old pump will be sold for $100 and a sunk cost of $1,200 on it will be revealed. The interest rate is 6%.

   (a) What error in equivalent annual costs will result if the municipality erroneously adds the sunk cost it has suffered to the cost of the new pump in making a comparison of the financial desirability of the two pumps?

   (b) Calculate the comparative-use value of the old pump.

4. A hydroelectric plant utilizing a continuous flow of 11 cubic feet of water per second with an absolute head of 860 feet was built 4 years ago. The 18-inch pipeline in the system cost $92,000 for pipe, installation, and right of way and has a loss of head due to friction of 81 feet. Additional water rights have been acquired which will result in a total of 22 cubic feet per second of water flow. The following plans are under consideration for utilizing the total flow. Plan A: Use the present pipeline. This will entail no additional expense but will result in a total loss of head due to friction of 346 feet resulting from the increased velocity of the water. Plan B: Add a second 18-inch pipeline at a cost of $68,000. The loss of head for this line will be 81 feet. Plan C: Install a 26-inch pipeline at a cost of $91,000 and remove the existing line, for which $3,800 can be realized. The loss of head due to friction for the 26-inch pipeline will be 63 feet.

   The energy of the water delivered to the turbine is valued at $64 per horsepower year, where horsepower $= h \times F \times 62.4 \div 550$. In this equation, h is the net head in feet, and F is the flow in cubic feet per second. Insurance and taxes amount to 2% of first cost. Operating and maintenance costs are essentially equal for all three plans. The interest rate is 10%. If all lines, including the one now in use, will be retired in 30 years with no salvage value, what is the comparable equivalent annual costs of the three alternatives?

5. A machine was purchased 3 years ago for $12,000. Its present value is $6,000 and its operating expenses are expected to continue at $1,000 a year. A second-hand machine costing $2,000 is available but its operating expenses are expected to be $1,700 per year. It is anticipated that the machines will be in service for 6 more years with $1,000 salvage value for the present machine and zero for the second-hand machine. Using the rate-of-return approach and a minimum attractive rate of return of 15%, find the best course of action.

6. Plot the data in Columns D, E, and F of Table 8.2. If an interest rate had been used the data would be slightly different. Indicate with a superimposed dashed line the change that interest would make.

7. At the end of the seventh year, replacement of the asset described in Table 8.1 is being considered. The salvage value of the asset has been determined to be zero at the end of the seventh year. What will be the average yearly cost to operate the asset 1, 2, or 3 years more?

8. At the end of the ninth year, replacement of the asset described in Table 8.3 is being considered. At that time, the salvage value of the asset was estimated to be zero. What will be the equivalent annual cost of operating the asset 1, 2, or 3 years more?

9. The maintenance cost of a certain machine is zero the first year and increases by $200 per year for each year thereafter. The machine costs $2,000 and has no salvage value at any time. Its annual operating cost is $1,000 per year. If the interest rate is zero what life will result in minimum average annual cost? Solve by trial and error showing yearly costs in tabular form.

10. The data is the same as in Problem 9 except that the interest rate is 10%. Solve using the tabular method.

11. A special milling machine is being installed at a first cost of $10,000. Maintenance cost is estimated to be $6,000 for the first year and will increase by 5% each year. If interest is neglected and the salvage value is $2,000 at any time, for what service life will the average annual cost be a minimum?

12. Use the same data given in Problem 11, but assume that the interest rate is 12% compounded continuously. Find the economic life of the machine.

13. A special-purpose machine is to be purchased at a cost of $20,000. The following table shows the expected annual operating costs maintenance costs and salvage values for each year of service. It interest is 10% what is the economic life for this machine?

Year of Service	Operation Cost for Year	Maintenance Cost for Year	Salvage Value at End of Year
1	$ 2,000	200	$10,000
2	3,000	300	9,000
3	4,000	400	8,000
4	5,000	500	7,000
5	6,000	600	6,000
6	7,000	700	5,000
7	8,000	800	4,000
8	9,000	900	3,000
9	10,000	1000	2,000
10	11,000	1100	1,000

14. Five years ago a conveyor system was installed in a manufacturing plant at a cost of $27,000. It was estimated that the system, which is still in good condition, would have a useful life of 20 years. Annual operating costs are $1,250. The number of parts to be transported have doubled and will continue at the higher rate for the rest of the life of the system. An identical system can be installed for $22,000, or a system with a 20 year life and double the capacity can be installed for $31,000. Annual operating cost is expected to be $2,800. The present system can be sold for $6,800. Either of the 3 systems will have a salvage value at retirement of 10% of original cost. The minimum attractive rate of return is 12%. Compare the two alternatives for obtaining the required services on the basis of equivalent annual cost over a 15-year study period, recognizing any unused value remaining in the systems at the end of that time.

15. Four years ago an ore-crushing unit was installed at a mine at a cost of $81,000. Annual operating costs for this unit are $3,540, exclusive of charges for interest and depreciation. This unit was estimated to have a useful life of 10 years and this estimate

still appears to be substantially correct. The amount of ore to be handled is to be doubled and is expected to continue at this higher rate for at least 20 years. A unit that will handle the same amount of ore and have the same annual operating cost as the one now in service can be installed for $75,000. A unit with double the capacity of the one now in use can be installed for $112,000. Its life is estimated at 10 years and its annual operating costs are estimated at $4,936. The present realizable value of the unit now in use is $26,000. All units under consideration will have an estimated salvage value at retirement age of 12% of the original cost. The interest rate is 10%. Compare the two possibilities of providing the required service on the basis of equivalent annual cost over a study period of 6 years, recognizing unused value remaining in the unit at the end of that time.

16. A small manufacturing company leases a building for machining of metal parts used in their final product. The annual rental of $10,000 is paid, in advance, on January 1. The present lease runs until December 31, 1976 unless terminated by mutual agreement of both parties. The owner wishes to terminate the lease on December 31, 1972 and offers the company $2,000 if it will comply with the request. If the company does not agree, the lease will remain in effect at the same rate until the 1976 termination date. The company owns a suitable building lot and has a firm contract for construction of a building for a total cost of $170,000 to be completed in one year. These figures are firm whether the building is constructed in 1972 or 1976.

If the company elects to stay in the leased building, it will spend $8,000, $6,500, and $7,000 in 1973, 1974, and 1975, respectively, on the facility with no salvage value resulting. It is estimated that operating expenses will be $3,500 less per year in the new building for a comparable level of output. Taxes, insurance, and maintenance will cost 3.5% of the first cost of the building per year. The life of the building is estimated to be 25 years with no salvage value and the interest rate is 6%. The decision is to be made on January 1, 1972 on the basis of the present worth of the two plans as of December 31, 1972.

(a) What is the present worth of the two plans as of December 31, 1972 if the study period is 25 years?

(b) What is the present worth of the two plans as of December 31, 1972 if the difference is ignored in the length of service provided?

(c) The company considers the privilege of waiting 4 years to build to have a present worth of $6,000. On the basis of this fact, and the results of part (a), which plan should be adopted?

17. Two bridge designs proposed for the crossing of a small stream are to be compared by the present-worth method for a period of 40 years. The wooden design has a first cost of $5,000 and an estimated life of 8 years. The steel design has a first cost of $10,000 and an estimated life of 20 years. Each structure has zero salvage value at the end of the given period and it is estimated that the annual expenditures will the same regardless of which design is selected. Using an interest rate of 6%, does the increased life of the steel design justify the extra investment? Make a present-worth comparison.

18. A manufacturer is considering the purchase of an automatic lathe to replace one of two turret lathes. The turret lathes were purchased 12 years ago at a cost of $3,400. The automatic lathe can be purchased for $14,800 and the turret lathe can be sold for $700. Other pertinent data are as follows:

	Turret Lathes	*Automatic Lathe*
Annual output, Part A	40,000 each	80,000
Annual use other than on Part A ..	400 hours each	0
Production, units of Part A per hour ..	34 each	82
Labor (one man per machine)	$5.10 per hour	$4.20 per hour
Estimated annual maintenance	$600 each	$1,400
Power cost per hour of operation	$0.12 each	$0.25
Taxes and insurance, 1.6% of	Present value	Original cost
Space charges per year	$250	$300

The turret lathes are in good mechanical condition and may be expected to serve an additional 10 years before maintenance rises appreciably. Their salvage value will probably never drop below $300 each. If one of the turret lathes is replaced by the automatic lathe, it is assumed that the automatic lathe will be used to produce the entire 80,000 units of Part A and that the remaining turret lathe will be used 800 hours per year. If the interest rate is 8%, how many years will be required for the automatic lathe to pay for itself?

19. Two years ago, a centrifugal pump driven by a direct-connected induction motor was purchased to meet a need for a flow of 2,000 gallons of water per minute for an industrial process. The unit cost $2,100 and had an estimated salvage value of $300 at the end of 6 years of use. It consumed electric power at the rate of 47 kw and was used 12 hours per day for 300 days a year. The process for which the water is required has been changed and in the future a flow of only 800 gallons per minute is needed 3 hours per day for 300 days a year. At the decreased flow both pump and motor are relatively inefficient and the current consumption is 36 kw.

A new unit of a capacity conforming to future needs will cost $1,180. This new unit will have an estimated life of 8 years with an estimated salvage value of $180 at the end of that time. Its current consumption will be 19 kw. The present unit can be sold for $600. The original estimates of useful life and salvage value are still believed to be reliable. Insurance, taxes, and maintenance are estimated at 6% of the original cost for both units. The cost of power is $0.022 per kilowatt-hour and the minimum attractive rate of return is 12%.

(a) Should the old unit be retained or should the new unit be purchased?

(b) At what number of hours per day for 300 days a year with a requirement of 800 gallons of water per minute will the future equivalent annual cost of the two units be equal?

20. A chemical processing plant secures its water supply from a well which is equipped with a 6-inch, single-stage centrifugal pump that is currently in good condition. The pump was purchased 3 years ago for $2,700 and has an expected life of 10 years. Due to design improvements, the demand for a pump of this type is such that its present value is only $1,000. It is anticipated that the pump will have a trade-in value of $400 seven years from now. An improved pump of the same type can now be purchased for $3,400 and will have an estimated life of 10 years with a trade-in value of $200 at the end of that time.

The pumping demand is 225 cubic feet per minute against an average head of 200 feet. The old pump has an efficiency of 75% when furnishing the demand above. The new pump has an efficiency of 81% when furnishing the same demand. Power costs $0.026 per horsepower-hour and either pump must operate 2,400 hours per year. Do the improvements made in design justify the purchase of a new pump if interest of 15% is required?

21. Three years ago a chemical processing plant installed at a cost of $20,000 a system to remove pollutants from the waste water that is discharged into a nearby river. The present system will cost $14,500 to operate next year with operating costs expected to increase thereafter. A new system has been designed to replace the existing system, and it is expected its installed cost will be $10,000. The new system is expected to have first-year operating costs of $9,000 with these costs increasing at a rate of $1,000 per year. The new system is estimated to have a useful life of 12 years. Because the original system and the new system are specially designed for this particular chemical process their salvage values are expected to be equal to their cost of removal. Should the company replace the existing pollution control system, if their minimum attractive rate of return is 12%? How long should the most economical system be kept in service?

22. A private hospital is considering the replacement of one of its artificial kidney machines. The machine being considered for replacement cost $35,000 4 years ago. If the present machine is kept one more year its operating and maintenance costs are expected to be $25,000. Operating and maintenance expenses in the second and third years are expected to be $27,000 and $29,000, respectively. The new machine, if purchased now, will cost $40,000 and will have an annual operating expense of $20,000. Its economic life is anticipated to be 5 years, and its salvage value at that time is estimated to be $10,000. The company selling the new machine will allow $9,000 on the old machine for trade-in. If the hospital delays its purchase for 1, 2, or 3 years the trade-in value on the existing machine is expected to decrease to $7,000, $5,000, and $3,000, respectively. If the interest rate is considered to be 15% what decision would be most economical? Use an annual cost comparison.

9

Break-Even and
Minimum Cost Analysis

In many situations encountered in engineering economic analysis, the cost of an alternative may be a function of a single variable. When two or more alternatives are a function of the same variable, it may be desirable to find the value of the variable that will result in equal cost for the alternatives considered. The value of such a variable is known as the *break-even* point.

If the cost of a single alternative is a function of a variable that may take on a range of values it may be useful to determine the value of the variable for which the cost of the alternative is a minimum. The value of such a variable is known as the *minimum cost* point. Multiple alternatives which depend upon the same variable can be compared on the basis of their minimum cost points. Several facets of the break-even and minimum cost aspect of engineering economic analysis will be presented in this chapter.

9.1. BREAK-EVEN ANALYSIS, TWO ALTERNATIVES

When the cost of two alternatives is affected by a common variable there may exist a value of the variable for which the two alternatives will incur equal cost. The costs of each alternative can be expressed as functions of the common independent variable and will be of the form

$$TC_1 = f_1(x) \text{ and } TC_2 = f_2(x)$$

where

$TC_1 =$ a specified total cost per time period, per project, or per piece applicable to Alternative 1;

$TC_2 =$ a specified total cost per time period, per project, or per piece applicable to Alternative 2;

$x =$ a common independent variable affecting Alternative 1 and Alternative 2.

Solution for the value of x resulting in equal cost for Alternative 1 and Alternative 2 is accomplished by setting the cost functions equal, $TC_1 = TC_2$. Therefore,

$$f_1(x) = f_2(x)$$

which may be solved for x. The resulting value for x yields equal cost for the alternatives considered and is, therefore, designated the break-even point.

Break-Even Point, Mathematical Solution. Where the cost of each alternative can be mathematically expressed as a function of a common variable the break-even point may be found mathematically. For example, assume that a 20-horsepower motor is needed to drive a pump to remove water from a tunnel. The number of hours that the pump will operate per year is dependent upon the rainfall and is, therefore, uncertain. The pump unit will be needed for a period of four years.

Two alternatives are under consideration. Proposal A calls for the construction of a power line and the purchase of an electric motor, at a total cost of $1,400. The salvage value of this equipment at the end of the four-year period is estimated at $200. The cost of current per hour of operation is estimated at $0.84, maintenance is estimated at $120 per year, and the interest rate is 10%. No attendant will be needed since the equipment is automatic. Let

$TC_A =$ total equivalent annual cost of Proposal A;
$D_A =$ equivalent annual cost of capital recovery and return
$$= (\$1,400 - \$200)(\overset{A/P\ 10,\,4}{0.3155}) + \$200(0.10) = \$399;$$
$M =$ annual maintenance cost $= \$120$;
$C =$ current cost per hour of operation $= \$0.84$;
$N =$ number of hours of operation per year.

Then

$$TC_A = D_A + M + NC.$$

Proposal B calls for the purchase of a gasoline motor at a cost of $550. The motor will have no salvage value at the end of the four-year period. The cost of fuel and oil per hour of operation is estimated at $0.42, maintenance is estimated at $0.15 per hour of operation, and the cost of wages chargeable to the engine when it runs is $0.80 per hour. Let

$TC_B =$ total equivalent annual cost of Proposal B;
$D_B =$ equivalent annual cost of capital recoverey and return
$$= \$550(\overset{A/P\ 10,\,4}{0.3155}) = \$174;$$

H = hourly cost of fuel and oil, operator, and maintenance
 = \$0.42 + \$0.80 + \$0.15 = \$1.37;
N = number of hours of operation per year.

Then

$$TC_B = D_B + NH.$$

There is a value of N for which the two alternatives will incur equal cost. This value may be found by setting $TC_A = TC_B$ and solving for N as follows:

$$D_A + M + NC = D_B + NH$$

$$N = \frac{D_B - (D_A + M)}{C - H}$$

Substituting

$$N = \frac{\$174 - (\$399 + \$120)}{\$0.84 - \$1.37} = 651 \text{ hours.}$$

That the total equivalent annual cost is equal for the two alternatives is shown as follows:

$$TC_A = TC_B$$

$$D_A + M + NC = D_B + NH$$

$$\$399 + \$120 + 651(\$0.84) = \$174 + 651(\$1.37)$$

$$\$1,066 = \$1,066.$$

For the cost data given, the annual cost of the two alternatives is calculated to be equal for 651 hours of operation per year. If the equipment is used less than 651 hours per year, selection of the gasoline motor is most economical; for more than 651 hours of operation per year, the electric motor is most economical. The total annual cost for each alternative, as a function of the number of hours of operation per year is shown graphically in Figure 9.1.

The difference in equivalent annual cost between the two alternatives may be calculated for any number of hours of operation. For example, suppose the equipment is to be operated 100 hours per year. Then

$$TC_A - TC_B = \Delta TC$$

$$D_A + M + NC - (D_B + NH) = \Delta TC.$$

Substituting

$$\Delta TC = \$399 + \$120 + 100(\$0.84) - \$174 - 100(\$1.37)$$

$$= \$292.$$

Break-Even Point, Graphical Solution. There are many situations in which setting up equations to represent the cost patterns of two alternatives is either too difficult or too time-consuming to be a feasible approach for the deter-

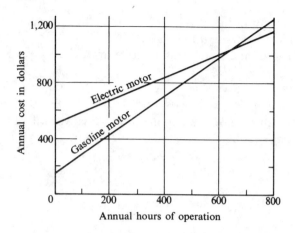

FIGURE 9.1. TOTAL ANNUAL COST AS A FUNCTION
OF THE NUMBER OF HOURS OF OPERATION PER YEAR.

mination of a break-even point. In situations for which the cost patterns of
two alternatives can be established by determining a number of points of the
pattern by calculation or experiment, the break-even point may be determined
graphically.

Consider the following example. Two methods, *A* and *B*, are under
consideration for packing different lengths of display material in paper
cartons of 24 inches in girth and 26, 36, 48, and 62 inches in length, respectively. The average times required to pack a number of cartons of the given
lengths by Method *A* and Method *B* were found by time studies. The results
are shown graphically in Figure 9.2. The length of carton resulting in an

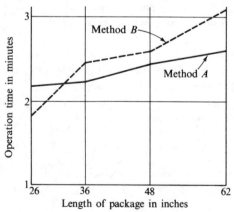

FIGURE 9.2. OPERATION TIME IN
MINUTES AS A FUNCTION OF CARTON
LENGTH.

equal operation time for the methods under consideration may be found by inspection from the graph.

9.2. BREAK-EVEN ANALYSIS, MULTIPLE ALTERNATIVES

In the examples considered thus far, break-even analysis has been applied in the case where only two alternatives confront the decision maker. This section illustrates the application of break-even analysis in cases where multiple alternatives are being considered. As an example consider an architectural engineering firm which has been asked to prepare preliminary plans for the construction of a three-story building. After careful analysis, three types of construction seem feasible.

In attempting to arrive at some quantitative basis for recommending a type of construction, the engineering firm developed fixed and variable cost data. These data are given below and are assumed to represent the cost of construction and operation for a building containing between 2,000 and 6,000 basic square feet.

Concrete and Brick
First cost per square foot .... $ 24
Annual maintenance $5,600
Annual climate control $2,400
Estimated life in years 20
Estimate salvage value is zero.

Steel and Brick
First cost per square foot .... $ 29
Annual maintenance $5,000
Annual climate control $1,500
Estimated life in years 20
Estimated salvage value is 3.2% of first cost.

Frame and Brick
First cost per square foot .... $ 35
Annual maintenance $3,000
Annual climate control $1,250
Estimated life in years 20
Estimated salvage value is 1.0% of first cost.

The total cost for each type of construction will be a function of the number of basic square feet enclosed by the building. For an interest rate of 8%, the total cost for each alternative is as follows:
Concrete and brick:

$$TC = \overset{A/P\ 8,\ 20}{\$24(A)(0.10189)} + \$8,000$$
$$= \$2.44(A) + \$8,000.$$

Steel and brick:

$$TC = \$29(A)(0.968)\overset{A/P\,8,\,20}{(0.10189)} + \$29(A)(0.032)(0.08) + \$6,500$$
$$= \$2.934(A) + \$6,500$$

Frame and brick:

$$TC = \$35(A)(0.99)\overset{A/P\,8,\,20}{(0.10189)} + \$35(A)(0.01)(0.08) + \$4,250$$
$$= \$3.658(A) + \$4,250.$$

Solution for the respective break-even points may be done mathematically by considering the alternatives in pairs. Or, by graphing the total cost of each alternative as a function of the area in square feet, it is possible to determine the break-even points by inspection. Each total cost function developed is graphed in Figure 9.3.

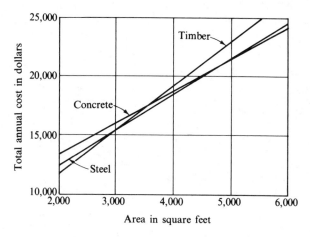

FIGURE 9.3. TOTAL ANNUAL COST AS A FUNCTION OF AREA IN SQUARE FEET.

Suppose that the client was considering a building measuring 40 feet by 100 feet. In this case, the engineering firm would recommend construction from steel and brick. However, if the client required a building of more than 5,000 square feet it would be most economical to use concrete and brick for the basic structure. For a building of less than 3,000 square feet, construction from frame and brick would be most economical. The extension of break-even analysis to cases where there are more than three alternatives follows the same reasoning.

9.3. MINIMUM COST ANALYSIS

An alternative may possess two or more cost components that are modified differently by a common variable. Certain cost components may vary directly with an increase in the value of the variable while others may vary inversely. When the total cost of an alternative is a function of increasing and decreasing cost components, most likely a value exists for the common variable that will result in a minimum cost for the alternative.

The general solution of the situation outlined above may be demonstrated for an increasing cost component and a decreasing cost component as follows:

$$TC = Ax + \frac{B}{x} + C$$

where

TC = a specified total cost per time period, per project, etc.;

x = a common variable;

A, B, and C = constants.

Taking the first derivative, equating the result to zero, and solving for x results in

$$\frac{dTC}{dx} = A - \frac{B}{x^2} = 0$$

$$x = \sqrt{\frac{B}{A}}.$$

The value for x found in this manner will be a minimum and is, therefore, designated the minimum cost point.

Minimum Cost Point, Mathematical Solution. A classical example of minimum cost analysis is given by the increasing and decreasing cost components involved in the choice of the cross sectional area of an electrical conductor. Since resistance is inversely proportional to the size of the conductor, it is evident that the cost of power loss will decrease with increased conductor size. However, as the size of the conductor increases, an increased investment charge will be incurred. For some given conductor cross section, the sum of the two cost components will be a minimum.

As an example, suppose a copper conductor is being considered to transmit the daily electrical load at a sub-station, and estimates call for transmission of 1920 amperes for 24 hours per day, respectively, for 365 days per year. The following engineering and cost data apply to the conductor installation: length of conductor, 140 feet; installed cost, $160 + $0.60 per pound of copper; estimated life, 20 years; salvage value, $0.50 per pound of copper.

Electrical resistance of a copper conductor 140 feet long and of 1 square inch cross section is 0.0011435 ohm, and the electrical resistance is inversely proportional to the area of the cross section. The energy loss in kilowatt-hours in a conductor due to resistance is equal to $I^2R \times$ number of hours $\div$ 1,000, where I is the current flow in amperes and R is the resistance of the conductor in ohms. Copper weighs 555 pounds per cubic foot. The energy lost is valued at $0.007 per kilowatt-hour; taxes, insurance, and maintenance are negligible; the interest rate is 6%.

The I^2R loss in dollars per year:

$$(1,920)^2(24)\left(\frac{365}{1,000}\right)\left(\frac{0.0011435}{A}\right)(\$0.007) = \frac{\$258.49}{A}$$

Weight of conductor in pounds:

$$\frac{[(140)(12)(A)(555)]}{1,728} = 539.6A$$

Capital recovery plus return in dollars per year:

$$[\$160 + (\$0.60 - \$0.50)(539.6)(A)]\overset{A/P\,6,\,20}{(0.08717)}$$
$$+ \$0.50(539.6)(A)(0.06) = \$20.89A + \$13.94.$$

The total cost per year:

$$TC = \$20.89A + \frac{\$258.49}{A} + \$13.94$$

$$\frac{dTC}{dA} = \$20.89 - \frac{\$258.49}{A^2} = 0$$

$$A = \sqrt{\frac{258.49}{20.89}} = 3.52 \text{ square inches.}$$

Therefore, the selection of a conductor with a cross-sectional area of 3.52 square inches will result in a minimum total cost. Note that this example exhibits a simple case involving an increasing and a decreasing cost component. The nature of these components may be tabulated from the expressions for I^2R loss cost and investment cost. These costs and the resulting total cost are given in Table 9.1. and graphed in Figure 9.4.

It was first indicated by Lord Kelvin that the most economical cross

Table 9.1. TOTAL ANNUAL COST AS A FUNCTION OF CROSS SECTIONAL AREA

Cost	Cross Sectional Area (in.²)				
	2	3	4	5	6
Investment cost	$ 55.72	$ 76.61	$ 97.50	$118.39	$139.28
I^2R loss cost	$129.37	$ 86.15	$ 64.68	$ 51.75	$ 43.12
Total annual cost..	$185.09	$162.76	$162.16	$170.14	$182.40

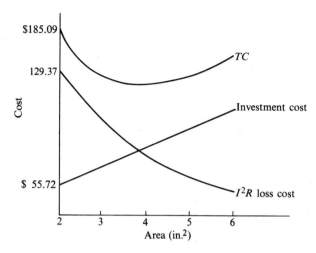

FIGURE 9.4. TOTAL ANNUAL COST CURVE.

sectional area for a conductor was one in which investment cost just equals the annual cost of lost energy. This is known as Kelvin's law.

Minimum Cost Point, Tabular Solution. There are many situations for which minimum cost points are sought but for which it is very difficult to set up equations that truly characterize the existing cost relationships. Since finding minimum cost points from cost equations involves differentiation, equations that only approximate true situations may result in grossly misleading results.

The graphical method for finding minimum cost points will be illustrated in the following example. The manufacturer of a pharmaceutical product plans to use an evaporative process. In this process one or several evaporators in multiple may be used. It is known that variable operating costs, of which the chief item is the cost of steam, will be approximately inversely proportional to the number of evaporators used in the installation. Fixed costs will be approximately in proportion to the number of evaporators used. Trial had demonstrated that a rather complex equation would be required to express accurately the cost relationships that existed, so a graphical solution was sought.

Estimates were made of the variable and the fixed operation costs that would be obtained if one, two, three, or four evaporators were used. Results from equipment manufacturers, experimental data, and a knowledge of the process formed the basis for estimating the costs. Table 9.2 summarizes the estimates based on 200 days' operation per year. It is clear that, on the basis of 200 days' operation per year, the lowest annual cost of $5,960 results when three evaporators are used.

Table 9.2. TOTAL ANNUAL COST AS A FUNCTION OF THE NUMBER OF EVAPORATORS USED

Costs	Number of Evaporators Used			
	1	2	3	4
Fixed costs (depreciation, insurance, taxes, etc.)	$ 860	$1,680	$2,350	$3,030
Variable costs (steam, labor, maintenance, etc.)	$7,850	$4,560	$3,610	$3,190
Total annual cost	$8,710	$6,240	$5,960	$6,220

9.4. MINIMUM COST ANALYSIS, MULTIPLE ALTERNATIVES

As an illustration of the application of minimum cost analysis where two alternatives are proposed, consider the following example. A double track railroad bridge is to be constructed for a 1,200 foot crossing. Two girder designs have been proposed. The first will result in the weight of the superstructure per foot to be $W_1 = 22(S) + 800$, where S is the span between piers. The second will result in a superstructure weight per foot of $W_2 = 20$ $(S) + 1,000$. Regardless of the girder design chosen, the piers will cost $220,000 each. The superstructure will be erected at a cost of $0.22 per pound. All other costs for the competing designs are the same.

In order to choose a girder design on the basis of minimum total cost for the superstructure and piers, it will be necessary to find the minimum cost pier spacing for each design. The general relationship that exists may be described as follows. As the number of piers increases, the amount of superstructure required decreases; and conversely, as the number of piers decreases, the amount of superstructure required increases. Therefore, this situation involves increasing and decreasing cost components, the sum of which will be a minimum for a certain number of piers.

The total cost for the superstructure and piers for girder Design 1 is as follows:

$$TC_1 = [22(S) + 800]($0.22)(1,200) + \left(\frac{1,200}{S} + 1\right)$220,000$$

$$= 5,810(S) + \frac{264,000,000}{S} + 431,000.$$

The minimum cost span between piers is

$$\frac{dTC}{dS} = 5,810 - \frac{264,000,000}{S^2} = 0$$

$$S = \sqrt{\frac{264,000,000}{5,810}} = 213 \text{ feet.}$$

And the minimum total cost for the superstructure and piers is

$$TC_1 = 5,810(213) + \frac{264,000,000}{213} + 431,000$$

$$= \$2,908,000.$$

The total cost for the superstructure and piers for girder Design 2 is as follows:

$$TC_2 = [20(S) + 1,000](\$0.22)(1,200) + \left(\frac{1,200}{S} + 1\right)\$220,000$$

$$= 5,280(S) + \frac{264,000,000}{S} + 484,000.$$

The minimum cost span between piers is

$$\frac{dTC}{dS} = 5,280 - \frac{264,000,000}{S^2} = 0$$

$$S = \sqrt{\frac{264,000,000}{5,280}} = 224 \text{ feet.}$$

And the minimum total cost for the superstructure and piers is

$$TC_2 = 5,280(224) + \frac{264,000,000}{224} + 484,000$$

$$= \$2,846,000.$$

On the basis of this analysis, the designer would choose girder Design 2 as the design that will result in a minimum total cost for the bridge. The pier spacing chosen for the design would be as approximate to 224 feet as possible. In this case, $1200 \div 224 = 5.36$ spans so 6 piers would be used for a span between piers of 240 feet. The nature of the increasing and decreasing cost components is given for various numbers of piers in Table 9.3.

Table 9.3. TOTAL COST AS A FUNCTION OF THE NUMBER OF PIERS FOR TWO COM-PETING BRIDGE DESIGNS

Piers	Cost of Piers	Superstructure Cost Design 1	Superstructure Cost Design 2	Total Cost Design 1	Total Cost Design 2
4	$880,000	$2,534,000	$2,376,000	$3,414,000	$3,336,000
5	1,100,000	1,953,600	1,848,000	3,053,600	2,948,000
6	1,320,000	1,605,120	1,531,000	2,925,120	2,851,000
7	1,540,000	1,372,800	1,320,000	2,912,800	2,860,000
8	1,760,000	1,210,176	1,172,160	2,970,176	2,932,160
9	1,980,000	1,082,400	1,056,000	3,062,400	3,036,000

The extension of minimum cost analysis to cases where there are more than two alternatives follows the same procedure.

PROBLEMS

1. A company may furnish a car for use by a service man for transportation between its properties, or the company may pay the service man for the use of his car at a rate of $0.13 per mile for this purpose. The following estimated data apply to company furnished cars: A car costs $3,000 and has a life of 4 years and a trade-in value of $500 at the end of that time. Monthly storage cost for the car is $20 and the cost of fuel, tires, and maintenance is $0.04per mile. What annual mileage must a service man travel by car for the cost of the two methods of providing transportation to be equal if the interest rate is 15%?

2. In considering the purchase of an automobile with air-conditioning, it is estimated that when the air-conditioning is off the mileage will be 14.0 miles per gallon and with air-condition on it will be 12.3 miles per gallon. The air-conditioning will add $400 to the first cost of the automobile and will increase its trade in value by $50 after a service life of 5 years. The air-conditioning will be used 20% of the time and it will add $60 per year to the maintenance of the automobile. The benefit of having air-conditioning when needed is considered to be worth $0.02 per mile. If gasoline costs $0.40 per gallon and if the interest rate is 8%, how many miles per year must be driven for the air-conditioner to pay for itself?

3. A coffee concern has a weigher for filling cans with coffee. Because of its lack of sensitivity the weigher must be set so that the average filled can contains $16\frac{1}{4}$ ounces in order to insure that no cans contain less than 16 ounces of coffee. The present weigher was purchased 10 years ago at a price of $1,500 and is expected to last an additional 10 years. A new weigher is being considered whose sensitivity is such that the average "overage" per pound can is $\frac{1}{8}$ ounce. This weigher will cost $6,400. $300 will be allowed for the present weigher on the purchase price of the new weigher. The salvage value of either weigher 10 years from now is considered to be negligible and their operating costs are estimated to be equal. The value of the coffee is $1.05 per pound. What is the minimum number of 1-pound packages packed annually for which the purchase of the new weigher is justified, if the interest rate is 10%?

4. Machine A costs $200, has zero salvage value at any time, and labor cost per unit of product on it is $1.14. Machine B costs $3600, has zero salvage value at any time, and labor cost per unit of product on it is $0.83. Neither machine can be used except make the product in question. The interest rate is 5%.
   (a) If the annual rate of production is 4,000 units, in how many years will the two machines break even in equivalent annual costs? Solve by trial and error.

5. An engineering consulting firm can purchase a small electronic computer for $25,000. It is estimated that the life and salvage value of the computer will be 6 years and $4,000, respectively. Operating expenses are estimated to be $50 per day, and maintenance will be performed under contract for $3,000 per year. As an alternative, sufficient computer time can be rented at an average cost of $120 per day. If the interest rate is 12%, how many days per year must the computer be needed to justify its purchase?

6. The cost of constructing a motel, including materials, labor, taxes, and carrying charges during construction and other miscellaneous items, is $980,000. The lots on which the motel is located cost $300,000. Furniture and furnishings cost $150,000.

Working capital of 30 days' gross income at 100% capacity is required. The investment in the furniture and furnishings should be recovered in 7 years and the investment in the motel should be recovered in 25 years. The land on which the motel is built is considered not to depreciate in value.

If the motel operates at 100% capacity, the gross annual income will be $1,400 per day for 365 days. The motel has fixed operating expenses (exclusive of capital recovery and interest) amounting to $115,000 a year and a variable operating cost of $78,000 for 100% capacity. Assume that the variable cost varies directly with the level of operation. If interest is taken at 8% compounded annually, at what percent of capacity must the motel operate to break even?

7. A certain assembly requires rods 0.10 square inches varying in length from 0.25 to 4 inches. The rods may be made from either brass or steel. The machining cost of brass rods is $0.0054 + $0.006L per piece, where L is the length of the rod in inches. For steel the machining cost may be expressed as $0.0072 + $0.010L per piece. Brass costs $0.70 per pound and steel costs $0.25 per pound. The weight of brass and steel is 0.309 and 0.283 pounds per cubic inch, respectively. What length will equalize the cost of producing the brass and steel rods? What type of rod should be used?

8. A certain area can be irrigated by piping water from a nearby river. Two competing installations are being considered for which the following engineering and cost data apply.

	Six-Inch System	Eight-Inch System
Size motor required	25	10
Energy cost per hour of operation........	$ 0.22	$ 0.07
Cost of motor installed	$ 360	$ 160
Cost of pipe and fittings	$2,050	$2,640
Salvage value at end of 10 years	$ 80	$ 100

On the basis of a 10-year life with an interest rate of 12%, determine the number of hours of operation per year for which the two systems will break even.

9. An electronic manufacturer is considering two methods for producing a required circuit board. The board can be hand wired at an estimated cost of $0.98 per unit and with an annual equipment cost of $200. A printed equivalent of the required circuit can be produced with an investment of $3,000 in printed circuit processing equipment which will have an expected life of 9 years and a salvage value of $100. It is estimated that labor cost will be $0.32 per unit and that the processing equipment will cost $150 per year to maintain. If all other costs are assumed equal, and if the interest rate is 10%, how many circuit boards must be produced each year for the two methods to break even in equivalent annual cost?

10. Two brands of a protective coating are being considered. Brand A costs $4.50 per gallon. Past experience with Brand A has revealed that it will cover 350 square feet of surface per gallon, will give satisfactory service for 3 years, and can be applied by a workman at the rate of 70 square feet per hour. Brand B, which costs $7.60 per gallon, is estimated to cover 400 square feet of surface per gallon and can be applied at the rate of 80 square feet per hour. The wage rate of the workman is $4.12 per hour. If an interest rate of 10% is used, how long should Brand B last to provide service at equal cost with that provided by Brand A?

11. A contractor is offered his choice of either a gasoline, diesel, or butane engine to

power a bulldozer he is to purchase. The gasoline engine will cost $1,600, will have an estimated maintenance cost of $200 per year, and will consume $3.60 worth of fuel per hour of operation. The diesel engine will cost $2,200, will cost an estimated $240 per year to maintain, and will consume $3.30 worth of fuel per hour. The butane engine will cost $2,500, will cost $315 per year to maintain, and will consume $2.90 worth of fuel per hour of operation. Since the salvage value of each engine will be identical, it may be neglected. All other costs associated with the 3 engines are equal and the interest rate is 12%. The service life of each engine is 4 years.

(a) Plot the total annual cost of each engine as a function of the number of hours of operation per year.

(b) Find the range of number of hours of operation for which it would be most economical to specify the gasoline engine; the diesel engine; the butane engine.

12. A 3-phase, 220-volt, squirrel cage induction motor is to be used to drive a pump which will fill a series of storage tanks. It is estimated that 300 horsepower-hours will be required each day for 365 days per year. Motors having the following characteristics may be leased:

Size (h.p.)	Lease Rate (Per Year)	Operating Cost (Per h.p.-hr.)
20	$128	$0.0044
40	$134	$0.0037
75	$142	$0.0031
100	$156	$0.0028
150	$192	$0.0025

Plot the total yearly cost as a function of the size of the motor and select the size that will result in a minimum cost per year.

13. An overpass is being considered for a certain railroad crossing. The superstructure design under consideration will be made of steel and will have a weight per foot depending upon the span between piers in accordance with $W = 32(S) + 1,850$. Piers will be made of concrete and will cost $185,000 each. The superstructure will be erected at a cost of $0.36 per pound. If the number of piers required is to be one less than the number of spans, find the number of piers that will result in a minimum total cost for piers and superstructure if $L = 1,275$ feet.

14. It has been found that the heat loss through the ceiling of a building is 0.13 Btu per hour per square foot of area per degree Fahrenheit. If the 2,200-square-foot ceiling is insulated, the heat loss in Btu per hour per degree temperature difference per square foot of area is taken as being equal to

$$\frac{1}{\dfrac{1}{0.13} + \dfrac{t}{0.27}}$$

where t is the thickness in inches. The in-place cost of insulation 1, 2, and 3 inches thick is $0.08, $0.11, and $0.15 per square foot, respectively. The building is heated to 75 degrees 3,000 hours per year by a gas furnace with an efficiency of 50%. The mean outside temperature is 45 degrees and the natural gas used in the furnace costs $0.60 per 1,000 cubic feet and has a heating value of 2,000 Btu per cubic foot. What thickness of insulation, if any, should be used if the interest rate is 10% and the resale value of the building 6 years hence is enhanced $200 if insulation is added, regardless of the thickness?

15. An hourly electric load of 1,600 amperes is to be transmitted from a generator to a transformer in a certain power plant. A copper conductor 150 feet long can be installed for $160 + $0.60 per pound, will have an estimated life of 20 years, and can be salvaged for $0.50 per pound. Power loss from the conductor will be a function of the cross-sectional area and may be expressed as $25,875 \div A$ kilowatt-hours per year. Energy lost is valued at $0.008 per kilowatt-hour; taxes, insurance, and maintenance are negligible; the interest rate is 8%. Copper weighs 555 pounds per cubic foot.
(a) Plot the total annual cost of capital recovery with a return and power loss cost for conductors for cross sections of 1, 2, 3, 4, and 5 square inches.
(b) Find the minimum cost cross section mathematically and check the result against the minimum point found in (a).

16. The government is considering charging a polluter's fee on the amount of pollutants discharged into our rivers in order to significantly reduce the pollution of our waterways. The polluter's fee is to be based on the amount of pollutant remaining in the effluent being discharged. The indicator to be used to measure the level of pollution of industrial wastes is biochemical oxygen demand, BOD.
A firm is presently discharging 4,000,000 pounds of BOD per year into a nearby river because it does not treat its waste products. The cost of installing and operating a waste treatment system that reduces the BOD output to a percent of its present level is shown below.

Percent BOD Remaining	Initial Investment	Operating Expenses
5%	$250,000	$40,000
10%	100,000	25,000
15%	85,000	15,000
20%	75,000	10,000
25%	75,000	5,000

It is expected that the treatment system will last 10 years and have zero salvage value at that time. The firm's minimum attractive rate of return is 15%.
(a) If the polluter's fee is $0.10 per pound of BOD remaining, what percent of BOD remaining will minimize the firm's annual costs?
(b) If the polluter's fee is $0.05 per pound of BOD remaining, what percent of BOD remaining will minimize the firm's costs?
(c) Plot a graph of the optimum level of BOD remaining as a function of the polluter's fee. Let the polluter's fee range from $0.02 per pound of BOD to $0.020. If you were charged with deciding the amount of the polluter's fee for this firm what amount would you choose?

17. The daily electrical load to be transmitted by a conductor in a power plant is 1,900 amperes per day for 365 days per year. Two conductor materials are under consideration, copper and aluminum. The following information is available for the competing materials.

	Copper	*Aluminum*
Length	120 ft.	120 ft.
Installed cost	150 + $0.60/lb.	150 + $0.30/lb.
Estimated life........................	10 yrs.	10 yrs.
Salvage value	$0.55/lb.	$0.10/lb.
Electrical resistance of conductor 120 feet by 1 sq. in. cross section	0.000982 ohms	0.001498 ohms
Density	555 lb./ft.^{3}	168 lb./ft.^{3}

The energy loss in kilowatt-hours in a conductor due to resistance is equal to I^2R times the number of hours divided by 1,000, where I is the current flow in amperes and R is the resistance in the conductor in ohms. The electrical resistance is inversely proportional to the area of the cross section. Lost energy is valued at $0.007 per kilowatt-hour.

(a) Plot the total annual cost of capital recovery and return plus power loss cost for each material for cross sections of 2, 3, 4, 5, 6, and 7 square inches if the interest rate is 15%.

(b) Solve mathematically for the optimum cross section of each material.

(c) Recommend the minimum cost conductor material and specify the cross sectional area.

18. Ethyl acetate is made from acetic acid and ethyl alcohol. Let x = pounds of acetic acid input, y = pounds of ethyl alcohol input, and z = pounds of ethyl acetate output. The relationship of output to input is

$$\frac{z^2}{(1.47x - z)(1.91y - z)} = 3.91.$$

(a) Determine the output of ethyl acetate per pound of acetic acid, where the ratio of acetic acid to ethyl alcohol is 2, 1.5, 1, 0.67, and 0.50, and graph the result.

(b) Graph the cost of material per pound of ethyl acetate for each of the ratios given and determine the ratio for which the material cost per pound of ethyl acetate is a minimum if acetic acid costs $0.10 per pound and ethyl alcohol costs $0.06 per pound.

10

The Evaluation of
Public Activities

The standards by which private enterprise evaluates its activities are markedly different from those that apply in the evaluation of public activities. In general, private activities are evaluated in terms of profit whereas public activities are evaluated in terms of the general welfare. The general welfare, as collectively and effectively expressed, is the primary basis for evaluating public activities. A basis for evaluating public activities is necessary for an understanding of the characteristics of the governmental agencies that sponsor them.

The government of the United States and its several subdivisions engage in innumerable activities—all predicated upon the thesis of promotion of the general welfare. So numerous are the services available to individual citizens, associations, and private enterprises that books are required to catalog them. This chapter will present concepts and methods of analysis applicable to the evaluation of such activities.

10.1. GENERAL WELFARE AIM OF GOVERNMENT

A national government is a super-organization to which all agencies of the government and all organizations in a nation, including lesser political subdivisions such as states, counties, cities, townships, and school districts as well as private organizations and individuals, are subordinate. In some of its aspects the government of the United States may be likened to a huge corporation. Its citizens play a role similar to that of stockholders. Each, if he chooses, may have a voice in the election of the policymaking group, the Congress of the United States, which may be likened to a board of directors.

In the United States the lesser political subdivisions, such as states, counties, cities, and school districts, carry on their functions in much the same way as does the United States, for in them each citizen may have a voice in determining their policy. Each of these lesser political subdivisions has certain freedom of action, although each is in turn subordinate to its superior organization. The subdivisions of the government are delineated for the most part as continuous geographical areas that are easily recognized.

It is a basic tenet that the purpose of government is to serve its citizens. The chief aim of the United States as stated in its Constitution is the *national defense* and the *general welfare* of its citizens. For convenience in discussion, these aims may be considered to be embraced by the single term—general welfare. This simply stated aim is, however, very complex. To discharge it perfectly requires that the desires of each citizen be fulfilled to the greatest extent and in equal degree with those of every other citizen.

Since the general welfare is the aim of the United States, the super-organization to which the lesser political subdivisions are subordinate, it follows that the latter's aims must conform to the same general objective regardless of what other specific aims they may have.

The General Welfare Aim as Seen by the Citizenry. Since each citizen may have a voice, if he will exercise it, in a government, the objectives of the government stem from the people. For this reason the objectives taken by the government must be presumed to express the objectives necessary for attainment of the general welfare of the citizenry as perfectly as they can be expressed. This must be so, for there is no superior authority to decide the issue.

Thus, when the United States declares war, it must be presumed that this act is taken in the interest of the general welfare. Similarly, when a state votes highway bonds, it also must be presumed to be in the interest of the general welfare of its citizens. The same reasoning applies to all activities undertaken by any political subdivision; for, if an opposite view is taken, it is necessary to assume that people collectively act contrary to their wishes.

Broadly speaking, the final measure of the desirability of an activity of any governmental unit is the judgement of the people in that unit. The exception to this is when a subordinate unit attempts an activity whose objective is contrary to that of a superior unit, in which case the final measure of desirability will rest in part in the people of the superior unit. Also, it must be clear that governmental activities are evaluated by a summation of judgments of individual citizens whose basis for judgment has been the general welfare as each sees it. The objectives of most governmental activities appear to be primarily social in nature, although economic considerations are often a factor. Public activities are proposed, implemented, and judged by the same group, namely, the people of the governmental unit concerned.

The situation of the private enterprise is quite different. Those in control of private enterprise propose and implement services to be offered to the public, which judges whether the services are worth their cost. To survive, a private business organization must, at least, balance its income and costs; thus profit is of necessity a primary objective. For the same reason a private enterprise is rarely able to consider social objectives except to the extent that they improve its competitive position.

The General Welfare Aim as Seen by the Individual. Public activities are evaluated by a summation of judgments of individual citizens, each of whose basis for judgment has been the general welfare as he sees it. Each citizen is the product of his unique heredity and environment; his home, cultural patterns, education, and aspirations differ from those of his neighbor. Because of this and the additional fact that human viewpoints are rarely logically determined, it is rare for large groups of citizens to see eye to eye on the desirability of proposed public activities.

The father of a family of several active children may be expected to see more point to expenditures for school and receational facilities than to expenditures for a street-widening program planned to enhance the value of downtown property. It is not difficult for a person to extol the value of aviation to his community if a proposed airport will increase the value of his property or if he expects to receive the contract to build it. Many public activities have no doubt been strongly supported by a few persons primarily because they would profit handsomely thereby.

But it is incorrect to conclude that activities are supported only by those who see in them opportunity for economic gain. For example, schools and recreational facilities for youth are often strongly supported by people who have no children. Many public activities are directed to the conservation of national resources for the benefit of future generations.

It is clear that the benefits of public activities are very complex. Some that are of great general benefit may spell ruin for some persons and vice versa. Lack of knowledge of the long-run effect of proposed activities is probably the most serious obstacle in the way of the selection of those activities that can contribute most to the general welfare.

10.2. THE NATURE OF PUBLIC ACTIVITIES

Governmental activities may be classified under the general headings of protection, enlightenment and cultural development, and economic benefits. Included under protection are such activities as the military establishments, police forces, the system of jurisprudence, flood control, and health services. Under enlightenment and cultural development are such services as the public school system, the Library of Congress and other publicly supported libraries, publicly supported research, the postal service, and recreation facilities. Economic benefits include harbors and canals, power development, flood control, research and information service, and regulatory bodies.

The list above, although incomplete, shows that there is much overlapping in classification. For example, the educational system is considered by many to contribute to the protection, the enlightenment, and the economic benefit of people. Consideration of the purposes of governmental activities as suggested by the classification above is necessary in considering the pertinency of economic analysis to public activities.

Multiple Purpose Projects. Many projects undertaken by govenment have more than one purpose. A good example of this arises in connection with the public lands, such as a forest reserve. For example, suppose that a new road is being planned for a certain section of a national forest. Since public land is managed under the multiple use concept, several benefits will result if the road is put into service. Among these are scenic driving opportunities, camping opportunities, improved fire protection, ease of timber removal, etc.

Justifying a public works project is normally easier if the project is to serve several purposes. This is especially true if the project is very costly and must rely upon the support from several groups. The forest road will probably appeal to persons who like to drive and camp, the U. S. Forest Service who is responsible for fire control, and the timber industry who will be granted contracts to harvest timber resources from time to time.

There are a number of problems which arise in connection with multiple use projects. Foremost among these is the problem of evaluating the aggregate benefit to be derived from the project. What is the benefit of a scenic drive or a camping trip? How can the benefit of improved accessibility for fire protection be measured? What is the benefit of easier timber removal? Each of these questions must be answered in quantitative monetary terms if an economic analysis of the project is to be performed. They must also be answered so that each group which benefits can share in the cost of the project.

A second problem arising from multiple purpose projects is the possibility for conflict of interest between the purposes. These conflicts frequently become political issues. A primary motive of every public servant is to get elected or re-elected. By demonstrating that direct benefits have been obtained for the parties concerned, a candidate obtains votes. Because the desire is to show that the direct benefits are not very costly, there is a tendency to allocate project costs to those benefit categories which are deemed essential by all. For example, a major portion of the cost of the road may be allocated to the U. S. Forest Service under the categories of improved fire protection. By so doing it is easy to show that the project is desirable to the general public due to its low cost in connection with scenic driving and camping opportunities.

Impediments to Efficiency in Public Activities. There are two major impediments to efficiency in public activities. First, the person who pays taxes has no practical way of evaluating what he receives in return for his tax payment. His tax payments go into a common pool and lose their identity. The taxpayer, with few exceptions, receives nothing in exchange at the time or place at which he pays his taxes on which to base a comparison of the worth of what he pays in and what benefits he will receive as the result of his payment.

Since governmental units are exclusive franchises, the taxpayer has no choice as to which unit he must pay taxes. Thus, he does not have an opportunity to evaluate the effectiveness of tax units on the basis of comparative performance nor an opportunity to patronize what he believes to be the most efficient unit.

The second deterrent to efficiency in governmental activities is the fact that recipients of the products of tax supported activities cannot readily evaluate the products in reference to what they cost. Where no direct payment is exchanged for products, a person may be expected to accept them on the basis of their value to the recipient only. Thus, the products of governmental activities will tend to be accepted even though their value to the recipient is less than the cost to produce them.

10.3. FINANCING PUBLIC ACTIVITIES

Funds to finance public activities are obtained through the assessment of various types of taxes and charges for services. Governmental receipts are derived chiefly from income, property, and excise taxes and duties on imports. Considerable income on some governmental levels is derived from fees collected for services. Examples of such incomes on the national level are incomes from postal services, and on the city level, incomes from supplying water service and from levies on property owners for sidewalks, pavements, and sewers adjacent to their property.

Two basic philosophies in the United States greatly influence the collection of funds and their expenditure by governmental subdivisions. These are collection of taxes on the premise of *ability to pay* and the expenditure of funds on the basis of *equalizing opportunity* of citizens. Application of the ability-to-pay viewpoint is clearly demonstrated in our income and property tax schedules. The equalization-of-opportunity philosophy is apparent in federal assistance to lesser subdivisions to help them provide improved educational and health programs, highway systems, old-age assistance, and the like.

Because of the two basic tenets of taxation on the basis of ability to pay and expenditure of tax funds on the basis of equalization of opportunity, there often is little relationship between the benefits that an individual receives and the amount he pays for public activities. This is in large measure true of such major activities as government itself, military and police protection, the highway system, and most educational activities.

Consideration of Interest. Since activities financed through taxation require payments of funds from citizens, the funds expended for public activities should result in benefits comparable with those which the same funds would bring if expended in private ventures. It is almost universal for individuals to demand interest or its equivalent as an inducement to invest their private funds. To maintain public and private expenditures on a comparative basis, it seems logical to consider interest in economic evaluation of public activities that are financed through taxation.

Some public activities are financed in whole or in part through the sale of services or products. Examples of such activities are power developments, irrigation and housing projects, and toll bridges. Many such services could be carried on by private companies and are in general in competition with private enterprise. Again, since private enterprise must of necessity consider interest, it seems logical to consider interest in the economic analysis of public activities that compete in any way with private enterprise in order that both may be on a comparable basis.

Expenditures for capital goods are made on the promise that they will ultimately result in more consumer goods than can be had for a present equal expenditure. Interest represents the expected difference. Not to consider interest in the evaluation of public activities is equivalent to considering a future benefit equal to a present similar benefit. This appears to be contrary to human nature.

The interest rate to use in an economy study of a public activity is a matter of judgment. The rate used should not be less than that paid for funds borrowed for the activity. In many cases, particularly where the activity is comparable or competitive with private activities, the rate used should be comparable with that used in private evaluations.

Consideration of Tax Loss. Many public activities result in loss of taxes through the removal of property from tax rolls or by other means, as, for example, the exemption from sales taxes. In a nation where free enterprise is a fundamental philosophy, the basis for comparison of the cost of carrying on activities is the cost for which they can be carried on by well-managed private enterprises. Therefore, it seems logical to take taxes into consideration in economy analysis, particularly where the activities are competitive with private enterprise.

The federal government agreed to pay $300,000 annually for 50 years to each of the states of Arizona and Nevada in connection with the Hoover Dam project. These payments are to partially compensate for tax revenue which would accrue to these states if the project had been privately constructed and operated. These payments are a cost and were considered in the economic justification of the project.

10.4. PUBLIC ACTIVITIES AND ENGINEERING

Engineering is a major factor in nearly all public activities because of the high levels of technological input required in most public ventures. It is evident that the complexity of systems and projects being undertaken by the federal and local governments is increasing instead of decreasing.

The contribution of engineering to the nation's space program, national defense, pollution control, urban renewal, and highway construction is well established. Engineering input to these activities involves engineers as employees of governmental agencies and as consultants to these agencies. Thus, public activities are a concern of all engineers as citizens and of many engineers as outlets for their talents.

The engineering process described in Chapter 1 involved the determination of objectives, identification of strategic factors, determination of means, evaluation of engineering proposals, and assistance in decision making. Each of these phases of the engineering process are applicable in public activities. The main modification required is the substitution of benefit or general welfare for profit.

For example, suppose that a municipality has under consideration two projects, one a swimming pool and the other a library. The municipality has resources for one or the other, but not for both. The selection cannot be made on the basis of profit, since no profit is in prospect for either venture. The selection must be made on the basis of which will contribute most to the general welfare as expressed by the citizens of the community, perhaps by a vote. There is no superior basis for evaluating the contribution of each alternative to the general welfare.

As a second example, consider the development of a weapons system to aid in national defense. Often several technically feasible weapons systems are under consideration. When this is the case it is desirable to evaluate the effectiveness of each with the thought that weapons system effectiveness and the general welfare are related. By considering the cost of each system in relation to its effectiveness, a basis for choice is established.

It should be noted that evaluations of public activities in terms of the general welfare encompass both the benefits to be received from and the cost of the proposed activity. No matter how subjective an evaluation of the contribution of an activity to the general welfare may be, its cost may often be determined quite objectively. It may be fairly simple to determine the immediate and subsequent costs for the swimming pool, the library, or the weapons systems. A knowledge of the costs in prospect for benefits to be gained may be expected to result in sounder selection of public activities in either the civil or the defense sector.

10.5. BENEFIT—COST ANALYSIS

Engineering economy studies directed to public projects almost always delineate the benefits to be derived from the project as a first step. This is in contrast to the consideration of profitability as a first step in evaluating an activity of private enterprise. When a public project is being considered, the question to be decided is: Will it result in the greatest possible enhancement of the general welfare in terms of economic, social, cultural, or other public satisfactions as judged by the people in the governmental unit concerned?

The second step in evaluating a public project involves an analysis of cost to the governmental agency. One common method for determining the cost of constructing and operating a public activity is to call for competitive bids from private organizations. The lowest bid received, after allowances for the bidder's ability to discharge the terms of his contract, is then a measure of cost. Except in unusual circumstances, there seems no justification for a governmental agency to undertake the construction and operation of a project unless it can do so for less than would be charged by private enterprise although there are times when governmental agencies carry on activities as a means of determining if costs of private enterprise are fair.

The Benefit-Cost Ratio. A popular method for deciding upon the economic justification of a public project is to compute the benefit-cost ratio. This ratio may be expressed as

$$\text{B-C ratio} = \frac{\text{Benefits to the Public}}{\text{Cost to the Government}}$$

where the benefits and the costs are present or equivalent annual amounts computed using the cost of money. Thus, the B-C ratio reflects the users' equivalent dollar benefits and the sponsors' equivalent dollar cost. If the ratio is 1, the equivalent benefits and the equivalent costs are equal. This represents the minimum justification for an expenditure by a public agency.

Considerable care must be exercised in accounting for the benefits and the costs in connection with benefit-cost analysis. Benefits are defined to mean all the advantages, less any disadvantages, to the users. Many proposals which embrace valuable benefits also result in inescapable disadvantages. It is the net benefits to the users which are sought. Similarly, costs are defined to mean all costs, less any savings, that will be incurred by the sponsor. Such savings are not benefits to the users but are reductions in cost to the government. It is important to realize that adding a number to the numerator does not have the same effect as subtracting the same number from the denominator of the

B-C ratio. Thus, incorrect accounting for the benefits and costs can lead to a ratio which may be misinterpreted.

As an example of the application of benefit cost analysis and the B-C ratio consider the following situation which is of interest to a state highway department. The accidents involving motor vehicles on a certain highway have been studied for a number of years. The calculable costs of such accidents embrace lost wages, medical expenses, and property damage. On the average there are 35 nonfatal accidents and 240 property damage accidents for each fatal accident. The average equivalent present cost of these three classes of accidents is calculated to be as follows:

Fatality per person	$84,000
Nonfatal injury accidents	5,400
Property damage accidents	1,200

From the data above the aggregate cost of motor-vehicle accidents per death may be calculated as follows:

Fatality per person	$84,000
Nonfatal injury accident $5,400 × 35	189,000
Property damage accident $1,200 × 240	288,000
Total	$561,000

The death rate on the highway in question has been 8 per 100,000,000 vehicle miles. A proposal to add a third lane is under consideration. It is estimated that the cost per mile will be $290,000, the service life of the improvement will be 30 years, and the annual maintenance will be 8 % of the first cost. The traffic density on the highway is 6,000 vehicles per day and the cost of money is 5%. It is estimated that the death rate will decrease to 4 per 100,000,000 vehicle miles. Although there are other benefits which will result from widening the highway, it is argued that the reduction in accidents is sufficient to justify the expenditure.

To verify the economic desirability of the widening project the highway department performs the following calculations. The equivalent annual benefit per mile to the public is

$$\frac{(8 - 4)(6,000)(365)(\$561,000)}{100,000,000} = \$49,144.$$

And the equivalent annual cost per mile to the highway department is

$$\$290,000(\overset{A/P\,5,\,30}{0.0660}) + \$290,000(0.08) = \$42,340.$$

This results in a benefit-cost ratio of

$$\text{B-C ratio} = \frac{\$49,144}{\$42,340} = 1.16.$$

Thus, widening the highway is justified based on the benefits to be derived from a reduction in accidents alone. Other benefits, such as the reduction in trip time, have not been included in the analysis and would increase the ratio.

Benefit-Cost Analysis for Multiple Alternatives. The example of the previous paragraphs illustrated a situation in which the sponsoring agency had the simple choice of widening the highway or leaving it as is. Usually, however, a sponsoring agency finds that different benefit levels and different costs result in meeting a specific objective. When this is the case, the problem of interpreting the corresponding benefit-cost ratios presents itself. A hypothetical situation and the correct method of analysis is presented in the following paragraphs.

Suppose that four mutually exclusive alternatives have been identified for providing recreational facilities in a certain urban area. The equivalent annual benefits, equivalent annual costs, and benefit-cost ratios are given in Table 10.1. Inspection of the B-C ratios might lead one to select Alternative B because the ratio is a maximum. Actually, this choice is not correct. The correct alternative can be selected by applying the principle of incremental return.

Table 10.1. BENEFIT—COST RATIOS FOR FOUR ALTERNATIVES

Alternative	Annual Benefits	Annual Costs	B-C Ratio
A	$182,000	$91,500	1.99
B	167,000	79,500	2.10
C	115,000	88,500	1.30
D	95,000	50,000	1.90

Table 10.2 exhibits the required computations. Alternative D is used as the initial base since it requires the minimum equivalent annual cost. Next, the incremental equivalent annual benefits and incremental equivalent annual costs are computed by comparing each acceptable alternative with the next alternative having the next higher equivalent annual costs. The resulting incremental B-C ratios are shown in the last column of Table 10.2. Alternative B is first compared to Alternative D. Since the B-C ratio for the difference between alternatives (B-D) exceeds 1, Alternative B is preferred to

Table 10.2. INCREMENTAL BENEFIT—COST RATIOS

Alternative	Incremental Annual Benefits	Incremental Annual Costs	Incremental B-C Ratio	Decision
D	$95,000	$50,000	1.90	Accept D
B-D	72,000	29,500	2.44	Accept B
C-B	−52,000	9,000	−5.78	Reject C
A-B	15,000	12,000	1.25	Accept A

Alternative D. Next, Alternative C is compared to Alternative B and the B-C ratio for this increment is less than 1. Thus Alternative B is preferred to Alternative C and Alternative B remains the "current best" solution. Finally, the B-C ratio is computed for the difference between Alternative A minus Alternative B. This ratio is greater than 1 indicating the Alternative A is more desirable than Alternative B. Since all the other alternatives have been eliminated Alternative A would be selected. Selection of this alternative will assure that the equivalent annual benefits less the equivalent annual costs are maximized and that its B-C ratio is greater than 1.

10.6. COST—EFFECTIVENESS ANALYSIS

Cost-effectiveness analysis had its origin in the economic evaluation of complex defense and space systems. Its predecessor, benefit-cost analysis had its origin in the civilian sector of the economy and may be traced back to the Flood Control Act of 1936. Much of the philosophy and methodology of the cost-effectiveness approach was derived from benefit-cost analysis and, as a result there are many similarities in the techniques. The basic concepts inherent in cost-effectiveness analysis are now being applied to a broad range of problems in both the defense and the civil sectors of public activities.

In applying cost-effectiveness analysis to complex systems three requirements must be satisfied. First the systems being evaluated must have common goals or purposes. The comparison of cargo aircraft with fighter aircraft would not be valid, but comparison with cargo ships would if both the aircraft and ships were to be utilized in military logistics. Second, alternate means for meeting the goal must exist. This is the case with cargo ships being compared with the cargo aircraft. Finally, the capability of bounding the problem must exist. The engineering details of the systems being evaluated must be available or estimated so that the cost and effectiveness of each system can be estimated.

The Cost-Effectiveness Approach. There are certain steps which constitute a standardized approach to cost-effectiveness evaluations.[1] These steps are useful in that they define a systematic methodology for the evaluation of complex systems in economic terms. The following paragraphs summarize these steps.

First, it is essential that the desired goal or goals of the system be defined. In the case of military logistics mentioned earlier, the goal may be to move a certain number of tons of men and supporting equipment from one point

---

[1] A. D. Kazanowski, "A Standardized Approach to Cost-Effectiveness Evaluations," Chapter 7 in J. Morley English, Editor, *Cost Effectiveness*, John Wiley and Sons, Inc., 1968.

to another in a specified interval of time. This may be accomplished by a few relatively slow cargo ships or a number of fast cargo aircraft. Care must be exercised in this step to be sure that the goals will satisfy mission requirements. Each delivery system must have the capability of delivering a mix of men and equipment that will meet the requirements of the mission. Comparison of aircraft that can deliver only men against ships that can deliver both men and equipment would not be valid in a cost-effectiveness study.

Once mission requirements have been identified, alternate system concepts and designs must be developed. If only one system can be conceived, a cost-effectiveness evaluation cannot be used as a basis for selection. Also, selection must be made on the basis of an optimum configuration for each system. In the previous chapter the minimum cost pier spacing was found for two competing bridge designs before a choice was made. A similar comparable basis must be established for a ship cargo system and an aircraft cargo system.

System evaluation criteria must be established next for both the cost and the effectiveness aspects of the system under study. Ordinarily, less difficulty exists in establishing cost criteria than in establishing criteria for effectiveness. This does not mean that cost estimation is easy. It simply means that the classifications and basis for summarizing cost are more commonly understood. Among the categories of cost are those arising throughout the system life cycle which include costs associated with research and development, engineering, test, production, operation, and maintenance. The phrase *life cycle costing* is often used in connection with cost determination for complex systems. Life cycle cost determination for a specific system is normally on a present or equivalent annual basis. The Department of Defense has adopted the policy of procuring systems on the basis of life cycle cost as opposed to first cost, as had been the practice in the past.

System evaluation criteria on the effectiveness side of a cost-effectiveness study are quite difficult to establish. Also, many systems have multiple purposes which complicates the problem further. Some general effectiveness categories are utility, merit, worth, benefit, and gain. These are difficult to quantify and, therefore, such criteria as mobility, availability, maintainability, reliability, and others are normally used. Although precise quantitative measures are not available for all of these evaluation criteria they are useful as a basis for describing system effectiveness.

The next step in a cost-effectiveness study is to select the fixed cost or the fixed effectiveness approach. In the fixed cost approach, the basis for selection is the amount of effectiveness obtained at a given cost. The selection criterion in the fixed effectiveness approach is the cost incurred to obtain a given level of effectiveness. When multiple alternatives which provide the same service are compared on the basis of cost, the fixed effectiveness approach is being used. This was the case with the two competing bridge designs presented in the previous chapter.

Candidate systems in a cost-effectiveness study must be analyzed on the basis of their merits. This may be accomplished by ranking the systems in order of their capability to satisfy the most important criterion. For example, if the criterion in military logistics is the number of tons of men and equipment moved from one point to another in a specified interval of time this criterion becomes the one of most importance. Other criteria, such as maintainability, would be ranked in a secondary position. Often this procedure will eliminate the least promising candidates. The remaining candidates can then be subjected to a detailed cost and effectiveness analysis. If the cost and the effectiveness for the top contender are both superior to the respective values for other candidates, the choice is obvious. If criteria values for the top two contenders are identical, or nearly identical, and no significant cost difference exists, either may be selected based on irreducibles. Finally, if system costs differ significantly, and effectiveness differs significantly, the selection must be made on the basis of intuition and judgment. This latter outcome is the most common in cost-effectiveness analysis directed to complex systems.

The final step in a cost-effectiveness study involves documentation of the purpose, assumptions, methodology, and conclusions. This is the communication step and it should not be treated lightly. No wise decision maker would base a major expenditure of capital on a blind trust of the analyst.

A Cost-Effectiveness Example. As an example of some aspects of cost-effectiveness analysis consider the goal of moving men and equipment from one point to another as discussed in the previous paragraphs. Suppose that only the cargo aircraft mode and the ship mode are feasible. Also suppose that some design flexibility exists within each mode, so that the effectiveness in tons per day may be established through design effort.

Assume that the Department of Defense has convinced Congress that such a military logistic system should be developed and that Congress has authorized a research and development program for a system whose present life cycle cost is not to exceed 1.2 billion dollars. The Department of Defense, in conjunction with a nonprofit research and engineering firm decided that three candidate systems should be conceived and costed. Table 10.3 shows the resulting present life cycle cost and corresponding effectiveness in tons

Table 10.3. COST AND EFFECTIVENESS FOR THREE SYSTEMS

System	Present cost in billions of $	Effectiveness in tons per day
Aircraft I	1.2	1,620
Ship	1.2	1,410
Aircraft II	1.0	1,410

per day for each system at maximim utilization. These data are also exhibited in Figure 10.1.

The three candidate systems were studied because of the following logic. First, using the fixed cost approach the study team projected the configuration of Aircraft System I and the Ship System and found that these systems would

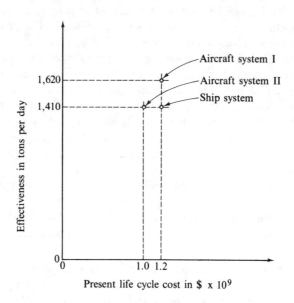

FIGURE 10.1.   COST AND EFFECTIVENESS FOR
THREE SYSTEMS.

have effectivenesses of 1,620 and 1,410 tons per day respectively. If the study had stopped here the conclusion would be to spend the 1.2 billion dollars for Aircraft System I. Realizing, however, that the Congress and the Department of Defense would welcome estimates of effectiveness for other levels of expenditure, the study team chose to determine the cost for Aircraft System II which would be designed to have an effectiveness equal to that of the Ship System costing 1.2 billion dollars. The cost of this system was estimated to be 1.0 billion dollars. This is an example of the fixed effectiveness approach.

At this point a decision must be made between Aircraft System I and Aircraft System II. The Ship System has a lower effectiveness of the same cost as Aircraft System I and a higher cost for the same effectiveness or Aircraft System II and so will not be a candidate in the decision process. In choosing between Aircraft System I and Aircraft System II, the Congress in conjunction with the Department of Defense must decide if the increase in effectiveness of 210 tons per day is worth a present life cycle cost of 0.2 billion dollars.

This example is quite simplified in its assumptions and analysis. Nothing has been said about how the study team was able to determine that a certain effectiveness would result from a system with a present life cycle cost of 1.2 billion dollars. Also, the example did not point out any secondary measures of effectiveness which would complicate the choice.

QUESTIONS AND PROBLEMS

1. Outline the function of government and the nature of public activities.
2. Contrast the criteria for evaluation of private and public activities.
3. Describe the impediments to efficiency in governmental activities.
4. Name the elements that should be considered in deciding on an interest rate to be used in the evaluation of public activities.
5. Describe the meaning of the general welfare objective as it relates to engineering economy analysis applied to public activities.
6. List some noneconomic indicators that could be useful in evaluating whether public activities are in the interest of the general welfare.
7. A state whose population is 4,000,000 has twelve state-supported colleges and universities with a total enrollment of 60,000 students and an annual budget for instructional programs approximating $120,000,000 per year.

    (a) Most students enrolled in the colleges state that prospective increased earnings is an important reason for attending. What percentage of the state's population do you believe desire to support the program of higher education through taxation for this reason?

    (b) What prospective annual increase in earnings must you achieve to recover the cost of your education? Assume the state provides $2,000 per student per year and that the interest rate is 8%.

    (c) Indicate any benefits that may accrue to the state from its investment in education.

8. List the factors that would have a significant effect on a city's decision to undertake a mass transportation system. Indicate which factors could be quantified and which factors would be considered nonquantifiable.

9. Analysis of accidents in one state indicates that increasing the width of highways from 20 ft. to 24 ft. may decrease the accident rate from 200 per 100,000,000 to 115 per 100,000,000 vehicle miles. Calculate the average daily number of vehicles that should use a highway to justify widening on the basis of the following estimates: Average loss per accident, $500; per mile cost of widening pavement 4 feet, $18,000; useful life of improvement, 25 years; annual maintenance, 3% of first cost; interest rate, 5%.

10. A suburban area has been annexed by a city which henceforth is to supply water service to the area. The prospective growth of the suburb has been estimated and on the basis of this the requirements for pipelines needed to meet the demand for water are as follows:

Years from Now	Pumping Cost Per Year	Pipe Diameter	First Cost
0	$18,000	10″	$200,000
10	$20,000	14″	$260,000
25	$22,000	18″	$300,000

From the standpoint of water-carrying capacity, a 14-inch pipe is equivalent to two 10-inch pipes and an 18-inch pipe is equivalent to three 10-inch pipes. On the basis of a minimum equivalent expenditure over the next 40 years, compare the following plans if the interest rate is 6% and if used pipeline can be sold for 20% of its original cost.

Years from Now	Plan A	Plan B	Plan C	Plan D	Plan E
0	Install 10″ Sell 10″	Install 10″	Install 14″	Install 14″	Install 18″
14	Install 14″ Sell 14″	Install 10″	....	.... Sell 14″	....
25	Install 18″	Install 10″	Install 10″	Install 18″	....

11. An inland state is presently connected to a seaport by means of a railroad system. The annual goods transported is 420 million ton miles. The average transport charge is 5 mills per ton mile. Within the next 20 years, the transport is likely to increase by 10 million ton miles per year.

It is proposed that a river flowing from the state to the seaport be improved at a cost of $700,000,000. This will make the river navigable to barges and will reduce the transport cost to 2 mills per ton mile. The project will be financed by 80% federal funds at no interest, and 20% raised by 7% bonds at par. There would be some side effects of the change-over as follows: (1) The railroad would be bankrupt and be sold for no salvage value. The right of way, worth about 25 million, will revert to the state; (2) 3,600 employees will be out of employment. The state will have to pay them welfare checks of $120 per month; (3) The reduction in the income from the taxes on the railroad will be compensated by the taxes on the barges.

(a) What is the benefit-cost ratio based on the next twenty years of operation?

(b) At what average rate of transport per year will the two alternatives be equal?

12. Let it be assumed that a certain state is contemplating a highway development embracing the construction of 8,000,000 square yards of pavement. Vehicle registration in the state is as follows:

Passenger cars	1,000,000
Light trucks	200,000
Medium trucks...........	40,000
Heavy trucks	10,000

The characteristics and pavements necessary to carry the vehicles are taken as follows:

Class of Vehicle	Pavement Thickness	Cost Per Sq. Yard
Passenger cars	5.5 inches	$10.50
Light trucks	6.0 inches	$11.80

Class of Vehicle	Pavement Thickness	Cost Per Sq. Yard
Medium trucks	6.5 inches	$12.50
Heavy trucks	7.0 inches	$13.00

On the assumption that paving costs should be distributed on the basis of the number of vehicles in each class and the incremental costs of paving required for each class of vehicle, what should be the taxes per vehicle for each vehicle class?

13. A toll road has been opened between two cities. The distance between the entrances of the two cities is 93 miles via the highway and 104 miles via the shortest alternate free highway. From the following data determine the economic advantage, if any, of using the toll road for the following conditions applicable to the operation of a light truck: toll cost, $2.50; driver cost, $5.10 per hour; average driving rate between entrances via toll road and free road respectively, 65 m.p.h. and 55 m.p.h.; estimated average cost of operating truck per mile via toll road, $0.14, via free road, $0.15.

14. The federal government is planning a hydroelectric project for a river basin. In addition to the production of electric power this project will provide flood control, irrigation and recreation benefits. The estimated benefits and costs that are expected to be derived from the three alternatives under consideration are listed below.

Alternatives	A	B	C
Initial Cost	$25,000,000	$35,000,000	$50,000,000
Annual benefits and costs:			
Power sales	$1,000,000	$1,200,000	$1,800,000
Flood control savings	250,000	350,000	500,000
Irrigation benefits	350,000	450,000	600,000
Recreation benefits	100,000	200,000	350,000
Operating and maintenance costs	200,000	250,000	350,000

The interest rate is 5% and the life of each of the projects is estimated to be 50 years.

(a) By comparing the benefit-cost ratios determine which project should be selected.

(b) Calculate the benefit-cost ratio for each alternative. Is the best alternative selected if the alternative with the maximum benefit-cost ratio is chosen.

(c) If the interest rate is 8% what alternative would be chosen?

15. Two sections of a city are separated by a marsh area. It is proposed to connect the sections by a four-lane highway. Plan A consists of a 2.4 mile highway directly over the marsh by the use of earth fill. The initial cost will be $9,500,000 and the required annual maintenance will be $11,000. Plan B consists of improving a 4.2 mile road skirting the swamp. The initial cost will be $4,200,000 with an annual maintenance cost of $5,600. A traffic survey estimates the traffic density to be as follows:

Years after Construction	Traffic Density in Vehicles Per Hour
1– 5	150
6–10	800
11–20	2200

The estimated average speed under these densities would be 60 m.p.h., 50 m.p.h., and 30 m.p.h., respectively. The traffic consists of 80% noncommercial vehicles with an operating cost of $0.12 per mile and 20% commercial vechicles with an operating cost of $0.16 per mile and $5.10 per hour. If Plan B is accepted the development of the property adjacent to the highway will result in an increase in tax revenue of $300,000 per year.

(a) Compare the alternatives on the basis of a 20-year period and 6% interest.

(b) At what traffic density are the plans equivalent for an interest rate of 6%?

16. In a certain city the fire insurance premium rate is $0.57 per $100 of the insured amount. There are approximately 8,000 dwellings of an average valuation of $20,000 in the city. It is estimated that the insured amount represents 70% of the value of the dwellings. The city commissioners have been advised that the fire insurance premium rate will be reduced to $0.50 per $100 if the following improvements are made to reduce fire losses.

Improvement	Cost	Life
Increase capacity of trunk water lines from pumping station	$80,000	40 years
Increase capacity of pumps	$ 8,000	20 years
Add supply tank to increase pressure in remote sections of town	$60,000	40 years
Purchase additional fire truck and related equipment	$50,000	20 years
Add two firemen	$40,000 per year	

Operation costs of added improvements are expected to be offset by decreased pumping cost brought about by the enlargement of trunk water lines. The city is in a position to increase its bonded indebtedness and can sell bonds bearing 7% interest at par. Should the above improvements be made on the basis of prospective savings in insurance premiums paid by the home owners? What is the benefit-cost ratio?

17. The federal government is presently considering a number of proposals for improving the speed of mail handling in large urban post offices. The measure of effectiveness to be used to evaluate these mail handling systems is the volume of mail processed per day. The cost of purchasing and installing these various systems, their resulting savings and their effectiveness are as follows.

System	Initial cost in millions of $	Annual savings in millions of $	Effectiveness in millions of letters processed per day
A	$1,400	$100	5
B	$2,200	$150	8
C	$2,600	$230	12
D	$3,700	$300	13
E	$5,100	$500	14

If the interest rate is 6% and the life of each system is estimated to be 10 years, plot the cost and effectiveness relationship for each proposal. Which alternatives can be dropped from further consideration? How would you decide between the remaining alternatives?

IV

ACCOUNTING, DEPRECIATION, AND INCOME TAXES

11

Accounting and
Cost Accounting

The accounting system of an enterprise provides a media for recording historical data arising from the essential activities employed in the production of goods and services. Engineering economic analysis provides a means for quantifying the expected future differences in the worth and cost of alternative engineering proposals. As compared with this function, accounting has the objective of providing summaries of the status of an enterprise in terms of assets and liabilities so that the condition of the enterprise may be judged at any point in time. Therefore, it is essential that the function of accounting and the function of engineering economy be clearly distinguished.

Accounting records are one of the most important sources of data for engineering economy studies. In them the analyst will find detailed quantitative data useful in estimating the future outcome of activities similar to those completed. In addition, the outcome of decisions based on economy studies will eventually be revealed in these records and may be used for post audit. For these reasons, it is desirable that the data provided by accounting systems be examined in relation to the requirements of engineering economy studies.

11.1. GENERAL ACCOUNTING

Two classifications of accounting are recognized; general accounting and cost accounting. Cost accounting is a branch of general accounting and is usually of greater importance in engineering economy studies than is general accounting. Cost accounting will be considered in the next section.

Basic Financial Statements. The primary purpose of the general accounting system is to make possible the periodic preparation of two basic financial statements for an enterprise. These are:

1. A *balance sheet* setting out the assets, liabilities, and net worth of the enterprise at a stated date.

2. A *profit and loss statement* showing the revenues and expenses of the enterprise for a stated period.

The accounts of an enterprise fall into five general classifications—assets, liabilities, net worth, revenue, and expense. Three of these—assets, liabilities, and net worth—serve to give the position of the enterprise at a certain date. The other two accounts—revenue and expense—accumulate profit and loss information for a stated period, which act to change the position of the enterprise at different points in time. Each of these five accounts is a summary of other accounts utilized as part of the total accounting system.

The balance sheet is prepared for the purpose of exhibiting the financial position of an enterprise at a specific point in time. It lists the assets, liabilities, and net worth of the enterprise as of a certain date. The monetary amounts recorded in the accounts must conform to the fundamental accounting equation

$$\text{Assets} - \text{Liabilities} = \text{Net worth.}$$

For example, the balance sheet for Company A shows these major accounts as of December 31, 19xx.

Assets		Liabilities	
Cash....................	$143,300	Notes payable	$ 22,000
Accounts receivable	7,000	Accounts payable	4,700
Raw materials	9,000	Accrued taxes	3,200
Work in process.........	17,000	Declared dividends	40,000
Finished goods	21,400		$ 69,900
Land	11,000	*Net Worth*	
Factory building	82,000	Capital stock	$200,000
Equipment	34,000	Profit for December	56,100
Prepaid services	1,300		256,100
	$326,000		$326,000

Balance sheets are normally drawn up annually, quarterly, monthly, or at other regular intervals. The change of a company's condition during the interval between balance sheets may be determined by comparing successive balance sheets.

Information relative to the change of conditions that have taken place during the interval between successive balance sheets is provided by a profit and loss statement. This statement is a summary of the income and expense for a stated period of time. For example, the profit and loss statement for Company A shows the income, expense, and net profit for the month of December, 19xx.

Gross income from sales		$251,200
Cost of goods sold		$142,800
Net income from sales		$108,400
Operating expense:		
Rent	$11,700	
Salaries	$28,200	
Depreciation	$ 4,800	
Advertising	$ 6,500	
Insurance	$ 1,100	$ 52,300
Net profit from operations		$ 56,100

The balance sheet and the profit and loss statement are summaries in more or less detail, depending on the purpose they are to serve. They are related to each other; the net profit developed on the profit and loss statement is entered under net worth on the balance sheet as shown above. This net profit is also used as a basis for computing the firm's income tax obligation shown on the balance sheet.

Together, the balance sheet and the profit and loss statement summarize the five major accounts of an enterprise. The major accounts are summaries of other accounts falling within these general classifications as, for example, cash, notes payable, capital stock, cost of goods sold, and rent. Each of these accounts is a summary in itself. For example, the asset item of raw material is a summary of the value of all items of raw material as revealed by detailed inventory records. In the search for data upon which to base economy studies, it will be necessary to trace each account back through the accounting system until the required information is found.

Profit as a Measure of Success for the Enterprise. Profit is the resultant of two components, one of which is the economy associated with income from the activity. It is obvious that some activities have greater profit potentialities than others. In fact some activities can only result in loss.

When profit is a consideration, it is important that activities be evaluated with respect to their effect on profit. The first step in making a profit is to secure an income. But, to acquire an income necessitates certain activities resulting in certain costs. Profit is, therefore, a resultant of activities which produce income and involve outlay which may be expressed as

$$\text{Profit} = \text{Income} - \text{Outlay}.$$

The total success of an organization is the summation of the successes of all the activities that it has undertaken. Also, the success of a major undertaking is the summation of the successes of the minor activities of which it is constituted. In Figure 11.1 each vertical block represents the income potentialities of a venture, the outlay incurred in seeking it, the outlay incurred in prosecuting it, and the net gain of carrying it on. In this conceptual scheme

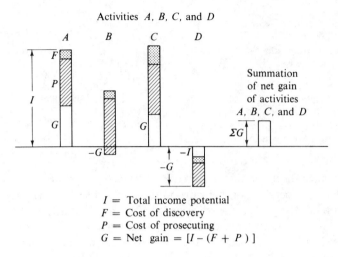

Activities *A*, *B*, *C*, and *D*

I = Total income potential
F = Cost of discovery
P = Cost of prosecuting
G = Net gain = $[I - (F + P)]$

FIGURE 11.1. AN ILLUSTRATION OF THE FINAL
OUTCOME OF SEVERAL ACTIVITIES.

the several quantities are considered to be measured in a single commensurable term, such as money.

From this figure it is apparent that the extent of the success of an activity depends upon its potentialities for income less the sum of the costs of finding it and carrying it on. It is also apparent that the success of an enterprise is the summation of the net success of the several actitivies undertaken during a period of time.

11.2. COST ACCOUNTING

Cost accounting is a branch of general accounting adapted to registering the costs for labor, material, and overhead on an item-by-item basis as a means of determining the cost of production. The final summary of this information is presented in the form of a cost of goods made and sold statement. It lists the costs of labor, material, and overhead applicable to all goods made and sold during a certain period. For example, the cost of goods made and sold statement for Company *A* during the month of December, 19xx is as follows:

Direct Material
In process Dec. 1, 19xx	$ 3,400	
Applied during the month	$39,500	
Total	$42,900	
In process Dec. 31, 19xx	$ 4,200	$ 38,700

Direct Labor

In process Dec. 1, 19xx	$ 4,300	
Applied during the month	$51,900	
Total	$56,200	
In process Dec. 31, 19xx	$ 5,700	$ 50,500

Overhead

In process Dec. 1, 19xx	$ 5,800	
Applied during the month	$60,100	
Total	$65,900	
In process Dec. 31, 19xx	$ 7,100	$ 58,800
Cost of Goods Made		$148,000
Finished goods Dec. 1, 19xx		$ 16,200
Total		$164,200
Finished goods Dec. 31, 19xx		$ 21,400
Cost of goods sold		$142,800

The cost of goods made and sold statement reflects summary data derived from the four accounts of materials in process, labor in process, overhead in process, and finished goods. The statement is subsidiary to the profit and loss statement since the "cost of goods sold" amount it develops is transferred to the profit statement. Similarly, the profit and loss statement was shown to be subsidiary to the balance sheet since the profit or loss amount developed thereon is transferred to the net worth section of the balance sheet.

The costs that are incurred to produce and sell an item of product are commonly classified as direct material, direct labor, factory overhead, factory cost, administrative cost, and selling cost. The first three are exhibited on the cost of goods made and sold statement, and give rise to the cost of goods sold entry on the profit and loss statement. Administrative and selling costs appear on the profit and loss statement under operating expense, and are subtracted from net income to arrive at the net profit amount. Each of these cost classifications will be considered in the paragraphs that follow.

Direct Material. The material whose cost is directly charged to a product is termed *direct material*. Ordinarily, the costs of principal items of material required to make a product are charged to it as direct material costs. Charges for direct material are made to the product at the time the material is issued, through the use of forms and procedures designed for that purpose. The sum of charges for materials that accumulate against the product during its passage through the factory constitutes the total direct material cost.

In the manufacture of many products, small amounts of a number of items of material may be consumed which are not directly charged to the product. These items are charged to factory overhead, as will be explained later. They are not directly charged to the product on the premise that the advantage to be gained will not be enough to offset the increased cost of record keeping.

Although perhaps less subject to gross error than records of other elements of cost, records of direct material costs should not be used in engineering economy studies without being questioned. Their accuracy in regard to quantity and price of material should be ascertained. Also, their applicability to the situation being considered should be established before they are used.

Direct Labor. *Direct labor* is labor whose cost is charged directly to the product. The source of this charge is time tickets or similar forms used to record the time and wages of workmen whose efforts are applied to a product during its journey through the factory. Unless the allocation of labor costs to products is very closely controlled, records of labor costs charged to specific products are likely to be in error.

As a result of either carelessness or a desire to conceal an undue amount of time spent on a job, some of the time applied to one job may be reported as being applied to another. Thus, direct-labor cost records should be carefully examined for accuracy and applicability to the situations under investigation before being used as data for engineering economy studies.

Various small amounts of labor may not be considered to warrant the record-keeping that is required to charge them as direct labor. Such items of labor became part of the factory overhead. The labor of personnel engaged in such activities as inspection, testing, or moving the product from machine to machine or in pickling, painting, or washing the product is often charged in this way.

Such items as social security, pension and insurance costs that are nearly proportional to direct wages are sometimes included in arriving at direct labor costs.

Factory Overhead. *Factory overhead* is also designated by such terms as factory expense, shop expense, burden, indirect costs, and on-cost. Factory overhead costs embrace all expenses incurred in factory production which are not directly charged to products as direct material or direct labor.

The practice of applying overhead charges arises because prohibitive costly accounting procedures would be required to charge all items of cost directly to the product.

Factory overhead costs embrace costs of material and labor not charged directly to product and fixed costs. Fixed costs embrace charges for such things as taxes, insurance, interest, rental, depreciation and maintenance of buildings, furniture and equipment, and salaries of factory supervision, which are considered to be independent of volume of production.

Indirect material and labor costs embrace costs of all items of material and labor consumed in manufacture which are not charged to the product as direct material or direct labor.

Factory Cost. The *factory cost* of a product is the sum of direct material, direct labor, and factory overhead. It is these items that are summarized on the cost of goods made and sold statement. This cost classification separates the manufacturing cost from administrative and selling costs thus giving an indication of production costs over time.

Administrative Costs. *Administrative costs* arise from expenditures for such items as salaries of executive, clerical, and technical personnel, office space, office supplies, depreciation of office equipment, travel, and fees for legal, technical, and auditing services that are necessary to direct the enterprise as a whole as distinct from its production and selling activities. Expenses so incurred are often recorded on the basis of the cost of carrying on subdivisions of administrative activities deemed necessary to take appropriate action to improve the effectiveness of administration.

In most cases, it is not practical to relate administrative costs directly to specific products. The usual practice is to allocate administrative costs to the product as a percentage of the product's factory cost. For example, if the annual administrative costs and factory costs of a concern are estimated at $10,000 and $100,000, respectively, for a given year, 10% will be added to the factory cost of products manufactured to absorb the administrative costs.

Selling Cost. The *selling cost* of a product arises from expenditures incurred in disposing of the products and services produced. This class of expense includes such items as salaries, commissions, office space, office supplies, rental and depreciation, operation of office equipment and automobiles, travel, market surveys, entertainment of customers, displays, and sales space.

Selling expenses may be allocated to various classes of products, sales territories, sales of individual salesmen, and so forth, as a means of improving the effectiveness of selling activities. In many cases it is considered adequate to allocate selling expense to products as a percentage of their production cost. For example, if the annual selling expense is estimated at $22,000 and the annual production cost is estimated at $110,000, 20% will be added to the production cost of products to obtain the cost of sales.

11.3. METHODS FOR ALLOCATING FACTORY OVERHEAD

There are four common methods of allocating factory overhead charges to the product being produced. These are the direct-labor-cost method, the direct-labor-hour method, the direct-material-cost method, and the machine-rate method. Each of these bases for allocation will be illustrated by reference to a hypothetical manufacturing concern which will serve to exhibit some aspects of cost accounting pertinent in engineering economy studies.

PLASCO Company is a small plastic manufacturing concern with plant facilities as is shown in Figure 11.2. Land for the plant cost $9,000 and the plant itself, constructed four years ago, cost $30,000. Two-thirds of each of these items, or $6,000 and $20,000, is attributed to the production department. Initial cost of Machine X and Machine Y was $18,000 and $30,000 respectively. In addition, other assets of the firm are factory furniture with a first cost of $2,000, small tools and dies which cost $2,100, and stores and stock inventory with a current value of $5,800.

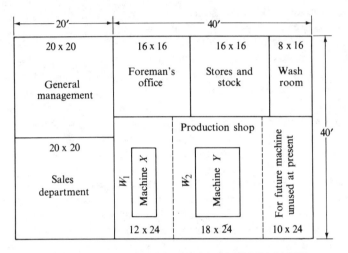

FIGURE 11.2. PLAN OF PLASCO COMPANY PLANT FACILITIES.

The factory salaries and wages during the current year are estimated to be as follows:

Foreman F supervises factory operations $12,000
Handyman H moves material, takes care of stock and stores, and does janitor work .... $ 5,200
Total indirect labor $17,200
Workman W_1 operates Machine X, $4.50/hr. × 2,000 hr. .... $ 9,000
Workman W_2 operates Machine Y, $3.60/hr. × 2,000 hr. .... $ 7,200
Total direct labor $16,200

The supplies to be used in the factory during the fifth year are estimated as follows:

Office and general supplies .... $ 600
Water (est. as $\frac{3}{4}$ of bill for entire building)............................ 75
Lighting current (est. as $\frac{2}{3}$ of bill for entire building) 200

Heating fuel (est. as $\frac{2}{3}$ of bill for entire building) 410
Electric power ($260 for Machine X and $400 for Machine Y) 660
Maintenance supplies ($100 for Machine X and $280 for Machine Y) .. 380
                                                                    $2,325

PLASCO Company makes three products L, M, and N. Estimated output, material cost, direct labor hours, and machine hours are given in Table 11.1

Table 11.1. ESTIMATED ACTIVITY DURING CURRENT YEAR

Pro-duct	Esti-mated output	Material cost		Direct-labor hours				Machine hours			
				Workman W_1		Workman W_2		Machine X		Machine Y	
		Each	Total	Each	Total	Each	Total	Each	Total	Each	Total
L	100,000	$0.10	$10,000	0.01	1,000	——	——	0.01	1,000	——	
M	140,000	0.08	11,200	——	——	0.01	1,400	——	——	0.01	1,400
N	80,000	0.12	9,600	0.0125	1,000	0.0075	600	0.0125	1,000	0.0075	600
			$30,800		2,000		2,000		2,000		2,000

The calculation of the PLASCO Company overhead rates during the current year may be summarized as follows:

A. *Overhead Items for Building*
  Depreciation, insurance, maintenance on building, ... $ 2,800
  Taxes on building and land, ... 1,000
  Interest on present value of building and land, ... 2,400
  Water, light, and fuel for factory, ... 685
    Total .... $ 6,885
B. *Miscellaneous Items of Overhead*
  Depreciation, taxes, insurance, and maintenance on factory furniture, $ 320
  Interest on present value of factory furniture, ... 140
  Depreciation, taxes, insurance, and interest on present value of small
    tools, ... 310
  Taxes, insurance, and interest on stores and stock inventory, ... 750
  Office and general supplies 600
    Total .... $ 2,120
C. *Indirect Labor and Labor Overhead*
  Salaries of indirect labor of F and H, ... $17,200
  Payroll taxes, ... 3,100
    Total $20,300
D. *Machine X Overhead Items*
  Depreciation, taxes, insurance, and maintenance on Machine X,.... $ 3,100
  Interest on present value of Machine X, ... 800
  Supplies for Machine X 100
  Power for Machine X 260
    Total .... $ 4,260

E. *Machine Y Overhead Items*

Depreciation, taxes, insurance, and maintenance on Machine Y, ..	$ 4,400
Interest on present value of Machine Y, ...	1,300
Supplies for Machine Y	280
Power for Machine Y	400
Total	6,380
Grand total, all factory overhead items	$39,945

On the basis of the information given, overhead allocation rates may be calculated as follows:

$$\text{direct-labor-cost rate} = \frac{\text{total factory overhead}}{\text{total direct labor wages}}$$

$$= \frac{\$39,945}{\$16,200} = 2.47.$$

$$\text{direct-labor-hour rate} = \frac{\text{total factory overhead}}{\text{total hours of direct labor}}$$

$$= \frac{\$39,945}{4,000} = \$9.99.$$

$$\text{direct-material-cost rate} = \frac{\text{total factory overhead}}{\text{total direct material cost}}$$

$$= \frac{\$39,945}{\$30,800} = 1.30.$$

Further analysis must be made before machine rates can be established for Machine X and Machine Y. In establishing machine rates, as many items of overhead as possible are directly allocated to each machine before their identity is lost by being charged to an overhead account.

Consider Item A. This item is equal to $6,885 and is equivalent to the rent of the factory building, which has a floor area of 1,600 square feet.

Annual cost per sq. ft. of floor area, $6,885 ÷ 1,600	$ 4.30
Space charge, Machine X, 288 × $4.30	$1,238
Space charge, Machine Y, 432 × $4.30	$1,858
Total space charged to Machines X and Y	$3,096
Balance of space cost to be allocated, $6,885 − $3,096	$3,789

This balance of Item A, together with Items B and C must be distributed to Machine X and Machine Y on some basis that is estimated to reflect actual conditions. In this example the sums of these items will be allocated equally to the two machines. One-half of the unallocated sum is equal to $\frac{1}{2}(\$3,789 + \$2,120 + \$20,300) = \$13,105$.

Machine X Overhead Charges and Machine Rate

Item *D*	$ 4,260
Space charge as calculated above	1,238
One-half of unallocated balance of Item A, Item B, and Item C, ...	13,105
Total	$18,603

$$\text{machine rate (Machine X)} = \frac{\text{overhead allocated to Machine X}}{\text{estimated annual hours of operation}}$$

$$= \frac{\$18,603}{2,000} = \$9.30 \text{ per hour.}$$

Machine Y Overhead Charges and Machine Rate

Item *E*	$ 6,380
Space charge as calculated above	1,858
One-half of unallocated balance of Item A, Item B, and Item C, ...	$13,105
Total	$21,343

$$\text{machine rate (Machine } Y) = \frac{\$21,343}{2,000} = \$10.67 \text{ per hour.}$$

The unit cost of Products *L*, *M*, and *N* may now be determined by each of the four methods of allocating factory overhead.

The Factory Cost of Product L

Direct-labor-cost method:

Direct material	$0.100
Direct labor, 0.01 × $4.50	0.045
Overhead, $0.045 × 2.47	0.111
	$0.256

Direct-labor-hour method:

Direct material	$0.100
Direct labor, 0.01 × $4.50	0.045
Overhead, 0.01 × $9.99	0.100
	$0.245

Direct-material-cost method:

Direct material	$0.100
Direct labor, 0.01 × $4.50	0.045
Overhead, $0.10 × 1.30	0.130
	$0.275

Machine-rate method:

Direct material	$0.100
Direct labor, 0.01 × $4.50	0.045
Overhead, 0.01 × $9.30	0.093
	$0.238

The Factory Cost of Product M

Direct-labor-cost method:

Direct material	$0.080
Direct labor, 0.01 × $3.60	0.036
Overhead, $0.036 × 2.47	0.089
	$0.205

Direct-labor-hour method:

Direct material	$0.080
Direct labor, 0.01 × $3.60	0.036
Overhead, 0.01 × $9.99	0.100
	$0.216

Direct-material-cost method:

Direct material	$0.080
Direct labor, 0.01 × $3.60	0.036
Overhead, $0.08 × 1.30	0.104
	$0.220

Machine-rate method:

Direct material	$0.080
Direct labor, 0.01 × $3.60	0.036
Overhead, 0.01 × $10.67	0.107
	$0.223

The Factory Cost of Product N

Direct-labor-cost method:

Direct material	$0.120
Direct labor, 0.0125 × $4.50 + 0.0075 × $3.60	0.083
Overhead, $0.083 × 2.47	0.205
	$0.408

Direct-labor-hour method:

Direct material	$0.120
Direct labor (calculated as above)	0.083
Overhead, 0.02 × $9.99	0.200
	$0.403

Direct-material-cost method:

Direct material	$0.120
Direct labor	0.083
Overhead, $0.12 × 1.30	0.156
	$0.359

Machine-rate method:

Direct material	$0.120
Direct labor	0.083
Overhead, 0.0125 × $9.30 + 0.0075 × $10.67	0.196
	$0.399

The cost of sales for the PLASCO Company is obtained by adding administrative and selling costs to the factory cost. Continuing with the example of the PLASCO Company, suppose that after careful analysis of expenditures, annual administrative costs have been estimated at $28,000 and annual selling costs at $16,500.

Annual direct material costs, direct labor costs, and factory overhead costs for PLASCO have been estimated previously as $30,800, $16,200, and $39,945, respectively. Therefore, the annual estimated cost of sales of PLASCO Company for the current year may be summarized as follows:

Estimated annual direct material cost	$30,800	
Estimated annual direct labor cost	16,200	
Estimated annual factory overhead cost	39,945	
Estimated annual factory cost	$86,945	
Estimated annual administrative cost	28,000	
Estimated annual production cost		$114,945
Estimated annual selling cost.....................................		16,500
Estimated annual cost of sales		$131,445

Difficulties Associated with Overhead Allocations. An examination of the factory costs obtained for Product L in the PLASCO Company example with the four methods of factory overhead employed shows considerable variation. Each of the three costs that make up the total is subject to estimating error. Direct material costs may not be accurate because of variations in pricing, charging a product with more material than is actually used, and the use of estimates. Similarly, and for much the same reasons, direct labor costs as charged will usually be inaccurate to some extent. However, with reasonably good control and accounting procedures, direct material costs and direct labor costs are generally reliable.

If attention is directed to factory overhead costs allocated to Product L, it will be observed that amounts allocated by the several methods range from $0.093 per unit to $0.130 per unit. The ratio of these amounts is approximately 1.4. The fact that the amounts shown for the several methods differ is evidence that significant differences can result from the choice of a method of overhead allocation.

In actual practice the item of factory overhead is a summation of a great number and variety of costs. It is therefore not surprising that the use of a single, simple method will not allocate factory overhead costs to specific products with precision. Although a particular method may be generally quite satisfactory, wide variations may result in some specific situations.

For example, consider the direct-labor-hour method as it applies to Product L and Product M. The assumption is that $9.99 will be incurred for each hour of direct labor, regardless of equipment used. Thus, the amount of overhead allocated to Product L and Product M is identical, even though it is apparent that the cost of operating the respective machines on which they are processed is quite different. Note that the machine rates of Machine X and Machine Y are $9.30 and $10.67, respectively.

If the direct-material-cost method is used, it is clear that overhead allocations to products will be dependent upon the unit cost as well as upon the quantity of materials used. Suppose, for example, that in the manufacture of tables a certain model may be made from either pine or mahogany. Processing might be identical but the amount charged for overhead might be several times as much for mahogany as for pine because of the difference in the unit cost of the materials used.

Effect of Changes in Level of Activity. In the determination of rates for the allocation of overhead, the activity of the PLASCO Company for the fifth year was estimated in terms of Products L, M, and N. This estimate served as a basis for determining annual material cost, annual direct labor cost, annual direct labor hours, and machine hours. These items then became the denominator of the several allocation rates.

The numerator of the allocation rates was the estimated factory overhead, totaling $39,945. This numerator quantity will remain relatively constant for changes in activity, as an examination of the items of which it is composed will reveal.

For this reason the several rates for allocating factory overhead will vary in some generally inverse proportion with activity. Thus, if the actual activity is less than the estimated activity, the overhead rate charged will be less than the amount necessary to absorb the estimated total overhead. The reverse is also true. When the under or overabsorbed balance of overhead becomes known at the end of the year, it is usually charged to profit and loss, or surplus.

In engineering economic analyses, the effect of activity on overhead charges and overhead rates is an important consideration. The total overhead charges of the PLASCO Company would remain relatively constant over a range of activity represented, for example, by 1,200 to 2,000 hours of activity for each of the two machines. Thus, after the total overhead has been allocated, the incremental cost of producing additional units of product will consist of direct material and direct labor costs.

11.4. USE OF ACCOUNTING DATA IN ECONOMY STUDIES

Since accounting data is the basis for many engineering economy studies, caution should be exercised in its use. An understanding of the relevance of accounting data is essential for its proper utilization. Two examples regarding the pertinence of cost accounting data will be presented in this section.

Cost Data Must be Applicable. It is a common error to infer that a reduction in labor costs will result in a proportionate decrease in overhead costs, particularly if overhead is allocated on a labor cost basis. In one instance a company was manufacturing an oil field specialty. An analysis revealed costs as follows:

Direct labor	$ 4.18
Direct material	1.84
Factory overhead, $4.18 × 2.30	9.61
	$15.63

The factory cost of $15.36 was slightly less than the price of the item in question. The first suggestion was to cease making the article. But after further analysis it became clear that the overhead of $9.61 would not be saved if the item was discontinued. The overhead rate used was based on heavy equipment required for most of the work in the department and on hourly earnings of workmen who averaged $3.56 per hour. For the job in question only a light drill press and hand tools were used. Little actual reduction in cost would have resulted from not using them in the manufacture of the article.

It has previously been calculated that items of overhead of the PLASCO Company for the building are equal to $4.30 per square foot of factory floor space. Figure 11.1 reveals a currently unused 10 ft. × 24 ft. space for future machines. The annual cost equivalent of this space is 10 ft. × 24 ft. × $4.30/ft.^{2} = $1,032.

The item need not be included as an item of cost attributable to a machine that may be purchased to occupy the space in question, since no actual additional cost will arise should the space be occupied. The $1,032 item has been entered as an overhead charge to be allocated to products made on the basis of one of several overhead rates. The addition of a new machine will probably result in changes in the overhead allocation rate used, but it will not result in a change in the overhead item pertaining to the building.

Average Costs are Inadequate for Specific Analysis. An important function of cost accounting, if not the primary one, is to provide data for decisions relative to the reduction of production costs and the increase of profit from sales. Cost data that are believed to be accurate may lead to costly errors in decisions. Cost data that give true average values and are adequate for over-all analyses may be inadequate for specific detailed analyses. Thus, cost data must be carefully scrutinized and their accuracy established before they can be used with confidence in engineering economy studies.

In Table 11.2 actual and estimated cost data realtive to the cost of three products have been tabulated. The actual production cost of Products *A*, *B*, and *C* are $9, $10, and $11, respectively, but owing to inaccuracies in overhead costs, the production costs of Proudcts *A*, *B*, and *C* and believed to be equal to $10 for each.

It should be noted that even though the average of a number of costs may be correct, there is no assurance that this average is a good indication of the

Table 11.2. ACTUAL AND ESTIMATED COST DATA

Product	Direct labor and material costs	Overhead costs, actual	Overhead costs, believed to be	Production cost, actual	Production cost, believed to be
A	$6.50	$2.50	$3.50	$ 9.00	$10.00
B	7.00	3.00	3.00	10.00	10.00
C	7.50	3.50	2.50	11.00	10.00
Average	7.00	3.00	3.00	10.00	10.00

cost of individual products. For this reason the accuracy of each cost should be ascertained before it is used in an economic analysis.

If, for example, the selling price of the products is based upon their believed production cost, Product A will be overpriced and Product C will be underpriced. Buyers may be expected to shun Product A and to buy large quantities of Product C. This may lead to a serious unexplained loss of profit. Average values of cost data are of little value in making decisions relative to specific products.

QUESTIONS AND PROBLEMS

1. What is the relationship between the balance sheet and the profit and loss statement?
2. Describe the difference in viewpoint in respect to time between a balance sheet and an economy study.
3. What is the function of general accounting; of cost accounting?
4. Describe the difference between general accounting activities and engineering economy studies?
5. What relationship between accuracy and cost determines the extent of expenditure that can be justified for the maintenance of a cost accounting system?
6. Name several bases for the distribution of overhead costs.
7. What precautions should be exercised in utilizing accounting data in engineering economic analysis?
8. A manufacturer makes 7,000,000 radio tubes per year. An assembly operation is performed on each tube for which a standard piece rate of $1.02 per hundred pieces is paid. The standard cycle time per piece is 0.624 minute. Overhead costs associated with the operation are estimated at $0.36 per operator hour.

   If the standard cycle time can be reduced by 0.01 of a minute and if one-half of the overhead cost per operator hour is saved for each hour by which total operator hours are reduced, what will be the maximum amount that can be spent to bring about the 0.01 of a minute reduction in cycle time? Assume that the improvement will be in effect for one year, that piece rates will be reduced in proportion to the reduction in cycle time, and that the average operator's time per piece is equal to the standard time per piece in either case.

9. The manufacturing costs of Products X and Y are believed to be $10 per unit. On the basis of this estimate and a desired profit of 10%, the selling price is set at $11 per unit.

   (a) What is the profit if 500 units of Product X and 1,500 units of Product Y are sold ?

   (b) If the actual manufacturing costs of Products X and Y are $9 and $11, respectively, what is the actual profit ?

10. A small factory is divided into four departments for accounting purposes. The direct labor and direct material expenditures for a given year are as follows:

Department	Direct Labor Hours	Direct Labor Cost	Direct Material Cost
A	800	4,500	19,000
B	890	4,900	6,800
C	1100	5,050	11,200
D	670	3,900	15,000

   Distribute an annual overhead charge of $24,000 to Departments A, B, C, and D on the basis of direct labor hours, direct labor cost, and direct material cost.

11. An automobile parts manufacturer produces batteries and distributor assemblies in his electrical products department. It is believed that the cost of manufacturing the batteries and the distributor assemblies is $18.40 and $23.20 per unit, respectively. These costs were derived on the basis of an equal distribution of overhead charges. A study of the firm's cost structure reveals that overhead would be more equitably distributed if overhead charges against the distributor assembly were 50% more than those against the battery. This conclusion was reached after careful consideration of the nature and source of a $220,000 annual overhead expenditure for these products.

   (a) Calculate the unit production cost applicable to the battery and to the distributor assembly if the annual production is made up of 24,000 batteries and 16,000 distributor assemblies.

   (b) What is the annual profit if the selling price is $21.80 and $28.90 for the battery and the distributor assembly, respectively ?

12. A factory producing lawn mowers works at 60% of its capacity and produces 18,000 mowers per year. The unit manufacturing cost is computed as follows:

Direct labor cost	$17.50
Direct material cost	12.50
Overhead cost	8.00
	$38.00

   The mowers are marketed through a factory distributor for $42.50 each. It is anticipated that the volume of production can be increased to 28,000 units per year if the price is lowered to $40 per unit. This action would not increase the present total overhead cost. Compute the present profit per year and the profit per year if the volume of production is increased.

12

Depreciation and
Depreciation Accounting

People satisfy their wants by the consumption of goods and services, the production of which is directly dependent upon the employment of large quantities of producer goods. But producer goods are not acquired without considerable investment. One characteristic of modern civilization is the large investment per worker in production facilities. Although this investment results in high worker productivity it should be recognized that this economy must be sufficient to absorb the reduction in value of these facilities as they are consumed in the production process.

Alternative engineering proposals will affect the type and quantity of producer goods required. As an essential ingredient in the process of want satisfaction, producer goods give rise to capital consumption and investment costs which must be considered in evaluating alternative engineering proposals. An understanding of the depreciation concept is essential if it is to be included as an integral part of engineering economy analysis.

12.1. CLASSIFICATIONS OF DEPRECIATION

Depreciation may be defined as the lessening in value of a physical asset with the passage of time. With the possible exception of land, this phenomenon is a characteristic of all physical assets. A common classification of the types of depreciation include (1) physical depreciation, (2) functional depreciation, and (3) accidents. Each of these types will be defined and explained in the paragraphs which follow.

Physical Depreciation. Depreciation resulting in physical impairment of an asset is known as *physical depreciation*. Physical depreciation manifests itself in such tangible ways as the wearing of particles of metal from a bearing and the corrosion of the tubes in a heat exchanger. This type of depreciation results in the lowering of the ability of a physical asset to render its intended service. The primary causes of physical depreciation are:

1. Deterioration due to action of the elements including the corrosion of pipe, the rotting of timbers, chemical decomposition, bacterial action, etc. Deterioration is substantially independent of use.
2. Wear and tear from use which subjects the asset to abrasion, shock, vibration, impact, etc. These forces are occasioned primarily by use and result in a loss of value over time.

Functional Depreciation. *Functional depreciation* results not from a deterioration in the asset's ability to serve its intended purpose, but from a change in the demand for the services it can render. The demand for the services of an asset may change because it is more profitable to use a more efficient unit, there is no longer work for the asset to do, or the work to be done exceeds the capacity of the asset. Depreciation resulting from a change in the need for the service of an asset may be the result of:

1. Obsolescence resulting from the discovery of another asset that is sufficiently superior to make it uneconomical to continue using the original asset. Assets also become obsolete when they are no longer needed.
2. Inadequacy or the inability to meet the demand placed upon it. This situation arises from changes in demand not contemplated when the asset was acquired.

Suppose that a manufacturer has a hand riveter in good physical condition, but he has found it more profitable to dispose of it and purchase an automatic riveter because of the reduction in riveting costs that the latter makes possible. The difference between the use value of the hand-operated riveter and the amount received for it on disposal represents a decrease in value due to the availability of the automatic riveter. This cause of loss in value is termed obsolescence. Literally, the hand riveter had become *obsolete* as a result of an improvement in the art of riveting.

As a second example, consider the case of a manufacturer who has ceased producing a certain item and finds that he has machines in good operating condition which he no longer needs. If he disposes of them at a value less than their former use value to him, the difference will be termed depreciation, or loss resulting from obsolescence. The inference is that the machine has become obsolete or out of date as far as the user is concerned.

Inadequacy, a cause of functional depreciation, occurs when changes in demand for the services of an asset result in a demand beyond the scope of the asset. For example, a small electric generating plant has a single 500-kilovolt-amperes (kva) generating unit whose capacity will soon be exceeded by the demand for current. Analyses show that it will cost less in the long run to replace the present unit with a 750-kva unit, which is estimated to meet the need for some time, than to supplement the present unit with a 250-kva unit.

In such a case the original 500-kva unit is said to be *inadequate* and to have been *superseded* by the larger unit. Any loss in the value of the unit below its use value is the result of inadequacy or supersession. The former term is preferred for designating this type of depreciation.

A disposition to replace machines when it becomes profitable to do so instead of when they are worn out has probably been an important factor in the rapid development of this nation. The sailing ship has given way to the steamship; in street transportation, the sequence of obsolescence has been horse cars, cable cars, electric cars, and automotive buses; one generation of computers is rapidly being replaced by the next. In power generation, nuclear plants are becoming an attractive alternative to fossil fuel plants. In manufacturing, improvements in the arts of processing have resulted in widespread obsolescence and inadequacy of equipment. Each technological advance produces improvements that result in the obsolescence of existing assets.

12.2. ACCOUNTING FOR THE CONSUMPTION OF CAPITAL ASSETS

An asset such as a machine is a unit of capital. Such a unit of capital loses value over a period of time in which it is used in carrying on the productive activities of a business. This loss of value of an asset represents actual piecemeal consumption or expenditure of capital. For instance, a truck tire is a unit of capital. The particles of rubber that wear away with use are actually small physical units of capital consumed in the intended service of the tire. In a like manner, the wear of machine parts and the deterioration of structural elements are physical consumptions of capital. Expenditures of capital in this way are often difficult to observe and are usually difficult to evaluate in monetary terms, but they are nevertheless real.

An understanding of the concept of depreciation is complicated by the fact that there are two aspects to be considered. One is the actual lessening in value of an asset with use and the passage of time, and the other is the accounting for this lessening in value.

The accounting concept of depreciation views the cost of an asset as a prepaid operating expense that is to be charged against profits over the life of the asset. Rather than charging the entire cost as an expense at the time the asset is purchased, the accountant attempts, in a systematic way, to spread the anticipated loss in value over the life of the asset. This concept of amortizing the cost of an asset so that the profit and loss statement is a more accurate reflection of capital consumption is basic to financial reporting and income tax calculation.

A second aim in depreciation accounting is to have, continuously, a monetary measure of the value of an enterprise's unexpended physical capital, both collectively and by individual units such as specific machines. This value can only be approximated with the accuracy with which the future life of the asset and the effect of deterioration can be estimated.

A third aim is to arrive at the physical expenditure of physical capital, in monetary terms, that has been incurred by each unit of goods as it is produced. In any enterprise, physical capital in the form of machines, buildings, and the like is used in carrying on production activities. As machines wear out in productive activities, physical capital is converted to value in the product. Thus the capital that is lost in wear by machines is recovered in the product processed on them. This lost capital needs to be accounted for in order to determine production costs.

12.3. THE VALUE-TIME FUNCTION

In considering depreciation for accounting purposes, the pattern of the future value of an asset should be predicted. It is customary to assume that the value of an asset decreases yearly in accordance with one of several mathematical functions. However, choice of the particular model that is to represent the lessening in value of an asset over time is a difficult task. It involves decisions as to the life of the asset, its salvage value, and the form of the mathematical function. Once a value-time function has been chosen, it is used to represent the value of the asset at any point during its life. A general value-time function is shown in Figure 12.1.

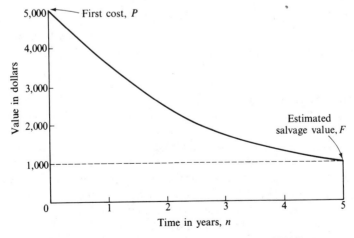

FIGURE 12.1. A GENERAL VALUE-TIME FUNCTION.

Accountants use the term *book value* to represent the original value of an asset less its accumulated depreciation at any point in time. Thus, a function similar to Figure 12.1 can represent book value.

The value at the end of any year is equal to the book value at the beginning of the year less the depreciation expense charged during the year. Table 12.1 presents the calculation of book value at the end of each year for an asset with a first cost of $12,000, an estimated life of 5 years, and a salvage value of zero with assumed depreciation charges.

Table 12.1. BOOK VALUE CALCULATION

End of Year	Depreciation charge during year	Book value at end of year
0	——	$12,000
1	$4,000	8,000
2	3,000	5,000
3	2,000	3,000
4	2,000	1,000
5	1,000	0

In determining book value the following notation will be used. Let

P = first cost of the asset;
F = estimated salvage value;
B_t = book value at the end of year t.
D_t = depreciation charge during year t.
n = estimated life of the asset.

A general expression relating book value and the depreciation charge is given by

$$B_t = B_{t-1} - D_t$$

where $B_0 = P$. The sections which follow are concerned with methods for determining D_t for $t = 1, 2, \ldots, n$.

12.4. STRAIGHT-LINE METHOD OF DEPRECIATION

The straight-line depreciation model assumes that the value of an asset decreases at a constant rate. Thus, if an asset has a first cost of $5,000 and an estimated salvage value of $1,000, the total depreciation over its life will be $4,000. If the estimated life is 5 years the depreciation per year will be $4,000 ÷ 5 = $800. This is equivalent to a depreciation rate 1 ÷ 5 or 20% per year. For this example, the annual depreciation and book value for each year is given in Table 12.2.

Table 12.2. THE STRAIGHT-LINE METHOD

End of year t	Depreciation charge during year t	Book value at end of year t
0	—	$5,000
1	$800	4,200
2	800	3,400
3	800	2,600
4	800	1,800
5	800	1,000

General expressions for the calculation of depreciation and book value may be developed for the straight-line method. Table 12.3 shows the de-

Table 12.3. GENERAL EXPRESSIONS FOR THE STRAIGHT-LINE METHOD

End of year t	Depreciation charge during year t	Book value at end of year t
0	—	P
1	$\dfrac{P-F}{n}$	$P - \left(\dfrac{P-F}{n}\right)$
2	$\dfrac{P-F}{n}$	$P - 2\left(\dfrac{P-F}{n}\right)$
3	$\dfrac{P-F}{n}$	$P - 3\left(\dfrac{P-F}{n}\right)$
t	$\dfrac{P-F}{n}$	$P - t\left(\dfrac{P-F}{n}\right)$
n	$\dfrac{P-F}{n}$	$P - n\left(\dfrac{P-F}{n}\right)$

preciation charge and book value expressions for each year. Thus, the depreciation in any year is

$$D_t = \frac{P-F}{n}$$

the book value is

$$B_t = P - t\left(\frac{P-F}{n}\right)$$

and the depreciation rate per year is $1/n$.

12.5. DECLINING-BALANCE METHOD OF DEPRECIATION

The declining-balance method of depreciation assumes that an asset decreases in value at a faster rate in the early portion of its service life than in the latter portion of its life. By this method a fixed percentage is multiplied times the book value of the asset at the beginning of the year to determine the depreciation charge for that year. Thus as the book value of the asset decreases through time so does the size of the depreciation charge. For an asset with a \$5,000 first cost, a \$1,000 estimated salvage value, an estimated life of 5 years and a depreciation rate of 30% per year the depreciation charge per year is shown in Table 12.4.

Table 12.4. THE DECLINING-BALANCE METHOD

End of year t	Depreciation Charge during year t	Book value at end of year t
0	—	\$5,000
1	(0.30)(\$5,000) = \$1,500	3,500
2	(0.30)(3,500) = 1,050	2,450
3	(0.30)(2,450) = 735	1,715
4	(0.30)(1,715) = 515	1,200
5	(0.30)(1,200) = 360	840

For a depreciation rate R the general relationship expressing the depreciation charge in any year for declining-balance depreciation is

$$D_t = R \cdot B_{t-1}.$$

From the general expression for book value shown in Section 12.3 it is seen that

$$B_t = B_{t-1} - D_t.$$

Therefore for declining-balance depreciation

$$B_t = B_{t-1} - RB_{t-1} = (1 - R)B_{t-1}.$$

Using this recursive expression it is possible to determine the general expressions for the depreciation charge and the book value for any point in time. These calculations are shown in Table 12.5 for an asset with first cost P.

Table 12.5. GENERAL EXPRESSIONS FOR THE DECLINING-BALANCE METHOD OF DEPRECIATION

End of year t	Depreciation charge during year t	Book value at end of year t
0	—	P
1	$R \times B_0 = R(P)$	$(1 - R)B_0 = (1 - R)P$
2	$R \times B_1 = R(1 - R)P$	$(1 - R)B_1 = (1 - R)^2 P$
3	$R \times B_2 = R(1 - R)^2 P$	$(1 - R)B_2 = (1 - R)^3 P$
t	$R \times B_{t-1} = R(1 - R)^{t-1} P$	$(1 - R)B_{t-1} = (1 - R)^t P$
n	$R \times B_{n-1} = R(1 - R)^{n-1} P$	$(1 - R)B_{n-1} = (1 - R)^n P$

Thus the depreciation in any year is

$$D_t = R(1 - R)^{t-1} P.$$

And the book value is

$$B_t = (1 - R)^t P.$$

If the declining-balance method of depreciation is utilized for income tax purposes the maximum rate that may be used is double the straight-line rate that would be allowed for a particular asset or group of assets being depreciated. Thus, for an asset with an estimated life of n years the maximum rate that may be used with this method is $2(1/n)$. Many firms and individuals choose to depreciate their assets using declining-balance depreciation with the maximum allowable rate. Such a method of depreciation is commonly referred to as the *double-declining-balance* method of depreciation.

In Table 12.4 the book value at the end of year 5 is \$840 which is less than the estimated salvage of \$1,000. If the salvage value for this example had been zero, it is observed that for this method of depreciation the book value would never reach zero regardless of the length of the time span over which the asset is depreciated. Thus, adjustments are necessary in order to rectify the differences between the estimated and calculated book value of the asset. In most situations these adjustments are made at the time of disposal of the asset when accounting entries are made to account for the difference between the asset's actual value and its calculated book value.

If it is desired to use a depreciation rate that will result in a particular book value at some point in time, it is possible to solve for the rate that should be used. Solving for R from the general expression

$$B_t = (1 - R)^t P$$

gives

$$R = 1 - \sqrt[t]{\frac{B_t}{P}}.$$

For the example previously considered the estimated salvage value is \$1,000 at the end of the fifth year and the asset's first cost is \$5,000. Thus,

$$R = 1 - \sqrt[5]{\frac{1,000}{5,000}} = 1 - 0.725 = 27.5\%.$$

This method of determining a depreciation rate for the declining-balance method is rarely used in practice. Usually the depreciation rate is selected with regard to its effect on income taxes, its effect on the profit and loss statement and its ease of calculation when assets are grouped for accounting purposes.

12.6. SUM-OF-THE-YEARS-DIGITS METHOD OF DEPRECIATION

The sum-of-the-years-digits depreciation model assumes that the value of an asset decreases at a decreasing rate. If an asset has an estimated life of 5 years, the sum of the years will be $1 + 2 + 3 + 4 + 5 = 15$. Thus, if the first cost of the asset is \$5,000, and the estimated salvage value is \$1,000, the depreciation during the first year will be ($5,000 - $1,000)\frac{5}{15} = $1,333.33. During the second year, the depreciation will ($5,000 - $1,000)\frac{4}{15} = $1,066.67, etc. These values are given in Table 12.6.

Table 12.6. AN EXAMPLE OF THE SUM-OF-THE-YEARS-DIGITS METHOD OF DEPRECIATION

End of year t	Depreciation charge during year t	Book value at the end of year t
0	—	\$5,000
1	$\frac{5}{15}$($4,000) = $1,333	3,667
2	$\frac{4}{15}$(4,000) = 1,067	2,600
3	$\frac{3}{15}$(4,000) = 800	1,800
4	$\frac{2}{15}$(4,000) = 533	1,267
5	$\frac{1}{15}$(4,000) = 267	1,000

The sum-of-the-years for any number of years n can be computed from the expression

$$\sum_{j=1}^{n} j = 1 + 2 + 3 + \ldots + n - 1 + n = \frac{n(n+1)}{2}.$$

Thus for an asset with a 5 year life the sum-of-the-years-digits is $(5)(5 + 1)/2 = 15$ as seen earlier.

The general expressions for the depreciation amount in each year and the book value at the end of each year are presented in Table 12.7. The first

Table 12.7. GENERAL EXPRESSIONS FOR THE SUM-OF-THE-YEARS-DIGITS METHOD OF DEPRECIATION

End of year t	Depreciation charge during year t	Book value at the end of year t
0	—	P
1	$\frac{n}{n(n+1)/2}(P - F)$	$P - \frac{(P-F)}{n(n+1)/2} n$
2	$\frac{(n-1)}{n(n+1)/2}(P - F)$	$P - \frac{(P-F)}{n(n+1)/2} [n + (n-1)]$
3	$\frac{n-2}{n(n+1)/2}(P - F)$	$P - \frac{(P-F)}{n(n+1)/2} [n + (n-1) + (n-2)]$
t	$\frac{n-t+1}{n(n+1)/2}(P - F)$	$P - \frac{(P-F)}{n(n+1)/2} \left[\sum_{j=n-t+1}^{n} j \right]$
n	$\frac{1}{n(n+1)/2}(P - F)$	$P - \frac{(P-F)}{n(n+1)/2} \left[\sum_{j=1}^{n} j \right]$

cost of the asset is P, while its estimated salvage value and estimated life are F and n respectively. Thus, the depreciation charge in any year can be expressed as

$$D_t = \frac{n - t + 1}{n(n + 1)/2}(P - F).$$

And the book value at the end of any year t is

$$B_t = P - \frac{(P - F)}{n(n + 1)/2}\left[\sum_{j=n-t+1}^{n} j\right].$$

However the expression for book value can be simplified when it is seen that

$$\sum_{j=n-t+1}^{n} j = \sum_{j=1}^{n} j - \sum_{j=1}^{n-t} j.$$

In other words if $n = 6$, $t = 4$ and $n - t + 1 = 3$ then

$$3 + 4 + 5 + 6 = 1 + 2 + 3 + 4 + 5 + 6 - [1 + 2].$$

Since it is known that $\sum_{j=1}^{n} j = n(n + 1)/2$ it follows that

$$\sum_{j=n-t+1}^{n} j = \frac{n(n + 1)}{2} - \frac{(n - t)[(n - t) + 1]}{2}.$$

Substituting back into the equation for book value at any time t

$$B_t = P - \frac{(P - F)}{n(n + 1)/2}\left[\frac{n(n + 1)}{2} - \frac{(n - t)(n - t + 1)}{2}\right]$$

$$= P - (P - F) + (P - F)\frac{(n - t)(n - t + 1)}{n(n + 1)}$$

$$= (P - F)\left(\frac{n - t}{n}\right)\left(\frac{n - t + 1}{n + 1}\right) + F.$$

The sum-of-the-years-digits method of depreciation produces larger depreciation charges in the early life of the asset with the smaller charges occurring later in the asset's life. In this method the depreciation rate decreases through time and this decreasing rate is multiplied times a fixed amount $(P - F)$. This calculation is in contrast to the double-declining method which multiplies a fixed rate times a decreasing book value. Both methods are similar in that the depreciation charges in the early portion of an asset's life are greater than in the later portion of its life.

The sum-of-the-years-digits method of depreciation is not computationally as simple as the double-declining balance method. However this method is used in practice because the rate at which it charges depreciation does produce a value-time curve that approximates the decrease in value of numerous categories of assets. That is, there are many assets that depreciate more rapidly during the early portion of their life and this fact is reflected by the sum-of-the-years-digits method which depreciates about three-fourths of the depreciable cost of an asset during the first half of its life.

12.7. SINKING-FUND METHOD OF DEPRECIATION

The sinking-fund depreciation model assumes that the value of an asset decreases at an increasing rate. One of a series of equal amounts is assumed to be deposited into a sinking fund at the end of each year of the asset's life. The sinking fund is ordinarily compounded annually and, at the end of the estimated life of the asset, the amount accumulated equals the total depreciation of the asset. Thus, if an asset has a first cost of $5,000, an estimated life of 5 years, an estimated salvage value of $1,000, and if the interest rate is 6%, the amount deposited into the sinking fund at the end of each year is

$$\overset{A/F\,6,\,5}{(\$5,000 - \$1,000)(0.1774)} = \$709.60.$$

The depreciation charge during any year is the sum of the amount deposited into the sinking fund at the end of the year and the amount of interest earned on the sinking fund during the year. For the conditions assumed, the capital recovered during the first year is $709.60; during the second year $709.60 + 0.06 \times \$709.60 = \$752.18$; during the third year $709.60 + 0.06 (\$709.60 + \$752.18) = \$797.31$; etc. These values are given in Table 12.8.

Table 12.8. AN EXAMPLE OF THE SINKING-FUND METHOD OF DEPRECIATION

End of year t	Depreciation charge during year t	Book value at end of year t
0	—	$5,000
1	$710	4,290
2	752	3,538
3	797	2,741
4	845	1,896
5	896	1,000

To develop the general expression for the amount of depreciation that will be charged each year for the sinking-fund method of depreciation it is first necessary to understand that the depreciation amount is the sum of two components. The first component is the amount deposited into the sinking fund at the end of each year so that its accumulated value at the end of the asset's life equals the depreciable value of the asset. The second component is the amount of interest earned on the accumulated value of the sinking fund at the beginning of the particular year in question. For an asset with first cost P, estimated salvage value F, life n and interest rate i, the first component of the depreciation charge each year is $(P - F)(\overset{A/F\,i,\,n}{\quad})$. The second component is different for each year and this is reflected by the second term in the second column of expressions in Table 12.9.

Table 12.9. GENERAL EXPRESSIONS FOR THE SINKING-FUND METHOD OF DEPRECIATION

End of year t	Depreciation charge at end of year t	Book value at end of year t
0	—	P
1	$(P-F)(\overset{A/F\,i,n}{\quad\quad}) + i(0)$	$P - (P-F)(\overset{A/F\,i,n}{\quad\quad})(\overset{F/A\,i,1}{\quad\quad})$
2	$(P-F)(\overset{A/F\,i,n}{\quad\quad}) + i\left[(P-F)(\overset{A/F\,i,n}{\quad\quad})(\overset{F/A\,i,1}{\quad\quad})\right]$	$P - (P-F)(\overset{A/F\,i,n}{\quad\quad})(\overset{F/A\,i,2}{\quad\quad})$
3	$(P-F)(\overset{A/F\,i,n}{\quad\quad}) + i\left[(P-F)(\overset{A/F\,i,n}{\quad\quad})(\overset{F/A\,i,2}{\quad\quad})\right]$	$P - (P-F)(\overset{A/F\,i,n}{\quad\quad})(\overset{F/A\,i,3}{\quad\quad})$
t	$(P-F)(\overset{A/F\,i,n}{\quad\quad}) + i\left[(P-F)(\overset{A/F\,i,n}{\quad\quad})(\overset{F/A\,i,t-1}{\quad\quad})\right]$	$P - (P-F)(\overset{A/F\,i,n}{\quad\quad})(\overset{F/A\,i,t}{\quad\quad})$
n	$(P-F)(\overset{A/F\,i,n}{\quad\quad}) + i\left[(P-F)(\overset{A/F\,i,n}{\quad\quad})(\overset{F/A\,i,n-1}{\quad\quad})\right]$	$P - (P-F)(\overset{A/F\,i,n}{\quad\quad})(\overset{F/A\,i,n}{\quad\quad})$

Thus for sinking-fund depreciation the depreciation amount for any year t can be determined from the general expression

$$D_t = (P - F)(\overset{A/F\,i,n}{\quad\quad}) + i\left[(P - F)(\overset{A/F\,i,n}{\quad\quad})(\overset{F/A\,i,t-1}{\quad\quad})\right].$$

By rewriting this expression

$$D_t = (P - F)(\overset{A/F\,i,n}{\quad\quad})\left[1 + i(\overset{F/A\,i,t-1}{\quad\quad})\right]$$

but $(\overset{F/A\,i,t-1}{\quad\quad})$ is $[(1 + i)^{t-1} - 1]/i$

so that

$$D_t = (P - F)(\overset{A/F\,i,n}{\quad\quad})[1 + (1 + i)^{t-1} - 1] = (P - F)(\overset{A/F\,i,n}{\quad\quad})(1 + i)^{t-1}$$

$$D_t = (P - F)(\overset{A/F\,i,n}{\quad\quad})(\overset{F/P\,i,t-1}{\quad\quad})$$

The general book value expression for sinking-fund depreciation is

$$B_t = P - (P - F)(\overset{A/F\,i,n}{\quad\quad})(\overset{F/A\,i,t}{\quad\quad}).$$

Although these expressions may appear complicated, the purpose of depreciation accounting is to charge the depreciable portion of the asset over the estimated life of the asset. The sinking-fund method of depreciation is just another systematic method of accomplishing this.

12.8. SERVICE OUTPUT METHOD OF DEPRECIATION

In some cases, it may not be advisable to assume that capital is recovered in accordance with a theoretical value-time model such as those considered previously. An alternative is to assume that depreciation occurs on the basis of service performed without regard to the duration of the asset's life. Thus, a trencher might be depreciated on the basis of pipeline trench completed. If the trencher has a first cost of $11,000 and a salvage value of $600, and if it is estimated that the trencher would dig 1,500,000 linear feet of pipeline trench in its life, the depreciation charge per foot of trench dug may be calculated as

$$\frac{\$11,000 - \$600}{1,500,000} = \$0.006933 \text{ per foot.}$$

The undepreciated capital at the end of each year is a function of the number of feet of trench dug during the year. If, for instance, 300,000 feet of pipeline trench were dug during the first year, the undepreciated balance at the end of the year would be $11,000 - 300,000(\$0.006933) = \$8,920.10$. This analysis would be repeated at the end of each year.

12.9. DEPLETION

Depletion differs in theory from depreciation in that the latter is the result of use and the passage of time while the former is the result of the intentional, *piecemeal removal* of certain types of assets. Depletion refers to an activity that tends to exhaust a supply and the word literally means emptying. When natural resources are exploited in production, depletion indicates a lessening in value with the passage of time. Examples of depletion are the removal of coal from a mine, timber from a forest, stone from a quarry, and oil from a reservoir.

In the case of depletion, it is clear that a portion of the asset is disposed of with each sale. But when a machine tool is used to produce goods for sale, a portion of its productive capacity is a part of each unit produced and, thus, is disposed of with each sale. A mineral resource has value only because the mineral may be sold and, similarly, the machine tool has value because what it can produce may be sold. Both depletion and depreciation represent decreases in value through the using up of the value of the asset under consideration.

There is a difference in the manner in which the capital recovered through depletion and depreciation must be handled. In the case of depreciation, the asset involved usually may be replaced with a like asset, but in the case of depletion such replacement is usually not possible. In manufacturing, the amounts charged for depreciation are reinvested in new equipment to continue operation. However, in mining, the amounts charged to depletion cannot be used to replace the ore deposit and the venture may sell itself out of business. The return in such a case must consist of two portions—the profit earned on the venture and the owners' capital which was invested. In the actual operation of ventures dealing with the piecemeal removal of resources, it is common to acquire new properties, thus enabling the venture to continue.

The *unit* method of depletion is similar to the service method of depreciation discussed in Section 12.8. That is, the depletion charge is based on the amount of the resource that is consumed and the initial cost of the resource. Suppose a reservoir containing an estimated 1,000,000 barrels of oil required an initial investment of $700,000 to develop. For this reservoir the unit depletion rate is $700,000/1,000,000 BBLS. = $0.70/BBL. If 50,000 BBLS. of oil are produced from this reservoir during a year the depreciation charge is (50,000 BBLS.) ($0.70/BBL.) = $35,000. As the estimates of the number of units of the resource remaining vary from year to year the unit depreciation rate is recalculated by dividing the unrecorded cost of the resource by the estimated units of the resource remaining.

For certain resources an optional method of calculating the depletion charge is provided by the United States income tax laws. This optional method is sometimes referred to as the *percentage* method of depletion. Percentage depletion allows a fixed percentage of the gross income produced by the sale of the resource to be the depletion charge. Thus over the life of such an asset the total depletion charges may exceed the initial cost of the asset. However, it is required that for any period the depletion charge may not exceed 50% of the net taxable income for that period computed without the depletion allowance. Some of the fixed percentages for depletion allowed by the tax laws for certain natural resources are oil and gas, uranium and sulfur, 22%; salt, 10%; and gravel 5%.

Using the oil reservoir example again suppose that the gross income from the property for the year is $190,000 and that the taxable income after taking all deductions except for depletion is $80,000. The depletion charge using the percentage method is 22% × $190,000 = $41,800 but that is more than $40,000 (50% × $80,000). Thus the depletion deduction that is allowed is $40,000.

12.10. CAPITAL RECOVERY WITH RETURN IS EQUIVALENT FOR ALL METHODS OF DEPRECIATION

It has been seen that the straight-line, declining-balance, sum-of-the-years-digits, and sinking-fund methods of depreciation all lead to different value-time functions for book value. It is therefore interesting to note that if the retirement of any asset takes place at the age predicted and the book value equals the estimated salvage value, the depreciation amount and interest on the undepreciated balance can be shown to be equivalent to the capital recovery with a return for any method of depreciation. Recall that capital recovery with a return is defined as $(P - F)(\overset{A/P\,i,\,n}{}) + Fi.$

If $(P - F) = A + B + C + \ldots + N$, where $A, B, C, \ldots, N$ are capital recovered amounts for the successive years, we get the following table:

End of Year	Depreciation at End of Year	Interest on Undepreciated Balance at End of Year
0	0	0
1	..................A	...............$Ai + Bi + \ldots + Ni + Fi$
2	..................B.	.................$Bi + \ldots + Ni + Fi$
n	..................N	.............................$Ni + Fi$

Interest, Fi, on the salvage value, F, will be equal for all methods of depreciation and need not be given further consideration. The quantity $B + Bi$ as of the end of Year 2 is equivalent to

$$(B + Bi) \times \frac{1}{(1 + i)} = B$$

as of the end of Year 1. Addition of this amount, B, to Bi results in a total $(B + Bi)$ as of the end of Year 1. This sum is in turn equivalent to B as of the end of Year 0. By similar calculations, quantities involving symbols A to N inclusive will be found to have a worth of $A, B, \ldots, N$ as of the end of Year 0 respectively. Since $(A + B + \ldots + N)$ equal $(P - F)$, and since $A, B, \ldots, N$ may be chosen to represent depreciation by any method, it may be concluded that the present-worth for i of the depreciation calculated by any method plus the interest on the undepreciated balance is equivalent to the present-worth of $(P - F)(\overset{A/P\ i,\ n}{\ })$ at the beginning of the depreciation period.

In the illustrative example on straight-line depreciation, an asset having a first cost of \$5,000 was depreciated to a salvage value of \$1,000 in a period of 5 years. The sum of *depreciation* and *interest on unrecovered balance* is shown in the second column of the following table. For example for Year 3, $\$1,004 = \$800 + 0.06(\$3,400)$.

End of Year	Sum of Depreciation and Interest on Undepreciated Balance		Single Payment Present Worth Factor		
1	\$1,100	$\times$	$P/F\ 6, 1$ (0.9434)	$=$	\$1,037.74
2	1,052	$\times$	$P/F\ 6, 2$ (0.8900)	$=$	936.28
3	1,004	$\times$	$P/F\ 6, 3$ (0.8396)	$=$	842.96
4	956	$\times$	$P/F\ 6, 4$ (0.7921)	$=$	757.25
5	908	$\times$	$P/F\ 6, 5$ (0.7473)	$=$	678.55
			Total Present Worth		\$4,252.78

The comparable figures calculated for sinking-fund depreciation are given in the second column of the following table.

Year No.	Sum of Depreciation and Interest on Undepreciated Balance		Single Payment Present Worth Factor		
1	$1,009.60	×	$P/F\,6,1$ (0.9434)	=	$ 952.46
2	1,009.60	×	$P/F\,6,2$ (0.8900)	=	898.54
3	1,009.60	×	$P/F\,6,3$ (0.8396)	=	847.66
4	1,009.60	×	$P/F\,6,4$ (0.7921)	=	799.70
5	1,009.60	×	$P/F\,6,5$ (0.7473)	=	754.47
			Total Present Worth		$4,252.83

The slight difference between the two resulting values, $4,252.78 and $4,252.83, results from using tables of too few decimal places.

There are ordinarily only two real transactions in an asset's depreciation. These are its purchase and its sale as salvage. In the above example the asset was purchased for $5,000 and its salvage five years later was presumed to have a value of $1,000. The present worth of these two amounts as of the time of purchase follows.

Present worth of $5,000 disbursement at time of purchase $5,000.00
Present worth of $1,000 received from sale of salvage value (a receipt is
$P/F\,6,5$
a negative disbursement), $1,000(0.7473) .... −747.30
Total present worth....................................... $4,252.70

Compare this with the two previous results. Also, note that capital recovery with return is

$$(\$5,000.00 - \$1,000.00)(\overset{A/P\,6,5}{0.2374}) + 1,000.00(0.06) = \$1,009.60$$

and that

$$\$4,252.70(\overset{A/P\,6,5}{0.2374}) = \$1,009.59.$$

Although this relationship between depreciation plus the interest on the undepreciated balances and capital recovery with a return does exist it is important for decision-making purposes that only the actual cash flows be considered. Depreciation charges by themselves are only accounting entries and therefore they do not represent an actual disbursement of cash.

12.11. DEPRECIATION AND ENGINEERING ECONOMY STUDIES

As was indicated earlier, depreciation is a cost of production. An asset is actually consumed in producing goods and thus its depreciation is a production cost. If the cost of capital consumption is neglected, profits will appear to be higher than they are by an amount equal to the depreciation that has taken place during the production period.

In economic analysis dealing with physical assets, it is necessary to compute the equivalent annual cost of capital recovered plus return, so that alternatives involving competing assets may be compared on an equivalent basis. Regardless of the depreciation model chosen to represent the value of the asset over time, the equivalent annual cost of capital recovered and return will be

$$(P - F)(\overset{A/P\ i,\ n}{}) + Fi.$$

Therefore, in making annual cost comparisons of alternative engineering proposals involving physical assets, this expression may be used to represent the equivalent annual costs of capital recovery for interest rate i.

In engineering economic analysis the primary importance of depreciation is its effect on estimated cash flows resulting from the payment of income taxes. Depreciation as an amortized cost influences profits as shown on a company's profit and loss statement since depreciation appears as an expense to be deducted from gross income. Income taxes are paid on the resulting net income figure and these taxes do represent actual cash flows although the depreciation charges are bookkeeping entries.

Depreciation is Based upon Estimates. It would be desirable to know the amount and pattern of an asset's depreciation at any point during its life in order that exact charges could be made against products as they are produced. Unfortunately, the depreciation of an asset cannot be known with certainty until after the asset has been retired from service.

Usually, it is impractical to defer calculation of depreciation costs until after an asset has been disposed of at the end of its life. In fact, depreciation costs of an asset should be taken into account prior to the purchase of the asset, as one of the factors to be considered in arriving at the desirability of purchasing it.

Since information on depreciation is needed on a current basis for making decisions, it has become a practice to estimate the amount of depreciation an asset will suffer and the pattern in which this depreciation will occur. This involves estimates of the service life, salvage value, and depreciation method. Usually the first cost of an asset is known with considerable accuracy. However, since the service life, salvage value, and pattern of depreciation refer to events in the future, they cannot be known with certainty. These estimates are usually based upon experience with similar assets and the judgment of the estimator. Certain aspects of the problem of estimating service life are presented in Appendix C.

QUESTIONS AND PROBLEMS

1. What is the difference between the accountant's concept of depreciation and depreciation as it is commonly understood by the general public?
2. How does depreciation affect the profit and loss statement of a company?
3. Does a depreciation charge represent an actual disbursement of funds? Explain.
4. Discuss the difference between physical and functional depreciation?
5. Describe the value-time function and name its essential components?
6. A man, who purchased a car for $2,500, was offered amounts for his car in succeeding years as follows:

Year	1	2	3	4	5	6
Offer	$1,900	$1,350	$925	$650	$425	$300

On the basis of the offers calculate the depreciation and the undepreciated value of the car for each year of its life. Is this the accountant's view of depreciation?
7. Calculate the book value at the end of each year using straight-line, sum-of-the-years-digits, double-declining balance and sinking-fund methods of depreciation for an asset with an initial cost of $50,000 and an estimated salvage value of $10,000 after 4 years. The interest rate is 12% and the results should be presented in tabular form.
8. An asset has a first cost of $60,000 with a salvage of $10,000 after 11 years. If it is depreciated by declining-balance method using a rate of 8% what will be the
   (a) depreciation charge in the second year?
   (b) depreciation charge in the eighth year?
   (c) book value in the eighth year?
9. An electronic calculator has a first cost of $1,600 with an estimated salvage value of $400 at the end of 6 years. The interest rate is 6%.
   (a) Graph the annual depreciation by the straight-line, sinking-fund, sum-of-the-years-digits and double-declining methods for each year.
   (b) Graph the book values obtained by the four methods for the end of each year.
10. A central air-conditioning unit is purchased for $22,000 and it has an expected life of 10 years. The salvage value for the unit at that time is expected to be $4,000. What will be the book value at the end of 8 years for (a) straight-line depreciation, (b) sum-of-the-years-digits depreciation, (c) double-declining balance depreciation, and (d) sinking-fund depreciation for 10%?
11. An apartment complex is purchased for 1.2 million dollars. It has an estimated life of 25 years and the salvage value at that time is expected to be $200,000. After five years what will be the total depreciation charged for (a) straight-line depreciation (b) sum-of-the-years-digits depreciation and (c) double-declining balance depreciation?

12. An asset was purchased 10 years ago for $2,400. It is being depreciated in accordance with the straight-line method for an estimated total life of 20 years and salvage value of $400. What is the difference in its book value and the book value that would have resulted if declining-balance depreciation at a rate of 9% had been used?

13. A special purpose machine is purchased for $200,000 with an expected life of 7 years. If its salvage value is expected to be zero, what will be the depreciation charge in the first and fifth years and what will be the book value at the end of the third year for (a) straight-line depreciation, (b) double-declining-balance depreciation, and (c) sum-of-the-years digits depreciation?

14. A company has purchased a numerically-controlled machine for $150,000. It is estimated that it will have a salvage value of $50,000 four years from now. What rate must be used with the declining-balance method of depreciation so that the book value of the machine will be equal to its salvage value at the end of its life? (a) Using the rate just calculated with declining-balance depreciation find the depreciation and book value for each year of the machine's life. (b) Compare those figures with similar figures for straight-line and sum-of-the-years-digits depreciation.

15. An asset has a first cost of $35,000 with an estimated life of 5 years. The salvage value at that time is estimated to be $5,000. What is the total accumulated depreciation charged during the first four years of the asset's life for (a) straight-line depreciation, (b) double-declining-balance depreciation, (c) sum-of-the-years-digits depreciation, and (d) sinking-fund depreciation for an interest rate of 15%?

16. An automatic control mechanism is estimated to provide 3,000 hours of service during its life. The mechanism costs $4,800 and will have a salvage value of zero after 3,000 hours of use. What is the depreciation charge if the number of hours the mechanism is used per year is (a) 700 (b) 1,100?

17. Filters costing $800 per set are used to separate particles from hot gases being discharged into the air. These filters must be replaced every 1,600 hours. If the plant operates 24 hours per day, 30 days per month, what is the depreciation charge per month for these filters?

18. A gold mine that is expected to produce 30,000 ounces of gold is purchased for $280,000. Gold can be sold for $16 per ounce to the federal government. If 3,500 ounces are produced this year what will be the depletion allowance for (a) unit depletion and (b) percentage depletion where the fixed percentage for gold is 22%?

19. An oil reservoir is estimated to have 200,000 barrels of oil. The cost of purchasing the lease and well is $450,000. If crude oil is selling for $3.20 per barrel and 30,000 barrels of oil per year is produced, what is the amount of the depletion allowance for (a) unit depletion and (b) percentage depletion at a fixed percentage of 22%?

20. A sulfur mining operation has a gross income of $400,000 in February from the production of sulfur. All expenses with the exception of depletion expenses for this month amount to $350,000. The fixed depletion rate for sulfur is 22%. What is the depletion allowance for this month?

21. A drilling rig was purchased for $3,800. One year later an offer of $4,000 was received for it. Two years later, the offer was $2,800 and three years later, a $1,800 offer was received.
(a) Determine the depreciation of the rig for each year based on the offers received.
(b) Determine the interest on the undepreciated balance for each of the 3 years using an interest rate of 7%.

(c) Determine the uniform year-end amount for the 3-year period which is equiva-lent to the sum of the depreciation and interest on the undepreciated balance found above.

(d) Determine the annual cost of capital recovery with a return for the 3-year period using the initial cost and the last offer. Compare with (c).

22. An asset has a first cost of $6,000 and an estimated salvage of $1,000 at the end of 4 years. The interest rate is 10%. Repeat steps a, b, and c of Problem 21 using 10% instead of 7% for straight-line and sum-of-the-years-digits methods of depreciation. Now compare the results of step c for both methods. Are they equal?

13

Income Taxes in
Economic Analysis

Income tax laws are the result of legislation over a period of time. Since they are man made, they incorporate many diverse ideas, some of which appear to be in conflict. These laws are expressed through a number of provisions and rules intended to meet current conditions. The provisions and rules are not absolute in application, but rather are subject to interpretation when applied.

Income taxes are levied by the federal government as well as by many individual states. State income tax laws will not be considered here because the principles involved are similar to those for federal tax laws, state income tax rates are relatively small, and there is a great diversity of state income tax law provisions.

13.1. RELATION OF INCOME TAXES TO PROFIT

In most cases, the desirability of a venture is measured in terms of differences between income and cost, receipts and disbursements, or some other measure of profit. It is the specific function of economic analysis to determine future profit potential that may be expected from prospective engineering proposals being examined. But income taxes are levies on profit that result in a reduction of its magnitude.

Regardless of how public-spirited a person may be or how clearly he may understand the government's need for income taxes, and even if he should place a high value upon the services he may receive in return for his income tax payments, income taxes are disbursements that differ from other disbursements associated with undertakings only in the manner in which their magnitudes are determined. Thus, in relation to engineering economy studies, income taxes are merely another class of expenditure, which require special treatment. Such taxes must be taken into account along with other classes of costs in arriving at the fruits accruing to sponsors of an undertaking.

Of the aggregate profit earned in the United States from such productive activities as manufacturing, construction, mining, lumbering and others in which engineering analysis is important, the aggregate disbursement for income taxes will be approximately 30 to 40% of net income. Their importance becomes clear when it is realized that in some cases income taxes may result in disbursements of as much as 70% of the profit to be expected from a venture.

13.2. INDIVIDUAL FEDERAL INCOME TAX

With few exceptions, every individual under 65 having a gross income of $1,700 or more during the year must file a tax return. For married persons under 65 years of age a gross income of $2,300 or more is required before a tax return must be submitted. Failure to pay the tax when it is due results in an interest charge at the rate of 6% per year. In addition if the failure to pay is due to willful neglect rather than some reasonable cause the penalty is $\frac{1}{2}$% of the unpaid tax for each month of delinquency.

An individual's income tax obligation is determined by applying a graduated tax rate to his net income from salaries, fees, commissions, and business activities less certain exemptions and deductions. For example, consider Mr. Doe who is unmarried, qualifies as the head of a household, and has four dependents. An outline of the steps necessary to compute Mr. Doe's tax obligation is given below.

Individual Federal Income Tax Outline [1]

1. Salary, wages, commission or other compensation received $12,000
2. Net income from business activities:
   This item consists of income received from the conduct of trade or business activities less the expense of carrying on such activities. These expenses include such items as rents, wages, fees, raw material, supplies, services, depreciation of facilities and losses incurred from the sale and exchange of other than capital assets. Business income less business expense 8,000
3. Income from capital gains:
   This item consists of 50% of the amount that net long-term (assets held more than six months) capital gains exceed net short-term (assets not held more than six months) capital losses. ($900 long-term gain less $300 short-term loss) $\times$ 0.5 300
4. Total adjusted gross income $20,300
5. Less deductions as follows:
   a. Contributions not in excess of 20% of adjusted gross income (plus 10% of adjusted gross income contributed directly to religious, educational organizations and

organizations providing medical care and hospitalization) $1,000
b. Interest paid 200
c. Taxes paid 1,600
d. Medical expenses in excess of 3% of adjusted gross
income 00
e. Loss from fire, storms, or other casualty not compensated
by insurance, etc. 00
f. Exemption[2]. The taxpayer is permitted an exemption of
$650 ($1,300 if 65 years old or older) (Blindness entitles
a taxpayer an additional exemption of $650) for himself
and $650 for each dependent. ($650 for a dependent
wife 65 years old or older and $650 additional if she is
blind.) A dependent is a person, 50% or more of whose
support is paid for by the taxpayer, and whose annual
earnings are less than $800 per year. The $800 limitation
on income does not apply to any child who is under
19 years of age or who is a student (Doe, and his four
dependents, five at $650) $3,250

Total deductions and exemptions $6,050   6,050

6. Taxable income $14,250

[1]Applicable to tax year 1971.
[2]Personal exemptions, $650 for 1971, $700 for 1972 and $750 for 1973 and later.

Table 13.1. TAX RATES FOR MARRIED TAXPAYERS FILING JOINT RETURNS AND
CERTAIN WIDOWS AND WIDOWERS (1971)

If the taxable income is:	The income is:
Not over $1,000	14% of taxable income
$ 1,000 to $ 2,000	$140 + 15% of excess over $ 1,000
2,000 to 3,000	290 + 16% of excess over 2,000
3,000 to 4,000	450 + 17% of excess over 3,000
4,000 to 8,000	620 + 19% of excess over 4,000
8,000 to 12,000	1,380 + 22% of excess over 8,000
12,000 to 16,000	2,260 + 25% of excess over 12,000
16,000 to 20,000	3,260 + 28% of excess over 16,000
20,000 to 24,000	4,380 + 32% of excess over 20,000
24,000 to 28,000	5,660 + 36% of excess over 24,000
28,000 to 32,000	7,100 + 39% of excess over 28,000
32,000 to 36,000	8,660 + 42% of excess over 32,000
36,000 to 40,000	10,340 + 45% of excess over 36,000
40,000 to 44,000	12,140 + 48% of excess over 40,000
44,000 to 52,000	14,060 + 50% of excess over 44,000
52,000 to 64,000	18,060 + 53% of excess over 52,000
64,000 to 76,000	24,420 + 55% of excess over 64,000
76,000 to 88,000	31,020 + 58% of excess over 76,000
88,000 to 100,000	37,980 + 60% of excess over 88,000
100,000 to 120,000	45,180 + 62% of excess over 100,000
120,000 to 140,000	57,580 + 64% of excess over 120,000
140,000 to 160,000	70,380 + 66% of excess over 140,000
160,000 to 180,000	83,580 + 68% of excess over 160,000
180,000 to 200,000	97,180 + 69% of excess over 130,000
over $200,000	110,930 + 70% of excess over 200,000

The amount of income tax due on the taxable income may be calculated from tabulated values such as those given in Tables 13.1 and 13.2. Table 13.2 is applicable to Mr. Doe's situation, therefore his tax obligation will be $2,980 + 0.28 × $250, or $3,050.

Table 13.2. TAX RATES FOR TAXPAYERS WHO QUALIFY AS HEAD OF HOUSEHOLD (1971)

If the taxable income is:	The income tax is:
Not over $1,000	14% of the taxable income.
$ 1,000 to $ 2,000	$ 140 + 16% of excess over $ 1,000
2,000 to 4,000	300 + 18% of excess over 2,000
4,000 to 6,000	660 + 19% of excess over 4,000
6,000 to 8,000	1,040 + 22% of excess over 6,000
8,000 to 10,000	1,480 + 23% of excess over 8,000
10,000 to 12,000	1,940 + 25% of excess over 10,000
12,000 to 14,000	2,440 + 27% of excess over 12,000
14,000 to 16,000	2,980 + 28% of excess over 14,000
16,000 to 18,000	3,540 + 31% of excess over 16,000
18,000 to 20,000	4,160 + 32% of excess over 18,000
20,000 to 22,000	4,800 + 35% of excess over 20,000
22,000 to 24,000	5,500 + 36% of excess over 22,000
24,000 to 26,000	6,220 + 38% of excess over 24,000
26,000 to 28,000	6,980 + 41% of excess over 26,000
28,000 to 32,000	7,800 + 42% of excess over 28,000
32,000 to 36,000	9,480 + 45% of excess over 32,000
36,000 to 38,000	11,280 + 48% of excess over 36,000
38,000 to 40,000	12,240 + 51% of excess over 38,000
40,000 to 44,000	13,260 + 52% of excess over 40,000
44,000 to 50,000	15,340 + 55% of excess over 44,000
50,000 to 52,000	18,640 + 56% of excess over 50,000
52,000 to 64,000	19,760 + 58% of excess over 52,000
64,000 to 70,000	26,720 + 59% of excess over 64,000
70,000 to 76,000	30,260 + 61% of excess over 70,000
76,000 to 80,000	33,920 + 62% of excess over 76,000
80,000 to 88,000	36,400 + 63% of excess over 80,000
88,000 to 100,000	41,440 + 64% of excess over 88,000
100,000 to 120,000	49,120 + 66% of excess over 100,000
120,000 to 140,000	62,320 + 67% of excess over 120,000
140,000 to 160,000	75,720 + 68% of excess over 140,000
160,000 to 180,000	89,320 + 69% of excess over 160,000
Over $180,000	103,120 + 70% of excess over 180,000

For a second illustration, suppose that Mr. Blue has a wife and three children, and he had items of income identical with those of Mr. Doe. The number of exemptions in both cases is five. Mr. and Mrs. Blue elect to file a joint return. The taxable income will remain at $14,250, and Mr. and Mrs. Blue may file a joint return. If this is done, rates in Table 13.1 apply and the tax to be paid is $2,260 + 0.25 × $2,250, or $2,822.50.

If Mr. and Mrs. Blue had divided the $14,250 between them and filed separate returns in accordance with the law, their total tax would have been $2,822.50, or the same as resulted from a joint return. However, if they had divided $14,250 income between them unequally, the tax to be paid would have been greater.

Item 2 in the outline above, *Net income from business activities*, consists essentially of business income less business expenses. Some items such as depreciation and allowed amortization must be determined by calculation. Such items are treated in subsequent sections as is the matter of capital gains appearing in Item 3 of the outline.

Adjusted Gross Income. The term adjusted gross income means net earnings as determined in accordance with the provisions of federal income tax law. The starting point is the taxpayer's total income. Certain items of income are excluded in whole or in part from the total income to determine gross income. Examples of excluded income are interest on certain federal, state or municipal obligations, health and accident insurance benefits, and annuities.

From the gross income certain deductions are made to arrive at the adjusted gross income. These deductions consist essentially of expenses incurred in carrying on a trade, profession, or business, reimbursement for expenses incurred in employment and certain longterm capital gains explained elsewhere.

Taxable Income. The individual's taxable income is his adjusted income less deductions he is permitted to take. Deductions up to 20% of adjusted gross income (30% under certain conditions) may be made for contributions to qualified individuals, fraternal organizations, governmental units of the United States, religious and educational organizations, and organizations providing medical care and hospitalization.

Interest paid by a taxpayer in carrying on business activities is taken into account in arriving at adjusted gross income. But interest paid on borrowed funds made for personal purposes is deductible under Item 5b in the above outline.

Practically all taxes, except federal income taxes, incurred and necessary to carry on business activity are taken into account in arriving at the adjusted gross income of an individual. In addition, individuals are permitted to make deductions for payment of most state and local taxes except estate, inheritance, and gift taxes, and for a few federal taxes with the exception of federal income, estate, and gift taxes.

Medical expenses in excess of 3% of adjusted gross income, incurred by a taxpayer for himself or his dependents, are deductible under Item 5d, if these expenses are not compensated for by insurance or other means. Cost of medicines and drugs in excess of 1% of adjusted gross income is included as a part of medical expense.

Losses sustained from fire, storm, theft or other casualties are deductible to the extent they are not compensated by insurance or similar means as shown under Item 5c.

The exemptions listed in Item 5f of the outline cover most of the provisions relative to this classification.

In lieu of itemized deductions 5a to 5c, the taxpayer may elect to take a standard deduction. The standard deductions on the separate return of an individual or the joint return of a husband and wife is $1,500 or 13% of the adjusted income, whichever is the least.[1] The maximum standard deduction on a separate return of a married person is $500.

13.3. CORPORATION FEDERAL INCOME TAX

The term *corporation*, as used in income tax law, is not limited to the artificial entity usually known as a corporation but may include joint stock associations or companies, some types of trusts, and some limited partnerships. In general, all business entities whose activities or purposes are the same as those of corporations organized for profit are taxed as such. This discussion will be based upon the tax requirements that apply to the usual business corporation.

The business corporation is subject to two taxes—the normal tax and the surtax. The normal tax is equal to 22% of taxable income and the surtax is equal to 26% of taxable income in excess of $25,000. In addition, a corporation may be liable for an accumulated earnings tax.

As a general rule, the taxable income of a corporation is computed in the same manner as that of an individual. There are, however, several provisions peculiar to corporations. Although corporations are in general entitled to the same deductions as individuals, deductions of a personal nature such as medical expenses, child care, alimony, or exemptions for the taxpayer and his dependents are excluded. Corporations are entitled to deductions for partially exempt interest received as income on certain government obligations, for dividends received, and for certain organizational expenses to which individuals are not entitled.

In general, a corporation's income tax is a levy on its net earnings, the difference between the income derived from and the expense incurred in business activity, with some exceptions. Because of the continuing nature of business activity and the fact of annual tax periods, net income must usually be a calculated amount.

The following is an outline of steps to be taken in computing the normal

---

[1]Applicable for 1971. For 1972 and 1973, these figures are ($2,000 or 14%) and ($2,000 or 15%), respectively.

income tax and the surtax of a corporation, illustrated by entering amounts applicable to a particular corporation:

1. Gross Income: This item embraces gross sales less cost of sales, dividends received on stocks, interest received on loans and bonds, rents, royalties, gains and losses from capital or other property, ... $700,000
2. Deductions: This item embraces expense not deducted elsewhere as, for example, in cost of sales. Includes compensation of officers, wages, and salaries, rent, repairs, bad debts, interest, taxes, contributions (not in excess of 5% of taxable income), losses by fire, storms, and theft, depreciation, depletion, advertising, contributions to employee benefit plans, and special deductions for partially exempt bond interest and partially exempt dividends 600,000
3. Taxable Income $100,000
4. Taxable income multiplied by normal tax rate, $100,000 × 0.22 .... $ 22,000
5. Taxable income $100,000
   Plus partially exempt interest and partially exempt dividends not subject to normal tax and not included in gross income ....
6. Total $100,000
   Less: $25,000 exemptions 25,000
7. Balance subject to surtax .... $ 75,000
8. Balance multiplied by surtax rate $75,000 × 0.26 .... $ 19,500
9. Total normal tax plus surtax (Items 4 and 8) .... $ 41,500

Corporate taxable income with some exceptions may be expressed as an equation applicable for taxable incomes from zero and up. This equation is

$$T = E_b \times \text{normal tax rate} + (E_b - \$25,000) \times \text{surtax rate}$$

where

T = total income tax;
E_b = taxable income.

Effective Income Tax Rates. Engineering economy studies involving income taxes can be simplified by the determination of applicable *effective income tax rates*. An effective income tax rate, as the term is used here, is a single rate which when multiplied by the taxable income of a venture under consideration will result in the income tax attributable to the venture. Effective income tax rates are essentially average rates that are applicable over increments of income.

The tax rate on any increment of taxable corporate income between $0 and $25,000 is of course 22%, and the tax rate for any increment between $25,000 and up is 48%. The previous statement does not hold to the extent that there are capital gains, exempt interest and dividends to which special rates apply.

The average tax rate, t_a, of any increment of income embracing taxable income less than $25,000 and greater than $25,000 may be calculated as follows: Find the difference of the tax payable on incomes corresponding to the upper and lower limits of the increments and divide this difference by the amount of the increment. For example, suppose a corporation wishes to find the average tax rate on the increment of income between \$24,000 and \$27,000. The tax payable on \$27,000 is equal to \$27,000 × 0.48 — \$25,000 × 0.26 = \$6,460, and the tax payable on \$24,000 is equal to \$24,000 × 0.22 = \$5,280. The difference is \$6,460 — \$5,280 = \$1,180. The average tax rate over the increment

$$t_a = \frac{\$1,180}{\$27,000 - \$24,000} = \frac{\$1,180}{\$3,000} = 0.393, \text{ or } 39.3\%.$$

Where there are capital gains, losses to be carried back or over, actual depreciation and depletion rates different from those allowed by tax schedules and other conditions, many factors may be involved in the determination of effective income tax rates. These determinations require careful study by individuals proficient in tax matters.

Once they are determined for a particular purpose, effective rates serve as a means for transmitting the thought and analyses that went into the determination in a single figure. By the use of effective rates the consideration of income taxes in economy studies is reduced to two distinct factors, namely, (1) the determination of effective tax rates applicable to particular activities or particular classes of activities and (2) the application of the effective tax rates in economy studies.

The Effect of Interest on Income Taxes. Interest paid for funds borrowed by an individual or a corporation to carry on a profession, trade, or business is deductible from income as expense of carrying on such activity. In addition, interest paid on funds borrowed for purposes not connected with a profession, trade, or business is deductible from the adjusted income in computing an individual's income tax.

Because interest paid is deductible as an expense the amount of borrowed funds may have a marked effect upon the amount of income tax that must be paid. For example, consider two corporations designated as Firm A and Firm B that are essentially identical except that in Firm A no borrowed funds are used and in Firm B an average of \$100,000 is borrrowed at the rate of 9% during a year. Assume that taxable income before interest payments in both cases is \$40,000. Then the taxable income in Firm A will be \$40,000 and taxable income in Firm B will be \$40,000 — (\$100,000 × 0.09) = \$31,000.

On the basis of a normal tax of 22% and a surtax of 26% the income tax for Firm A will be \$40,000 × 0.22 + (\$40,000 — \$25,000) × 0.26 = \$12,700. And the income tax for Firm B will be \$31,000 × 0.22 + (\$31,000 — \$25,000) × 0.26 = \$8,380. This is a difference of \$4,320. Thus the net cost of borrowing the \$100,000 for one year is \$9,000 — \$4,320 = \$4,680. This is equivalent to an interest rate of 4.68%.

Capital Gains and Losses. Capital gains and losses are recognized as being short-term if they apply to assets held not more than six months. Long-term gains and losses apply to assets held more than six months.

If the aggregate of short-term and long-term activities results in a net loss for a year, a corporation may not deduct such loss from current income. But such loss may be carried forward for five subsequent years, being considered as a short-term capital loss, and offset against capital gains during that period.

If the aggregate of short-term and long-term activities results in a net gain, it is added in full as an item of income. It will then be subject to the normal tax rate and the surtax rate.

The maximum tax that a corporation need pay on the excess of long-term capital gain over short-term capital loss is such excess multiplied by a tax rate of 30%. For an individual as long as the net long-term gain over the net short-term loss is $50,000 or less the net gain is taxed at a rate of 25%.

Research and Experimental Expenditures. Research and experimental activities result in new knowledge which has little value in itself, but is valuable only in application. If knowledge is used it may be presumed to result in increased taxable income of the activity in which it is used, and, therefore, in increased income taxes from this source. This viewpoint seems to be borne out by a provision which became effective for tax years beginning after December 31, 1953, and permits expenditures for research and experimentation to be deducted in the year in which they were incurred. The taxpayer also has the option of treating such expenditures as deferred expenses chargeable to capital account and deducting it evenly over a period of 60 or more months. In the latter case, the activity is treated essentially the same as an asset that depreciates.

The provision permitting expenditures for research and experimental activities to be deducted in the year in which they are incurred is indeed an encouragement to carry on such activities when the income tax rate is high. For example, suppose that the effective tax rate of a taxpayer applicable to the consideration of research expenditures is 0.45 and that such expenditures amount to $100,000 during a year. If the research results in no increase in income, its cost to the taxpayer will be $100,000(1 − 0.45), or $55,000. Thus, in a sense, research expenditures are in part underwritten by the government until they result in increased income, if ever.

13.4. DEPRECIATION AND INCOME TAXES

Since the amount of taxes to be paid during any one year is dependent upon deductions made for depreciation, the latter is a matter of consideration for the Bureau of Internal Revenue of the United States Treasury Department and state taxing agencies. Directives are issued by governmental agencies as guides to the taxpayers in properly handling depreciation for tax purposes.

In general, an asset must be used for the purpose of producing an income, whether or not an income actually results from its use, in order that a deduction may be made for its depreciation. In cases where an asset such as an automobile is used both as a means for earning income that is taxable and for personal use, a proportional deduction is allowable for depreciation. Intangible property such as patents, designs, drawings, models, copyrights, licenses, and franchises may be depreciated.

There are restrictions which limit the percentage of the first cost of an asset that may be depreciated during the initial years of life. Depreciation models that yield depreciation amounts in early years which are in excess of that permitted by the Bureau of Internal Revenue may not be used for tax purposes. For example, where the declining-balance model is used the rate of depreciation must not exceed twice the allowable straight-line rate. If a five year life is considered reasonable for an asset whose installed cost is $1,000 and there is no salvage value, its straight-line depreciation rate would be 20% and its annual depreciation would be $200 per year. The corresponding maximum allowable declining balance rate would be 40%. This would result in a deduction for depreciation of $1,000 × 0.40 = $400 the first year, ($1,000 − $400) × 0.40 = $240 the second year, etc.

Effect of Method of Depreciation on Income Taxes. A taxpayer has considerable choice in the method of depreciation he may use for tax computation. If the effective tax rate remains constant and the operating expense is constant over the life of the equipment, the depreciation method used will not alter the total of the taxes payable over the life of the equipment. But, methods providing for high depreciation and consequent low taxes in the first years of life will be of advantage to the taxpayer because of the time value of money.

For illustration of the comparative effect of the straight-line and the fixed percentage on a diminishing balance method, consider the following example. Assume that a taxpayer has just installed a machine whose first cost is $1,000, whose estimated life is 10 years, and whose salvage value is nil. The machine is estimated to have a constant operating income before depreciation and income taxes of $200 per year. The taxpayer estimates the applicable effecive income tax rate for the life of the machine to be 40%. He considers money to be worth 6% and wishes to compare the effect of the straight-line method and the fixed percentage on a diminishing balance method. Let the first method be represented by Alternative *A* and the second method be represented by Alternative *B*. These alternatives are summarized in the following tabulations.

The total income tax paid during the life of the equipment is equal to $400 with either alternative, but the present worths of the taxes paid differ. For Alternative A, the present worth of taxes paid as of the beginning of year No. 1 is equal to $294.40. The corresponding figure for Alternative B is $275.60. The difference in favor of the declining balance method is $18.80, or 6.4% of the Alternative A tax.

ALTERNATIVE A

Income Taxes for Straight-Line Method of Depreciation and 10 Year Life							
Year end no. A	First cost B	Income before depr. and income tax C	Annual book depr. D	Total book depr. to date ΣD E	Income less depr. (taxable income) $C - D$ F	Income tax rate G	Income tax $F \times G$ H
0	$1,000						
1		$ 200	$100	$ 100	$100	0.4	$ 40
2		200	100	200	100	0.4	40
3		200	100	300	100	0.4	40
4		200	100	400	100	0.4	40
5		200	100	500	100	0.4	40
6		200	100	600	100	0.4	40
7		200	100	700	100	0.4	40
8		200	100	800	100	0.4	40
9		200	100	900	100	0.4	40
10		200	100	1,000	100	0.4	40
		$2,000					$400
Present worth of income taxes $= \$40(\overset{P/A \, 6,\, 10}{7.3601}) = \$294.40$							

By examining the book value of an asset over its life for the various methods of depreciation, one can easily see which methods produce larger depreciation deductions in the early portion of the asset's life. The time-value functions of the four depreciation methods discussed in Chapter 12 are presented in Figure 13.1. Since the book value of an asset is the first cost of the asset less accumulated depreciation charges, the curves with the lowest values in their early life are those which are most favorable to the taxpayer.

From Figure 13.1 it is seen that double-declining balance and sum-of-the-years-digits depreciation have faster depreciation write-offs than straight-line or sinking-fund depreciation. Thus for tax purposes double-declining balance and sum-of-the-years-digits depreciation have found wide acceptance since they effectively postpone the payment of taxes to a later date. Such a postponement is favorable to the taxpayer because money has a time value and a dollar paid in taxes at the present is worth more than

ALTERNATIVE B

		Income before depr. and income tax	Annual book depr.	Total book depr. to date	Income less depr. (taxable income)	Income tax rate	Income tax
Income Taxes for Percentage on Diminishing Balance Method of Depreciation (20%) and 10 Year Life							
Year end no.	First cost			ΣD	$C - D$		$F \times G$
A	B	C	D	E	F	G	H
0	$1,000						
1		$ 200	$200	$ 200	$ 0	0.4	$ 0
2		200	160	360	40	0.4	16
3		200	128	488	72	0.4	29
4		200	102	590	98	0.4	39
5		200	82	672	118	0.4	47
6		200	66	738	134	0.4	54
7		200	52	790	148	0.4	59
8		200	42	832	158	0.4	63
9		200	34	866	166	0.4	66
10		200	134	1,000	66	0.4	27
		$2,000					$400

$$\text{Present worth of income taxes} = \sum_{n=1}^{10} [\text{Col. } H \times (\overset{P/F\,6,\,n}{\quad})] = \$275.60$$

a dollar paid in taxes at some future time. For the same reasons the sinking-fund method of depreciation is the least appealing of the four methods shown in Figure 13.1.

Effect of Estimated Life on Income Taxes. There is often little connection between the life of an asset that will be realized and that which a taxpayer may use for tax purposes. If the applicable effective tax rate remains constant through the realized life of an asset, the use of a shorter life for tax purposes will usually be favorable to a taxpayer. The reason is that use of short estimated lives for tax purposes results in relatively high annual depreciation, low annual taxable income and consequently, low annual income taxes during the early years of an asset's life. Even though early low annual income taxes will result in correspondingly higher annual income taxes· during the later years of an asset's life, the present worth of all income taxes during the asset's life will be less.

If the conditions in Alternative A in the previous section had permitted the use of an estimated life of five years for tax purposes, the annual income tax for the first five years would have been nil, and during the second five years of life, $80 per year. The present worth of these payments at 6% as of the beginning of the first year would be equal to

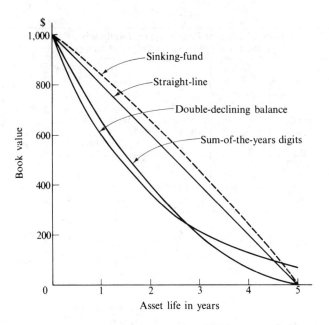

FIGURE 13.1. BOOK VALUE COMPUTED BY FOUR
METHODS OF DEPRECIATION.

$$\overset{P/A\,6,\,5\quad P/F\,6,\,5}{\$80(\,4.212\,)(0.7473)} = \$252.78.$$

The difference in favor of the shorter life is $42.62.

The matter of useful life may be settled by entering into an agreement with the Treasury Department as to useful lives of assets. Such agreements are binding on both parties and will not be changed unless facts not previously contemplated arise.

If challenged in regard to the estimated life selected for assets, the burden of proof rests upon the taxpayer. Changes of rate during the life of an asset must be based upon facts related to the use of the asset. For example, mark-edly increased use of a machine over that on which a depreciation rate was initially established would generally be considered an acceptable reason for using an increased rate of depreciation.

The lessening in value due to extraordinary obsolescence ordinarily cannot be anticipated; consequently, it will usually be taken into account by an adjustment in the remaining life of an asset. When it becomes apparent that the value of an asset is being adversely affected by revolutionary inventions, radical economic changes, and abnormal needs that will result in retirement of the asset at a shorter life than originally estimated, the estimated life may be decreased to reflect the new conditions.

Special provisions are made for the rapid amortization of facilities that have been certified by proper authority as necessary in the national interest. Deductions for tax purposes may be made for such assets on the basis of amortization over a 60-month period regardless of their estimated useful life. This provision has essentially the same effect as permission to use an estimated life of five years for the assets involved.

Effect of a Gain or Loss on Disposal on Income Taxes. When a depreciable asset used in business is exchanged or sold for an amount greater or smaller than its book value, this "gain" or "loss" on disposal has an important effect on income taxes. If the asset has been held for more than 6 months and it is disposed of for an amount greater than the asset's book value at the time of disposal the "gain" is treated as a long-term capital gain. If the asset is sold for less than its book value the "loss" on disposal is treated as a long-term capital loss. The disposal of an asset for an amount equal to the asset's estimated salvage value or book value at the time of disposal incurs no taxes.

For example, suppose an asset purchased for $7,000 has a life of 3 years and produces revenue of $4,000 per year. If the asset is depreciated by the straight-line method and it is estimated that the salvage value of the asset will be $1,000, the annual depreciation charge is ($7,000 − $1,000)/3 = $2,000. The calculations of taxes for an effective tax of 40% and long-term capital gain tax of 30% are presented in Table 13.3, where it is assumed that $1,500 is received from the disposal of the asset.

Table 13.3. TAXES FOR AN ASSET WHEN ITS ACTUAL SALVAGE VALUE EXCEEDS ITS ESTIMATED SALVAGE VALUE

End of Year	Before Tax Cash Flow	Deprecia-tion	Book Value	Taxable Income		Taxes
				Ordinary Taxable Income	Capital Gains or Loss	
0	−$7,000	—	$7,000	—		—
1	4,000	$2,000	5,000	$2,000		$800
2	4,000	2,000	3,000	2,000		800
3	4,000+ 1,500	2,000	1,000	2,000	$500	800 + 150

The taxable income for Year 3 consists of two elements. The first element consists of the ordinary revenue received less the annual depreciation or $4,000 − $2,000 = $2,000. The second element reflects the fact that the actual amount received on disposal exceeds the estimated salvage value at the time of disposal. Thus a capital gain of $1,500 − $1,000 = $500 is realized. For corporations this capital gain is taxed at a rate of 30% while the other income is taxed at the effective rate. If the amount realized on the sale of the asset had been less than the estimated salvage value, the difference would appear as a capital loss. This loss could then be used to offset other capital gains resulting from other activities in Year 3. If there are no other capital gains during Year 3, the loss can be carried forward for five years and offset against capital gains during that period.

13.5. DEPLETION AND INCOME TAXES

Natural resources such as minerals, oil, gas, timber, and certain others are found in deposits or stands of definite quantities. They are exploited by removal. As quantities are removed to produce income, the amount remaining in a deposit or stand is reduced.

In arriving at taxable income, depletion is deducted from gross income. The basis for determining the amount of depletion for a given year is the total cost of the property, the total number of units in the property, and the number of units sold during the year in question. This may be expressed by the equation:

$$\text{depletion for year} = \frac{\text{cost of property}}{\text{total units in property}} \times \text{units sold during year.}$$

Oil, gas, and mineral properties are also depleted for tax purposes on the basis of a percentage of gross income, provided that the amount allowed for depletion is not greater than 50% of the taxable income of the property before depletion allowances. In general, percentage depletion will be elected if it results in less tax than does depletion based on cost.

Typical of percentage depletion allowances for mineral and similar resources are the following:

Oil and gas wells	22%
Sulphur, uranium, asbestos, bauxite, graphite, mica, antimony, bismuth, cadmium, cobalt, lead, manganese, nickel, tin, tungsten, vanadium, zinc	22%
Various clays, diatomaceous earth, dolomite, feldspar, and metal mines if not in 22% group	14%
Coal, lignite, sodium chloride	10%
Brick and tile clay, gravel, mollusk shells, peat, pumice, and sand	5%

Suppose that a property estimated to have 1,000,000 barrels of oil was purchased for $600,000 or at a cost of $0.60 per barrel. During the tax year 50,000 barrels were pumped and sold from the property for a gross income of $150,000. Total operating expenses during the year, deductible from gross income to obtain taxable income, amounted to $70,000.

If the taxpayer did not use the 22% depletion allowance, his taxable income would be calculated as follows:

Gross income, 50,000 barrels sold at $3 .... $150,000
Deductible operating expense .... $ 70,000
Depletion, 50,000 barrels at $0.60 30,000
                                                                    $100,000   100,000
Taxable income .... $ 50,000

If the taxpayer uses the 22% depletion allowance, his taxable income would be calculated as follows:

Gross income, 50,000 barrels sold at $3 ....$150,000
Deductible operating expense 70,000
Taxable income before depletion .... $ 80,000
Depletion allowance, $150,000 × 0.22 (since this is less than 50% of the taxable income before depletion, this method of depletion may be used) 33,000
Taxable income .... $ 47,000

Timber may not be depleted on a percentage basis. In general, timber is depleted for tax purposes on the basis of its cost at a specific date, the total number of units of timber on the property and the units of timber removed during the tax year. Special rules are applicable for determining cost and for revising the number of units on the property from time to time.

Table 13.4. APPLICATION OF COST AND PERCENTAGE DEPLETION TO A MINERAL DEPOSIT

Year	Units produced	Gross income ($B \times \$2$)	Operating cost	Net income before depletion and income tax	50% of net income ($E \times 0.5$)	Allowable cost depletion at $0.40 per unit	Allowable percentage depletion at 22% ($C \times 0.22$)	Taxable income (E less F, H, or I)	Income tax (tax rate of 0.3)($J \times 0.3$)
A	B	C	D	E	F	H	I	J	K
1	20,000	$40,000	$20,000	$20,000	$10,000	$8,000	$8,800‡	$11,200	$3,360
2	16,000	32,000	16,000	16,000	8,000	6,400	7,040‡	8,960	2,688
3	12,000	24,000	14,000	10,000	5,000*	4,800	5,280	5,000	1,500
4	8,000	16,000	8,000	8,000	4,000	3,200	3,960‡	4,040	1,212
5	4,000	8,000	6,000	2,000	1,000	1,600†	1,760	400	120
									$8,880

*Deduct this amount for depletion as full percentage depletion will not be allowed and because it is greater than cost depletion.

†Deduct this amount for depletion because it is greater than the allowable percentage depletion that may be used.

‡Deduct this amount for depletion because it is full percentage depletion allowances and it is greater than allowable cost depletion.

Effect of Depletion Method on Income Taxes. Most resources subject to depletion may be depleted on either a cost basis or a percentage basis in computing income taxes. Consider an example of a mineral deposit estimated to contain 60,000 units of ore and purchased at a cost of $24,000 including equipment. The depletion rate is $24,000 ÷ 60,000, or $0.40 per unit of ore.

This property is also subject to percentage depletion at rate of 22% per year, but not in excess of 50% of the net taxable income of the property before depletion. An analysis of this situation is presented in Table 13.4.

By taking advantage of the two methods of depletion, the income taxes on the property above total $8,880. If only cost depletion had been used, income taxes in successive years would have been $3,600, $2,880, $1,560, $1,440 and $120 for a total of $9,600.

13.6. ECONOMY CALCULATIONS INVOLVING INCOME TAXES

The introduction of income taxes into economy studies requires special consideration because of three factors in the nature of income taxes. One factor is that income taxes are dependent upon net income. What interests a taxpayer is the net return of a venture after income taxes. Net earning before income taxes, which determines the amount of income taxes, is a difference of income and cost which makes net earning subject to the joint effect of errors in estimates of the latter two quantities. A second factor is that in computing income taxes only interest on borrowed money may be considered a cost. A third factor is that the method of depreciation used in computing income taxes must be considered in analyses.

A Basic Tax Equation. The following symbolism will be used and will apply to annual quantities:

E_b = net profit before income taxes;
E_a = net profit after income taxes;
G = estimated gross income from the activity under consideration;
C = estimated annual costs, all not included elsewhere;
D = estimated annual depreciation;
D_t = estimated annual depreciation allowed for tax purposes;
I = interest paid on borrowed funds;
t = effective applicable income tax rate;
T = income tax payable.

A general equation for annual profit before income taxes is derived as follows:

$$E_b = G - (C + D + I) \text{ and } E_a = E_b - T$$

and
$$T = [G - (C + D_t + I)]t.$$
Then
$$E_a = [G - (C + D + I)] - [G - (C + D_t + I)]t.$$

This equation may be simplified by assuming that $D_t = D$, and substituting D for D_t. This may be justified by the fact that although actual and allotted annual depreciations may differ, the total actual depreciation should equal the total allowed depreciations over the life of an asset. If this substitution is made
$$E_a = E_b(1 - t) = [G - (C + D + I)](1 - t).$$

Notice should be taken of the fact demonstrated in this equation that income after income taxes of a venture is equal to its gross income times $(1 - t)$ less its costs times $(1 - t)$.

The Effect of Taxes on Alternatives with Equal Income. When two alternatives have equal income, their comparison may be based only upon costs. In many cases, gross incomes of alternatives can be assumed to be identical even though their amount may not be known. For example, when either of two machines can satisfactorily handle an equal flow of production, their contribution to the total income of the enterprise may be assumed to be equal even though the amount of the contribution of each to income may not be known. Assume that Machine A and Machine B result in gross incomes of G and G', respectively. Then income taxes of the alternatives can be taken into account as follows:

Machine A: $E_a = G(1 - t) - (C + D + I)(1 - t)$

Machine B: $E'_a = G'(1 - t) - (C' + D' + I')(1 - t).$

But by the conditions given, $G = G'$
and,
$$E_a - E'_a = [-(C + D + I) + (C' + D + I')](1 - t).$$

If the costs of Machine A are less than the costs of Machine B, the annual "saving" in operating costs before income taxes will amount to $(C' + D' + I') - (C + D + I)$. Thus the advantage of Machine A over Machine B after income taxes will be equal to this "saving" multiplied by $(1 - t)$.

After-Tax Cash Flows for Comparison of Alternatives. The cash flows that represent the actual receipts and disbursements associated with an investment alternative can be either before or after-tax cash flows. Since taxes constitute a substantial portion of the disbursements that are related to an alternative it is sound decision making to compare the after-tax cash flows of investment alternatives.

To determine the after-tax cash flows of an investment alternative a tabular method may be used. Tabular methods have the advantage that they can be made to reflect complex situations with simple mathematics. They are also easy for the layman to understand.

Suppose a firm has $30,000 available for investment. Three possible alternatives have been suggested and they are considered to be mutually exclusive. Alternative A1 and Alternative A2 differ only in the method of depreciation used in the calcualtion of their after-tax cash flows. Alternative A1 requires sum-of-the-years-digits depreciation while Alternative A2 uses the straight-line method. The net before-tax cash flows for Alternative A1 and Alternative A2 are presented in Column B of Tables 13.5 and 13.6, respectively. The lives of these alternatives are 5 years and the estimated salvage value is zero. The effective rate is 40% and the MARR *after taxes* is considered to be 10%.

Table 13.5. TABULAR METHOD FOR ALTERNATIVE A1.

End of Year	Before-tax Cash Flow	Depreciation Charges	Taxable Income $B - C$	Taxes $0.4 \times D$	After-Tax Cash Flow $B - E$
A	B	C	D	E	F
0	− $30,000				− $30,000
1	10,000	$10,000	$ 0	$ 0	10,000
2	10,000	8,000	2,000	800	9,200
3	10,000	6,000	4,000	1,600	8,400
4	10,000	4,000	6,000	2,400	7,600
5	10,000	2,000	8,000	3,200	6,800

Table 13.6. TABULAR METHOD FOR ALTERNATIVE A2.

End of Year	Before-Tax Cash Flow	Depreciation Charges	Taxable Income $B - C$	Taxes $0.4 \times D$	After Tax Cash Flow $B - E$
A	B	C	D	E	F
0	− $30,000				− $30,000
1	10,000	$6,000	$4,000	$1,600	8,400
2	10,000	6,000	4,000	1,600	8,400
3	10,000	6,000	4,000	1,600	8,400
4	10,000	6,000	4,000	1,600	8,400
5	10,000	6,000	4,000	1,600	8,400

Alternative A3 is somewhat different from A1 and A2 in that it requires an initial investment of $50,000. Since the firm has only $30,000 available it must borrow the additional $20,000 required at an interest rate of 8% per year in order to undertake Alternative A3. This alternative has a life of five years, similar to A1 and A2, but the estimated salvage at the end of its life is expected to be $25,000. Column B of Table 13.7 presents the before-tax cash flow of the alternative exclusive of the loan and its associated interest of payments. The cash flow related to the loan is shown in Column C of Table 13.7.

Table 13.7. TABULAR METHOD FOR ALTERNATIVE A3

End of Year	Before-Tax Cash Flow	Cash Flow for Loan	Cash Flow After Loan Payment	Deprecia- tion Charges	Taxable Income	Taxes	After-Tax Cash Flow
A	B	C	D	E	F	G	H
0	$-$50,000	$20,000	$-$30,000				$-$30,000
1	12,000	$-1,600$	10,400	$5,000	$5,400	$2,160	8,240
2	12,000	$-1,600$	10,400	5,000	5,400	2,160	8,240
3	12,000	$-1,600$	10,400	5,000	5,400	2,160	8,240
4	12,000	$-1,600$	10,400	5,000	5,400	2,160	8,240
5	12,000	$-1,600$	10,400	5,000	5,400	2,160	8,240
5	25,000*	$-20,000$	5,000				5,000

*Salvage Value

Now that the after-tax cash flows are available for each of the three alternatives the computation of the present-worth amounts for these alternatives is straightforward. For an after-tax MARR of 10% the present-worth on total investment for each alternative is computed.

$$PW_{A1}(10) = -\$30,000 + \left[\$10,000 - \$800(\overset{A/G\ 10,\,5}{1.8101})\right](\overset{P/A\ 10,\,5}{3.7908}) = \$2,419$$

$$PW_{A2}(10) = -\$30,000 + \$8,400(\overset{P/A\ 10,\,5}{3.7908}) \qquad = \$1,843$$

$$PW_{A3}(10) = -\$30,000 + \$8,240(\overset{P/A\ 10,\,5}{3.7908}) + \$5,000(\overset{P/F\ 10,\,5}{0.6209}) = \$4,340.69$$

Economically, the most desirable alternative is $A3$. Although Alternative $A1$ and Alternative $A2$ are economically the same on a before-tax basis, Alternative $A1$ is favored over Alternative $A2$ on an after-tax basis. This difference between these two alternatives arises because of the effect the different methods of depreciation have on income taxes. This subject was discussed earlier in Section 13.4.

Suppose that Alternative $A3$ is implemented and it is now five years later. An examination of the receipts and disbursements produced by Alternative $A3$ over the last five years reveals that the predictions about the future regarding this alternative were accurate with one exception. Instead of the salvage value being $25,000 as predicted the actual amount received was $30,000. Thus the after-tax cash flow resulting from the salvage value is the estimated salvage value plus the after-tax gain realized on disposal. That is, the after-tax contribution from the $30,000 received on disposal is

$$\$25,000 + (\$30,000 - \$25,000)(1 - 0.30) = \$28,500.$$

However since $20,000 of that amount is used to repay the loan, the net benefit derived from the $30,000 salvage is $8,500. That amount now replaces the $5,000 amount shown for year 5 in Column H of Table 13.7. Thus if one is interested in the after-tax rate of return actually earned on the investment made in Alternative $A3$ all that is required it to solve for i in the equation that follows.

$$0 = -\$30,000 + \$8,240(\overset{P/A\ i,\ 5}{\quad}) + \$8,500(\overset{P/F\ i,\ 5}{\quad})$$

By trial and error

$$i^{*}_{A3} = 17.3\%$$

Thus in retrospect, Alternative $A3$ can be considered a good investment since it has produced a return greater than the MARR of 10%.

PROBLEMS

1. A self-employed book salesman is 32 years old, married, and has two children. Last year he earned $2,000 in commissions in connection with which he had $600 traveling and other expenses. From the operation of a technical book shop he had $66,000 gross income and had expenses of $43,400, exclusive of those expenses listed below which should properly be charged to the book shop. In addition, he sold a trailer for $1800 that he purchased nine months ago for $1400. During the year he made contributions of $400 to his church, Red Cross, etc., paid $500 and $300 in interest respectively on mortgages on his home and his book shop. To keep abreast of the fields in which he sells books he is a member of several technical societies and pays annual dues of $80 per year. Last year he paid $1400 in federal income tax. On his home and bookshop, he paid, respectively, $200 and $400 in state, county, and municipal taxes. Medical expenses were $120. Water from a leaky pipe caused $140 damage to books in his shop of which $120 was compensated by insurance.
   (a) Determine adjusted income.
   (b) Determine taxable income.
   (c) Determine the income tax to be paid, using the proper table in the text and assuming that the salesman and his wife file a joint return.
2. From data in Problem No. 1 determine the effective income tax rate applicable to the salesman's (a) taxable income, (b) adjusted income.
3. The salesman is considering giving up selling on commission.
   (a) What would have been his net income after taxes if he had done so in the year covered in Problem 1 ?
   (b) What is the effective income tax rate pertinent to the increment of income from the activity he is considering giving up ?
4. Mr. Brown, who is in business for himself, is interested in an activity which will necessitate the borrowing of $3,000 for 9 months at 6% interest. The new activity is expected to increase his taxable income by $600, bringing his total taxable

income to $17,000. If Mr. Brown is single and qualifies as head of a household, what will be his net income after income taxes?

(a) If he undertakes the new activity and it turns out as estimated?

(b) If he does not undertake the new activity?

5. To develop a new product a corporation is considering an investment of $1,500,000 in research and development. It is estimated that the increase in taxable income from marketing the new product in the coming year will be $6,000,000. Currently, the taxable income for the firm is $30,000,000 per year.

(a) What will be the increase in the net income after taxes if the research and marketing program is successful?

(b) What will be the decrease in the net income after taxes if the research and marketing program is unsuccessful?

6. A corporation's taxable income for next year is estimated at $400,000. It is considering an activity for next year which it is estimated will result in an additional taxable income of $40,000. Normal and surtax rates are respectively 22% and 26% on excess over $25,000. Calculate effective income tax rate applicable.

(a) To the present estimated $400,000 taxable income.

(b) To the $40,000 estimated increment of income.

(c) If the profit to the new venture is $40,000 before income taxes, what will it be after income taxes are paid?

7. The same conditions exist as in Problem 6, except that a new activity has been undertaken but instead of resulting in an increase of $40,000 in taxable income as estimated, it has resulted in a decrease in the taxable income of $20,000. Calculate

(a) Net income after income taxes on a taxable income of $400,000.

(b) Net income after income taxes if the new venture results in a loss of $20,000.

(c) The loss in income after income taxes that would be caused by the $20,000 loss.

8. A corporation is considering a study to develop a new process to replace a present method of manufacturing transistors which, if successful, is estimated to result in a saving of $300,000 in fulfilling a contract that will be completed during the current tax year. The study is estimated to cost $25,000 during the same tax year. If the new process is not successful, the corporation taxable income will be $5,000,000.

(a) If the study is successful, what will be the increase in income after taxes?

(b) If the study is unsuccessful, what will be the decrease in income after taxes?

9. A corporation has a taxable income of $21,000 during a year. It is considering a venture that will result in an additional income of $10,000 during next year. The normal tax rate is 22% and the surtax rate on taxable income in excess of $25,000 is 26%.

(a) What is the effective income tax rate if earnings remain at $21,000?

(b) What is the effective income tax rate if the new venture is undertaken and turns out as estimated?

(c) What is the effective income tax rate applicable to the increment of income due to the new venture?

10. Corporation A has a total investment of $1,000,000, uses no borrowed funds, and has a taxable income of $100,000.

(a) How much will it have for payment of dividends after income taxes if the normal tax rate is 22% and the surtax rate applicable to income in excess of $25,000 is 26%?

(b) If it is capitalized at $1,000,000 what would be the rate of return it earned for its stockholders?

Corporation B is identical with A except that it is capitalized at $800,000, uses $200,000 of funds borrowed at 9% and its taxable income before payment of interest on borrowed funds is $100,000.

(c) How much will it have for payment of dividends after income taxes?

(d) What rate of return did it earn for its stockholders?

11. The applicable effective tax rate of a taxpayer is 0.45. He has purchased a machine for $2,000 whose life and salvage value for tax purposes is 4 years and zero respectively. The estimated annual income of the machine is $1,200 per year before income taxes. Compare the present worth of the income taxes that will be payable if the straight-line method, the sum-of-the-years depreciation and the double-declining balance methods are elected, and if the interest rate is 15%.

12. A corporation purchases an asset for $60,000 with an estimated life of 5 years and a salvage value of zero. The gross income per year will be $20,000 before depreciation and taxes. If the tax rate applicable to this activity is 40% and if the interest rate is 12%, calculate the present worth of the income taxes if the straight-line method of depreciation is used; if the sinking-fund method of depreciation is used; if the double-declining balance method of depreciation is used; if the sum-of-the-years digits is used.

13. An asset with an estimated life of 10 years and a salvage value of $5,000 is purchased by a firm for $60,000. The firm's effective tax rate is 50% and the minimum attractive rate of return is 10%. The gross income from the asset before depreciation and taxes will be $8,500 per year. Calculate the present-worth of the taxes for the straight-line and sum-of-the-years digits methods of depreciation.

(a) Assume that the firm is profitable in its other activities.

(b) Assume that the firm has no profits in its other activities.

14. A corporation purchases a $60,000 asset with an estimated life of 4 years and expected salvage value of $30,000. If after 4 years the asset is sold for $20,000, what are the capital gains or losses for the straight-line depreciation method and the double-declining balance depreciation method in the fourth year? Assume all accounting adjustments between actual and estimated salvage value occur at the time the asset is sold. What are the capital gain tax effects of these two alternative methods of depreciation?

15. The income tax rate of a corporation is 44%. Four months ago it purchased a warehouse for $22,000. It has just received an offer to sell the warehouse for $28,000 for a short-term gain of $6,000. What selling price two to eight months later will be equivalent to the present $28,000 offer in terms of income after taxes if the corporation has no other short-term gains in prospect?

16. A contractor purchased $400,000 worth of equipment to build a dam. For tax purposes, he was permitted to depreciate the equipment by the straight-line method on the basis of no salvage value and a five year life. After two years and the completion of the dam, the contractor sold the equipment for $300,000. The contractor attributed sale of the equipment at an amount higher than the "book value" to the excellence of maintenance. He estimated that he spent $16,000 per year more for maintenance than necessary "just to keep the equipment running," and by so doing reduced his operating costs exclusive of maintenance on the contract by $12,000

per year. The contractor's income tax rate is 42% and the long-term capital gain tax rate is 30%. The contractor averaged $48,000 taxable income for the two years required to complete the contract. There were no short-term capital losses.

(a) How much capital gains tax would the contractor pay as a result of selling his equipment?

(b) What was his total income tax including capital gains tax for the two-year period, and what was his total income after income taxes?

(c) What would have been his total income tax for the two-year period if he had spent $12,000 per year less maintenance and as a result his operating cost had been increased by $16,000 per year and his equipment had been sold for its book value? What was his total income after income taxes for the two-year period?

(d) How much was his total income after income taxes for the two-year period increased by his maintenance policy if his estimates were correct?

17. A mineral deposit discovered by a mining company is estimated to contain 100,000 tons of ore. The company has made an initial investment of $5,000,000 to recover the ore which sells for $210 per ton. The company has a minimum attractive rate of return of 15% and an effective tax rate of 46%. The fixed percentage depletion rate for this mineral is 14%. If the ore is produced and sold at a rate of 20,000 tons per year and the operating expenses exclusive of depletion expenses are $1,000,000 per year, what is the present-worth of the after-tax cash flow for this company if (a) unit depletion is used? (b) percentage depletion is used?

18. An oil lease is purchased for $200,000 and it is estimated that there are 400,000 barrels of oil on this lease. It costs $60,000 per year exclusive of depletion charges to recover the oil and the oil can be sold for $3.20 per barrel. The effective tax rate is 40% and 40,000 barrels of oil are produced each year. If the interest rate is 12%, what is the present worth of the after-tax cash flow for this lease for (a) unit depletion; (b) percentage depletion?

19. A prospector acquired a mine containing 300,000 tons of tungsten ore for $540,000. Annual operating costs of the mine are $132,000, and 40,000 tons of ore are being sold at the rate of $14.50 per ton. The prospector wishes to compare the results of using cost depletion and percentage depletion (22%). Determine the income tax for each method assuming that the applicable income tax rate is 0.35.

20. An investment proposal has the following estimated cash flow before taxes.

End of Year	0	1	2	3	4
	−$50,000	$30,000	$10,000	$5,000	$2,000 + $20,000 (salvage value)

The effective tax rate is 40%, the tax rate on capital gains is 30%, and the interest rate is 10%. Assume the company considering this proposal is profitable in its other activities. Find the after-tax cash flows and the present-worth of those cash flows for (a) the straight-line method of depreciation (b) the double-declining method of depreciation, and (c) the sum-of-the-years digits depreciation.

21. An investment proposal is described by the following tabulation:

Year	1	2	3
Gross income at end of year ..	$ 8,000	$16,000	$20,000
Investment at beginning year	14,000	10,000	0
Operating cost	2,000	3,000	4,000
Depreciation charge	8,000	8,000	8,000

The company considering this proposal is profitable in its other activities. Thus any depreciation not use to off-set income for this proposal can be used to reduce taxes on the income from the other activities. Find the rate of return earned by this proposal (a) before taxes, (b) after taxes. The effective income tax rate is 40%.

22. An asset is being considered whose first cost, life, salvage value, and annual operating expenses, respectively, are estimated at $15,000, 10 years, zero, and $800. The asset will be depreciated by the straight-line method of depreciation. The effective income tax is 40%. Determine the equivalent annual cost at 12% for the after-tax cash flow if

(a) the initial investment is made from equity funds

(b) the initial investment is borrowed at 8% with repayment of principal and interest in 10 equal annual amounts

(c) the initial investment is borrowed at 8% with repayment of interest at the end of each period and the loan principal repaid at the end of 10 years.

V

ESTIMATES,
UNCERTAINTY,
AND IMPLEMENTATION

14

Treatment of Estimates
in Economy Studies

The approach to estimating in the physical environment approximates certainty in many applications. Instances are the pressure that a confined gas will develop under a given temperature, the current flowing in a conductor as a function of the voltage and resistance, and the velocity of a falling body at a given point in time. It was noted, however, that much less is known with certainty about the economic environment with which the engineering process is concerned. Economic laws depend upon the behavior of people and are unlike physical laws which depend upon well-ordered cause and effect relationships.

A large portion of creative engineering activity has as its objective the search for activities with profit potential that is high in relation to the risk involved. This search necessitates the estimation of pertinent facets of the anticipated economic outcome. In this chapter attention is directed to the process of estimating and to elementary techniques useful in allowing for inaccuracies in estimates.

14.1. THE ELEMENTS TO BE ESTIMATED

Any activity that is undertaken requires an input of thought, effort, material, and other elements for its performance. In a purposeful activity, an input of some value is surrendered in the hope of securing an output of greater value. The terms input and output as used here have the same meaning as when they are used to designate, for instance, the number of heat units that are supplied to an engine and the number of energy units it contributes for a defined purpose.

Success of a venture in terms of economy is determined by considering the relationship between the input and output of the venture through time, with the time value of money taken into consideration. But anticipated success can be imaginary if certain elements are omitted. Thus, an important task in an economy study is the delineation of inputs and outputs associated with the proposed activity.

Outputs. Activities are normally proposed in response to a need or requirement. Outputs, therefore, should be considered first and in conjunction with the need. Benefit, worth, effectiveness, and other terms are used to describe outputs in relationship to requirements.

The outputs of commercial organizations and governmental agencies are endless in variety. Commercial outputs are differentiated from governmental outputs by the fact that it is usually possible to evaluate the former accurately but not the latter. A commercial organization offers its products to the public. Each item of output is evaluated by its purchaser at the point of exchange. Thus the monetary values of past and present outputs of commercial concerns are accurately known item by item.

Since engineering is concerned with the future, it is concerned with future output. Generally, information on two subjects is needed to come to a sound conclusion. One of these is the physical output that may be expected from a certain input. This is a matter for engineering analysis. The second is a measure of output that may be expressed in terms of monetary income.

Monetary income is dependent upon two factors: one is the volume of output, in other words the amount that will be sold; the other is the monetary value of the output per unit. The determination of each of these items for the future must of necessity be based upon estimates. Market surveys and similar techniques are widely used for estimating the future output of commercial concerns. In the case of large scale systems or projects the monetary income to the contractor is determined through contractual agreement.

Internal intermediate outputs of commercial organizations are determined with great difficulty and are usually estimated as dictated by judgment. For example, the value of the contribution of an engineer, a production clerk, or a foreman to a final output is rarely known with reasonable accuracy to either himself or to his superior. Similarly, it is difficult to determine the value of the contributions of most intermediate activites to the final result.

The outputs of many governmental activities are distributed without regard to the amount of taxes paid by the recipient. When there is no evaluation at the point of exchange, it appears utterly impossible to evaluate many governmental activities. John Doe may recognize the desirability of the national military establishment, the U. S. Forest Service, or Public Health activities, but he will find it impossible to demonstrate their worth in monetary terms. However, some government outputs, particularly those which are localized, such as highways, drainage, irrigation, and power projects, may be fairly accurately evaluated in monetary terms by calculating the reduction in cost or the increase in income they result in for the user.

Inputs. The inputs of private and governmental enterprise involve a wide variety of items. A number of these common to most organizations are presented in the following paragraphs.

A most important item of input is services of people for which salaries and wages are paid. In a commercial organization the total input of human services, as measured by the cost for a given period of time, is ordinarily reflected quite accurately. Where involuntary service is secured, as in military services of governments, it will be clear that the input of human services is not necessarily reflected by amounts paid out as wages. In a commercial organization the input of human services may be classified under the headings of direct labor, indirect labor, investigation and research. Of these, direct labor is the only item whose amount is known with reasonable accuracy and whose identity is preserved until it becomes a part of output.

Input devoted to investigation and research is particularly hard to relate to particular units of output. Much research is conducted with no particular specified goal in mind and much of it results in no appreciable benefit that can be associated with a particular output. Expenditures for people for investigation and research may be made for some period of time before this type of service has a concrete effect upon output. Successful research of the past may continue to affect output for a long time in the future. The fact that expenditures for this type of service do not parallel in time the benefits provided makes it very difficult to relate them to an organization's output.

Indirect labor, supervision, and management have characteristics falling between those of direct labor and of investigation and research. The input of indirect labor and, to a lesser extent, that of supervision, parallel the output fairly closely in time, but their effects can ordinarily be identified only with broad classes of output items.

Management is associated with the operations of an organization as a whole. Its important function of seeking out desirable opportunities is similar in character to research. Input in the form of management effort is difficult to associate with output either in relation to time or to classes of product.

As difficult as it may be to associate certain inputs with a final measurable output, such as of products sold on the market, input can ordinarily be identified closely with intermediate ends that may or may not be measurable in concrete terms. For example, the cost of the input of human effort assigned to the engineering department, the legal department, the labor relations department, and the production department are reflected with a high degree of accuracy by the payrolls of each. However, the worth of the output of the personnel assigned to these departments may and usually does defy even reasonably accurate measurement.

A second major category of input is that of material. Many items of material are acquired to meet the objectives of commercial and govenmental enterprise. For convenience, material items may be classified as direct material, indirect material, equipment, land and buildings.

Inputs of direct material are directly allocated to final and measurable outputs. The measure of material items of input is their purchase price plus costs for purchasing, storage, and the like. This class of input is subject to reasonably accurate measurement and may be quite definitely related to final output, which in the case of commercial organizations is easily measurable.

Indirect material and power inputs are measurable in much the same way and with essentially the same accuracy as are direct material and power inputs. One of the important functions of accounting is to allocate this class of input in concrete terms to items of output or classes of output. This may ordinarily be done with reasonable accuracy.

An input in the form of an item of equipment requires that an immediate expenditure be made, but its contribution to output takes place piecemeal over a period of time in the future which may vary from a short time to many years, depending upon the use life of the equipment. Inputs of equipment are accurately measurable and can often be accurately allocated to definite output items except in amount. The latter limitation is imposed by the fact that the number and kinds of output to which any equipment may contribute are often not known until years after many units of the product have been distributed. The function of depreciation accounting is to allocate equipment inputs to outputs.

Inputs of land and buildings are treated in essentially the same manner as inputs of equipment. They are somewhat more difficult to allocate to output because of their longer life and because a single item, such as a building, may contribute simultaneously to a great many output items. Allocation is made with the aid of depreciation and cost accounting techniques and practices.

Allocation of inputs of indirect materials, equipment, buildings, and land rests finally upon estimates or judgments. Although this fact is often obscured by the complexities of and the necessary reliance upon accounting practices for day-to-day operations, it should not be lost sight of when economy studies are to be made.

Capital in the form of money is a very necessary input, although it must ordinarily be exchanged for producer goods in order for it to make a contribution to output. Interest on money used is usually considered to be a cost of production and so may be considered to be an input. Its allocation to output will necessarily be related to the allocation of human effort, services, material, and equipment in which money has been invested.

Taxes are essentially the purchase of governmental service required by private enterprise. Since business activity cannot be carried on without the payment of taxes, they comprise a necessary input. There are many types of taxes, such as ad valorem, excise, sales, and income. The amounts may be precisely known and, therefore, are accurately called inputs. However, it is often difficult to allocate taxes to outputs especially in the case of income taxes which are levied after the profit is derived.

14.2. ESTIMATING OUTPUTS AND INPUTS

The success of an activity as a whole may be estimated. Thus, if the establishment of a construction department is under consideration by a corporation, it might be directly estimated that the department will yield a return to the firm equivalent to a certain per cent per year on the amount invested. This return will be a resultant of a number of prospective receipts and disbursements, which may be classified as income, operating expenses, depreciation, interest, and taxes. It is a rare intellect that can combine accurately four complex items to obtain their resultant without resort to paper and pencil, even when the items are clearly known. Thus for best results in estimating the final outcome of an undertaking, it will almost always be found advantageous to begin with detailed estimates of income and costs as they may be expected to originate in the future. The detailed estimates are then combined mathematically to obtain their results.

Prospective receipts and disbursements are a secondary result of prospective activities. Estimates of receipts and disbursements should, therefore, be based on these prospective activities. It follows that receipts and disbursements often arise from the same data. For example, if the construction department is to take charge of all plant expansion and new plant construction, this activity may be expected to result in both income and expense.

Estimating Income. If there is a demand for goods or service, an income can be derived from supplying that demand. If the cost of supplying the service is less than the income received, a profit can be made. The first step toward a profit is an income. Thus, in estimating the desirability of a prospective undertaking, it seems logical to estimate income as a first step. As used here, income also embraces the possibility for making a saving.

Estimates of a result will usually be more accurate if they are based upon estimates of the factors having a bearing on the result than if the result is estimated directly. For example, in estimating the volume of a room, it will usually prove more accurate to estimate the several dimensions of the room and calculate the volume than to estimate the volume directly. Similar reasoning applies to estimates of economic factors in prospect.

If, for example, the problem under consideration is the saving that may result from automating a given operation, the amount of saving will depend upon the number of units processed in a given time and the saving per unit. The first step is to bring all possible information to bear upon estimating the number of units expected to be processed during each year of the future period to be considered. In this connection, use should be made of such information as the records of past sales, present sales trends, the product's relation to the building trades, general business activity, and anything else that may be useful in arriving at the most accurate estimate of future sales.

In estimating the savings per unit, all items of saving—direct labor, direct material, overhead items, storage, inspection, and any others—should be estimated separately and totaled in preference to estimating the total of the savings of the items directly. The total estimated saving is then determined as a product of the number of units processed and the saving per unit. If more than one product is to be processed on a machine, the savings for each product should be estimated as above and totaled.

Under some circumstances—for example, if the income is represented by the saving resulting from an improvement in a process for manufacturing a staple product made at a constant rate—an estimate of income is easily made. But estimating income for new products with reasonable accuracy may be very difficult. Extensive market surveys and even trial sales campaigns over experimental areas may be necessary to determine volume. When work is done on contract, as is the case with much construction work, for example, the necessity for estimating income is eliminated. Under these circumstances the income to be received is known in advance with certainty from the terms of the contract.

Estimating Operating Expense. Operating expense is used here to embrace both direct and indirect costs pertinent to systems in the operational state. Operating expense originates as expenditures for such items as fuel, water, electric current, materials, supplies, wages, taxes, and insurance.

Some of these items will be based on the same facts that determine income. For instance, the income derived from the manufacture and sale of a given number of units of a product will be based upon the number and the income derived per unit. The number of units sold will also be a factor in estimating the needed materials, labor, power, and other inputs needed in their production. In general, operating expenses should be estimated item by item rather than as a whole.

Operating expenses commence after a system has been produced or constructed. The magnitude of these expenses are a result of original design as well as environmental factors. Since engineers have direct control over design more attention should be given to the expenses of operation and maintenance which might be reduced through proper design.

Estimating Depreciation. Depreciation of machines often parallels income. The period and extent of use of single-purpose machines are dependent upon the number of units to be processed, which will have been estimated in arriving at income. Where depreciation is dependent upon wear and tear, the extent of use will be the determining factor; experiences in the past with like or similar machines may be helpful. If weathering or other causative factors of depreciation associated with the passage of time are the determining factors in estimating depreciation, mortality tables may prove to be of aid.

A portion of an investment in a physical asset is lost through physical or functional depreciation. If the loss resulting from depreciation is offset by an equal income from the use of the asset, that portion of the investment has been recovered. The actual undepreciated balance of the asset represented by its salvage value has not been lost and so need not be recovered. The salvage value of the asset is merely reconverted into another medium, usually money, by selling what remains of the asset.

In the recovery of an investment, attention is focused upon realizing an income to offset value lost through depreciation. And since depreciation is taken to be the difference between first cost and the amount realized from salvage, the prime problem in capital recovery is the securing of income equal to depreciation.

A depreciation estimate should be considered to be a summation of four estimates covering (1) installed cost, (2) service life in years, (3) salvage value at the end of service life, and (4) pattern of depreciation during service life.

Little difficulty is ordinarily experienced in making a reasonable estimate of the installed cost of an asset. The pattern of depreciation is ordinarily selected by considering two factors; its effect on income taxes and its effect on corporate profit over the life of the asset. This leaves estimates of service life and salvage value. Of these, the estimate of service life is most important and perhaps the most difficult to make. The longer the service life of an asset, the less is the significance of error in estimates of salvage value.

It was pointed out earlier that the equivalent annual cost of an asset is not affected by the depreciation method. Thus, estimates of the service life and salvage value are of primary importance in a decision concerning the desirability of an asset.

Estimating Interest. In comparative evaluations of proposed activities it is believed to be sound to take the viewpoint that interest is an item of cost. By considering interest as an expense, the interest rate can be determined more or less objectively. An enterprise that borrows money for its operations may use the rate that it is paying for funds. A concern investing its own funds is justified in using the rate that it can receive for funds for purposes similar to the one under consideration.

Regardless of the method used in arriving at the interest rate, it will be necessary to estimate the total investment to which the rate will be applied. This may be broken down into the return expected on capital unrecovered in physical assets and the return expected on funds used for operating expense. In either case, the return will enter the final analysis as an interest cost and, life operating expense, depreciation and taxes, will be deducted from gross income yielding a net income. The final magnitude of the estimated net income will depend upon the magnitude of the estimated interest.

As an item of expense, interest for most activities will be relatively small. Thus, the interest rate estimated will have a minimum effect. It is desirable that the rate selected be used in all studies to be compared.

Estimating Taxes. Income taxes are difficult to estimate in regard to a specific project that is an element of a larger activity for the reason that the income tax on the project is partly determined by net income of the activity as a whole. Also, tax schedules are subject to the fiscal needs of the nation. Income taxes of a specific project may be based upon the estimated net income of the project and an estimated effective income tax rate.

An estimate of the income taxes of a project may be made on a cost basis as a per cent of the investment in the project. The applicable per cent is the ratio of the income taxes of the total activity and the investment in the total activity and was called the effective tax rate.

14.3. AN EXAMPLE OF A DECISION BASED ON ESTIMATES

An example will be used to illustrate some aspects of estimating, treating estimated data, and arriving at an economic decision. To simplify the discussion of these subjects, the same example will be used throughout the balance of this chapter. Income taxes will not be considered and it will be assumed that all funds invested in the project are equity funds.

The purchase of a machine for etching a printed circuit, now performed in another manner, is considered likely to result in a saving. It is known with absolute certainty that the machine will cost $1,000 installed. All other factors pertinent to the decision are unknown and must be estimated.

Income Estimate. From a study of available data and the result of judgment, it has been estimated that a total of 3,000 units of the circuit board will be made during the next six years. The number to be made each year is not known; but, since it is believed that production will be fairly well distributed over the six-year period, it is believed that the annual production should be taken as 500 units. A detailed consideration of materials used, time studies of the methods employed, wage rates, and the like, have resulted in an estimated saving of $1.04 per unit, exclusive of the costs incident to the operation of the machine if the machine is used. Combining the estimated production and the estimated unit saving results in an estimated saving (income) of 500 × $1.04, or $520 per year.

Capital Recovery Estimate. The machine is a single-purpose machine and no use is seen for it except in processing the product under consideration. Its service life has been taken to be 6 years to coincide with the estimated production period of 6 years. It is believed that the salvage value of the machine will be offset by the cost of removal at retirement. Thus the estimated net receipts at retirement will be zero.

Interest is considered to be an expense in this evaluation, and the rate of interest has been estimated at 5%. The next step is to combine the estimates of first cost, service life, and salvage value to determine the estimated annual capital recovery with a return. For the first cost of $1,000 a service life of 6 years, a salvage value of zero, and an interest rate of 5%, the resultant estimate of the annual cost of capital recovery will be:

$$\overset{A/P\,5,\,6}{(\$1,000 - \$0)(0.1970)} + \$0(0.05) = \$197.$$

Operation Cost Estimate. Operating costs will ordinarily be made up of several items such as fuel, maintenance, supplies, and labor. Consider for simplicity that the operating expense of the equipment in this example consists of items a, b, c, and d. Each of these items is estimated on the basis of the number of units of product that it is estimated are to be processed per year. Assume that these items have been estimated as follows:

$$\text{Item } a = \$90 \text{ per year};$$
$$\text{Item } b = \$60 \text{ per year};$$
$$\text{Item } c = \$40 \text{ per year};$$
$$\text{Item } d = \$10 \text{ per year}.$$

The estimated income and cost items of the example may be summarized as follows:

Estimated annual capital recovery and return, $1,000(0.1970)$^{A/P\,5,\,6}$	$197
Estimated annual operating cost, $90 + $60 + $40 + $10	200
Estimated total annual cost	$397
Estimated total annual income	520
Estimated net annual profit for venture	$123

This final statement means that the venture will result in an equivalent annual profit of $123 per year for a period of 6 years if the several estimates prove to be accurate.

The resultant equivalent annual profit is itself an estimate, and experience teaches that the most certain characteristic of estimates is that they nearly always prove to be inaccurate, sometimes in small degree and often in large degree. Once the best possible estimates have been made, however, whether they eventually prove to be good or bad, they remain the most objective basis on which to base decision. It should be realized that decision making can never be an entirely objective process.

14.4. ALLOWANCE FOR INACCURACIES IN ESTIMATES

The success of the scientific approach depending upon a cause and effect relationship in the physical realm has carried over into other realms. The idea that the future can be predicted if sufficient knowledge is available is now generally accepted. Great emphasis is placed on securing sufficient data and applying it carefully in arriving at estimates that are representative of actualities to the highest possible extent. The better the estimates, the less allowance need be made for error. It should be realized at the outset that allowances for errors (factors of safety) do not make up for deficiency of knowledge in the sense that allowances correct errors. Allowances are merely a means of eliminating some consequences of error at a cost.

As an example of the effect of an allowance in an economic undertaking, consider the following illustration. A contractor has estimated the cost of a project on which he has been asked to bid at $100,000. If he undertakes the job, he wishes to profit by 10% or $10,000. How shall he make allowance for errors in his cost estimate? If he makes an allowance of 10% for errors in his estimates, comparing to the very low factor of safety of 1.10, his estimated cost becomes $110,000. To allow for his profit margin, he will have to enter a bid of $121,000. But the higher his bid; the less the chance that he will be the successful bidder. If he is not the successful bidder, his allowance for errors in his estimate may have served to insure not only that he did not profit from the venture, but that he was left with a loss equal to the cost of making the bid. This illustration serves to emphasize the necessity for considering the cost of making allowances for errors in estimates.

Allowing for Error in Estimates by High Interest Rates. A policy common to many industrial concerns is to require that prospective undertakings be justified on the basis of a high minimum acceptable rate of return, say 25%. One basis for this practice is that there are so many opportunities which will result in a return of 25% or more that those yielding less can be ignored. But since this is a much greater return than most concerns make on the average, the high rate of return represents an allowance for error. It is hoped that if undertaking of ventures is limited to those that promise a high rate of return, none or few will be undertaken that will result in a loss.

Returning to the example of the etching machine given previously, suppose that the estimated income and the estimated cost of carrying on the venture when the interest rate is taken at 25% are as follows:

Estimated annual income (for six years)		$520
Estimated annual capital recovery and return, $1,000($\overset{A/P\ 25,\ 6}{0.3388}$)	$339	
Estimated annual operating cost...........................	200	
Estimated total annual operating cost		539
Estimated net annual profit of venture		−$ 19

If the calculated loss based on the high interest rate is the deciding factor, the venture will not be undertaken. Though the estimates as given above, except for the interest rate of 25%, might have been correct, the venture would have been rejected because of the arbitrary high interest rate taken, even though the resulting rate of return would be 22.5%.

Suppose that the total operating costs had been estimated as above but that annual income had been estimated at $600. On the basis of a policy to accept ventures promising a return of 25% on investment, the venture would be accepted. But if it turned out that the annual income was, say, only $150, the venture would result in loss regardless of the calculated income with the high interest rate. In other words, an allowance for error embodied in a high rate of return does not prevent a loss that stems from incorrect estimates if a venture is undertaken that will result in loss.

Allowing for Error in Estimates by Rapid Payout. The effect of allowing for error in estimates by rapid payout is essentially the same as that of using high interest rates for the same purpose. Let it be assumed that a policy exists that equipment purchases must be based upon a three-year payout period when the interest rate is taken at 5%.

Returning to the example of past paragraphs, suppose that the estimated income and the estimated cost of carrying on the venture when a three-year payout period is taken is as follows:

Estimated annual income (Estimated for six years but taken as being for three years to conform to policy)		$520
Estimated annual capital recovery and return, $1,000 ($\overset{A/P\ 5,\ 3}{0.3672}$)..	$367	
Estimated annual operating cost...........................	200	
Estimated total annual operating cost		567
Estimated net annual profit for venture............................		−$ 47

Under these conditions the venture would not have been undertaken. The effect of choosing conservative values for the components making up an estimate is to improve the certainty of a favorable result, if the outcome results in values that are more favorable than those chosen.

14.5. LEAST FAVORABLE, FAIR, AND MOST FAVORABLE ESTIMATES

A plan for the treatment of estimates considered to have some merit is to make a least favorable estimate, a fair estimate, and a most favorable estimate of each situation. The *fair estimate* is the estimate that appears most reasonable to the estimator after a diligent search for and a careful analysis of data. This estimate might also be termed the most likely estimate.

The *least favorable estimate* is the estimate that results when each item of data is given the least favorable interpretation that the estimator feels may reasonably be realized. The least favorable estimate is definitely not the very worst that could happen. This is a difficult estimate to make. Each element of each item should be considered independently in so far as this is possible. The least favorable estimate should definitely not be determined from the fair estimate by multiplying the latter by a factor.

The *most favorable estimate* is the estimate that results when each item of data is given the most favorable interpretation that the estimator feels may reasonably be realized. Comments similar to those made in reference to the least favorable estimate, but of reverse effect, apply to the most favorable estimate. The use of the three estimates will be illustrated by application to the example of previous paragraphs.

Items Estimated	Least Favorable Estimate	Fair Estimate	Most Favorable Estimate
Annual number of units	300	500	700
Savings per unit	$0.80	$1.04	$1.15
Annual saving	$240	$520	$750
Period of annual savings, n	3	6	10
Capital recovery and return, $A/P\ 5, n$ $1,000(\quad)$	$367	$197	$130
Operating cost			
Item A	$ 60	$ 90	$120
Item B	60	60	70
Item C	50	40	60
Item D	20	10	2
Estimated total of capital recovery, return and operating items	$557	$397	$382
Estimated net annual saving in prospect for n years	−$317	$123	$368

An important feature of the least favorable, fair, and most favorable estimate plan of comparison is that it provides for bringing additional information to bear upon the situation under consideration. Additional information results from the estimator's analysis and judgment in answering two questions relative to each item. These questions are: "What is the least favorable value that this item may reasonably be expected to have?" and the reverse, "What is the most favorable value that this item may reasonably be expected to have?"

Judgment should be made item by item, for a summation of judgments can be expected to be more accurate than a single judgment of the whole. A second advantage of the three-estimate plan is that it reveals the consequences of deviations from the fair or most likely estimate. Even though the calculated consequences are themselves estimated, they show what is in prospect for different sets of conditions. It will be found that the small deviations in the direction of unfavorableness may have disastrous consequences in some situations. In others even a considerable deviation may not result in serious consequences. Studies of this type are called sensitivity analyses.

The results above can be put on other bases for comparison. The present-worth basis has merit in this instance because of the variation in the number of years embraced by the above three estimates. The payout period and saving per unit of product can easily be calculated for other viewpoints. The estimated present-worth of savings is calculated as follows:

Items Estimated	Least Faborable Estimate	Fair Estimate	Most Favorable Estimate
Net annual saving	−$317	$123	$368
Period of annual savings, n	3	6	10
Present worth factor, ($P/A\ 5,n$)	2.723	5.076	7.722
Present worth of saving	−$863	$624	$2,842

And, the estimated rate or return is:

Items Estimated	Least Favorable Estimate	Fair Estimate	Most Favorable Estimate
Annual saving	$240	$520	$750
Annual operating cost, items A, B, C, and D	190	200	252
Difference applicable to capital recovery and return at 5%	50	320	498
Capital recovery period, years	3	6	10
Rate of return	minus	22.5%	48.9%

The net estimated saving per unit of product is:

Items Estimated	Least Favorable Estimate	Fair Estimate	Most Favorable Estimate
Net annual saving	−$317	$123	$368
Annual number of units	300	500	700
Net saving per unit of product	−$1.057	$0.246	$0.526

There are many who feel that it is an aid to judgment to have several bases on which to compare a single situation. Since the cost of making extra calculations is usually insignificant in comparison with the worth of even a small improvement in decision, the practice should be followed by all who feel they benefit from the additional information. But there are limits beyond which further calculations can serve no useful purpose. This occurs when calculations are made that are beyond the scope of the data used. In the example above, it would seem that no useful purpose would be served, for instance, by averaging the results calculated from the three estimates, least favorable, fair, and most favorable.

QUESTIONS

1. Contrast the effectiveness of estimating in the physical environment and the economic environment.
2. Explain why engineering economy studies must rely heavily upon estimates.
3. List each item in the classification of outlays involved in an activity.
4. List the common items of input encountered in the production of goods and services.
5. Explain why an estimate of a result will probably be more accurate if it is based upon estimates of the factors that have a bearing on the result, than if the result is estimated directly.
6. Describe the estimating procedure for operating expense, depreciation, interest, and taxes.
7. Why should an estimate of the income of a venture be made before estimating the factors that have a bearing on the result?
8. Relate the usual "factor of safety" concept in engineering design to allowances for error in engineering economy studies.
9. Discuss the value of allowing for error in estimates by high interest rates and rapid payout.
10. Discuss the value of making least favorable, fair, and most favorable estimates.

15

Dealing with Uncertainty
in Economy Studies

Because economic decision making is primarily concerned with future events, rather than past events, there is usually an element of uncertainty associated with the alternatives under consideration. Not only are the estimates of future economic conditions problematical, but in addition the anticipated future economic effects of most projects are known with only some degree of assurance. It is this lack of certainty about the future that makes economic decision making one of the more difficult and challenging tasks faced by individuals, industry, and government.

This chapter begins with an introduction to probability theory and its application to decision making. A number of techniques are then presented for utilizing the concepts of probability to quantify uncertainty associated with economic alternatives.

15.1. PROBABILITY AND DECISION MAKING

To formally incorporate uncertainty about future events into a logical decision process it is necessary to utilize probability theory. Probability theory consists of an extensive body of knowledge concerned with the quantitative treatment of uncertainty. Since probability theory is well developed and rigorously defined it is appropriate to apply it to decision problems involving the mathematical consideration of uncertainty.

By using probability theory it is possible to uniquely define events so that no ambiguities exists and so that each statement made within the theory is explicit and clearly understood. Probability theory allows uncertainty to be represented by a number so that the uncertainty of different events can be directly compared. In addition the structure of probability theory prevents the introduction of extraneous notions without full knowledge of the decision maker.

The probability that an event will occur may be expressed by a number that represents the likelihood of the occurence. This likelihood may be determined by examining all available evidence related to the occurrence of the event. Thus, probability can be viewed as a state of mind since it represents our belief about the likelihood of an event occurring.

To illustrate, suppose you were presented with a normal looking coin and you were asked to guess the probability of the coin landing with heads up after it is tossed. Since you have never tossed this particular coin you might answer that the probability of heads occurring is approximately one half. This answer is based on all your past experience of tossing two sided coins which has lead you to believe that the probability of heads occurring is one half. Now suppose you were allowed to toss the coin in question one hundred times and heads occurred 60% of the time. Your belief that this coin will yield heads 50% of the time is somewhat shaken. If you tossed the coin one million times and heads occurred 70% of the time you would be convinced that this coin is not "fair" and that the likelihood of heads occurring is approximately 70%. As your knowledge about the frequency of heads for this coin has increased, your feeling about the likelihood of getting a heads from a toss of this coin has been substantially modified. Thus your information leads you to believe that the probability of a heads occurring when *this* coin is tossed is approximately 0.7.

Accordingly, any decision involving a wager on the basis of a toss of this coin will certainly be affected by your feeling about the likelihood of heads occurring. Since our beliefs about the occurrence of future events are the bases for making investment decisions, the concept of probability as a state of mind is quite useful in the analysis of economic alternatives. In this book it will be assumed that the subjective probabilities just discussed follow the laws of classical probability theory.

15.2. BASIC PROBABILITY RELATIONSHIPS

The three axioms of probability are:

1. For any event A, the probability of A, $P(A) \geq 0$
2. $P(S) = 1$ for the certain event S
3. If $AB = 0$, then $P(A + B) = P(A) + P(B)$

That is, for two mutually exclusive events the probability that event A or B occurs is equal to the sum of their probabilities.

The conditional probability of event A occurring given that event B has occurred is described as the probability of A given B, $P(A/B)$

$$P(A/B) = \frac{P(AB)}{P(B)}$$

where $P(AB)$ equals the probability of both event A and event B occurring. When events A and B are independent

$$P(AB) = P(A)P(B).$$

Thus, the probability of the event (A and B) equals the probability of event A multiplied by the probability of event B. When events A and B are independent, the conditional probability of event A occurring given that event B has occurred equals the probability that event A occurs expressed as

$$P(A/B) = \frac{P(AB)}{P(B)} = \frac{P(A)P(B)}{P(B)} = P(A).$$

Probability Distribution Functions. A *random variable* is a function which assigns a value to each event included in the set of all possible events. For instance if a coin is to be tossed twice a random variable describing the number of heads occurring can have the values 0, 1, or 2. When a random variable is discrete as in the coin tossing example a *probability mass function* is used to describe the probability of the random variable being equal to a particular value.

If the probability of a coin showing heads is considered to be 0.5 the probability mass function for the random variable "number of heads" for two tosses of the coin is found in the following manner. The possible outcomes of the two tosses can be described by the events H_1, heads on 1st toss; H_2, heads on 2nd toss; T_1, tails on 1st toss; and T_2, tails on 2nd toss. The possible outcomes of tossing the coin twice are shown in Column A of Table 15.1. The probability of each outcome is shown in Column B of Table 15.1. These probabilities are easy to calculate since the outcome on the first toss is considered to be independent of the outcome of the second toss. Thus the probabilities in Column B consist of the probability of the given event occurring on the first toss multiplied times the probability of the given event occurring on the second toss. The value of the random variable "number of heads" is shown in Column C.

Table 15.1. RANDOM VARIABLE "NUMBER OF HEADS" FOR TWO TOSSES OF A COIN

Possible Outcomes A	Probability of Occurrence B	Number of Heads C
$H_1\ H_2$	$P(H_1)P(H_2) = (0.5)(0.5) = 0.25$	2
$H_1\ T_2$	$P(H_1)P(T_2) = (0.5)(0.5) = 0.25$	1
$T_1\ H_2$	$P(T_1)P(H_2) = (0.5)(0.5) = 0.25$	1
$T_1\ T_2$	$P(T_1)P(T_2) = (0.5)(0.5) = 0.25$	0

The *probability mass function* can be directly described from the information in Table 15.1. The probability that the random variable "number of heads" has the value 2 is 0.25. The probability of having no heads occurring for two tosses is also 0.25. Since the events H_1T_2 and T_1H_2 are mutually exclusive it is necessary to sum the probabilities of those events to determine the probability that the random variable "number of heads" is equal to 1. Thus the probability of getting one head on two tosses of the coin is 0.5. The probability mass function describing the probability of there being a particular number of heads after two tosses of a coin is shown in Figure 15.1.

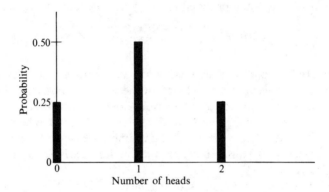

FIGURE 15.1. PROBABILITY MASS FUNCTION.

For any probability mass function the sum of the probabilities for all possible outcomes must total to one and the probability of each possible outcome must be greater than or equal to zero and less than or equal to one. Thus for the random variable x

$$0 \leq P(x) \leq 1 \qquad \sum_x P(x) = 1$$

where $\sum_x$ indicates summing over all possible values of x.

When a random variable is continuous a *probability density function* is used to relate the probability of an event to a value or range of values for the random variable. A probability density function for a random variable x is shown in Figure 15.2.

The probability of an event occurring is described by the area under the probability density function for those values of x included in the event. For example in Figure 15.2 the dark area under the curve for $1 \leq x \leq 3$ represents the probability that the random variable x will occur within that range. In order to satisfy the axioms of probability a probability density function must have the following properties.

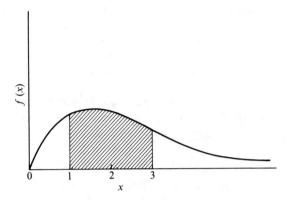

FIGURE 15.2. PROBABILITY DENSITY
FUNCTION.

$$0 \le f(x) < \infty \qquad \text{and} \qquad \int_{-\infty}^{\infty} f(x)\, dx = 1$$

The probability of an event in the range a to b is defined as

$$P(a \le x \le b) = \int_{a}^{b} f(x)\, dx.$$

Mean and Variance. There are two parameters of a probability distribution function that help characterize the function. The first of these parameters is the *mean* or *expected value* of the random variable. For a discrete random variable x the expected value is defined as

$$E(x) = \sum_{x} x P(x).$$

For a continuous random variable x the expected value is defined as

$$E(x) = \int_{-\infty}^{\infty} x f(x)\, dx.$$

The mean of a probability distribution is a measure of central tendency or concentration of mass for the distribution.

The second parameter is a measure of the spread of the probability distribution and it is called the *variance*. The variance for any random variable x is defined as

$$\text{Var}\,(x) = E\{[x - E(x)]^2\} = E(x^2) - [E(x)]^2.$$

For discrete random variables

$$E(x^2) = \sum_{x} x^2 P(x)$$

and

$$[E(x)]^2 = \left[\sum_{x} x P(x) \right]^2.$$

For continuous random variables

$$E(x^2) = \int_{-\infty}^{\infty} x^2 f(x)\, dx$$

and

$$[E(x)]^2 = \left[\int_{-\infty}^{\infty} x f(x)\, dx\right]^2.$$

Calculation of the mean and variance for the probability mass function in Figure 15.1 for the random variable "number of heads" N yields

$$E(N) = \sum_N NP(N) = 0(0.25) + 1(0.50) + 2(0.25) = 1$$

$$\text{Var}\,(N) = E(N^2) - [E(N)]^2 = 0^2(0.25) + 1^2(0.50) + 2^2(0.25) - 1^2 = 0.5$$

To see how these parameters can be useful for decision making consider the following game. A coin will be tossed twice. If no heads occur $100 will be lost, if one head occurs $40 will be won and if two heads appear $80 will be won. What is the expected value of the random variable G, "profit from game"?

$$E(G) = -\$100(0.25) + \$40(0.50) + \$80(0.25) = \$15$$

If one was able to participate in a large number of such games the expected or average winnings per bet over the long run will be approximately $15. Since the expected value reflects the long run expectation from uncertain events it is widely used as a basis for comparison of investment alternatives.

15.3. DECISION MAKING USING KNOWN PROBABILITIES

In many decision making problems the decision maker feels he has sufficient information to reasonably describe the nondeterministic problem parameters by probability distributions. For example a decision to purchase an asset can be dependent on the length of its life, its first cost, its receipts, its salvage value and the interest rate used to consider the time value of money. Usually the values of these parameters are not known with certainty until the termination of the project. If for similar assets there is substantial historical data pertaining to these parameters the decision maker may feel justified in utilizing this data to describe probability distribution functions for these random variables. The technique of decision making where it is possible to assign probabilities to uncertain future events is frequently referred to as decision making under *risk*.

Expected Value Decision Making. It has been shown in Section 15.2 that it is possible to calculate an expected value for a random variable if its probability mass function or probability density function is known. If such probability distributions are used to describe the profit or cost of an investment as a random variable, the expected values of these parameters can provide a reasonable basis for comparing alternatives. That is, the expected profit or cost of an investment proposal reflects the long-term profit or cost that would be realized if the investment were repeated a large number of times and its probability distribution remained unchanged. Thus where large numbers of investments are made it may be reasonable to make decisions based on the average or long-term effects of each proposal. Of course it is necessary to recognize the limitations of using the expected value as a basis for comparison on unique or unusual projects where the long-term effects are less meaningful.

Because most industries and governments are generally long lived the expected value as a basis for comparison seems to be a sensible method for evaluating investment alternatives under risk. The long term objectives of such organizations may include the maximization of expected profits or minimization of expected costs. If it is desired to include the effect of the time value of money where risk is involved all that is required is to state expected profits or costs as expected present-worths, expected annual equivalents, or expected future-worths.

Suppose a firm is planning to introduce a new product which is similar to an existing product. After a detailed study by the marketing and production departments estimates are made of possible future cash flows as related to varying market conditions. The estimates shown in Table 15.2 indicate that this new product will produce cash flow A as demand decreases, cash flow B if demand remains constant and cash flow C if demand increases. Based on past experience and projections of future economic activity it is believed that the probabilities of decreasing, constant, and increasing demand will be 0.1, 0.3, and 0.6, respectively.

Table 15.2. PROBABILITY OF OCCURRENCE FOR NEW PRODUCT PROPOSAL CASH FLOWS

	Probability of Occurrence for Cash Flow		
	A	B	C
Year	$P(A) = 0.1$	$P(B) = 0.3$	$P(C) = 0.6$
0	$-\$30,000$	$-\$30,000$	$-\$30,000$
1	11,000	11,000	4,000
2	10,000	11,000	7,000
3	9,000	11,000	10,000
4	8,000	11,000	13,000

This firm generally uses the present-worth criterion to make such decisions and the minimum attractive rate of return is 10%. By calculating the present-worth for each level of demand the firm develops a probability mass function for the present-worth amount. Because this firm makes numerous decisions where risk is involved they use the expected value of the present-worth amount to determine the acceptability of investment proposals. From Table 15.2 the expected present-worth may be calculated as

$$E[PW(10)] = (0.1)PW_A(10) + (0.3)PW_B(10) + (0.6)PW_C(10)$$

$$= (0.1)\left[-\$30,000 + \$11,000(\overset{P/A\ 10,\ 4}{3.170}) \right.$$

$$\left. - \$1,000(\overset{A/G\ 10,\ 4}{1.3812})(\overset{P/A\ 10,\ 4}{3.170}) \right]$$

$$+ (0.3)\left[-\$30,000 + \$11,000(\overset{P/A\ 10,\ 4}{3.170}) \right]$$

$$+ (0.6)\left[-\$30,000 + \$4,000(\overset{P/A\ 10,\ 4}{3.170}) \right.$$

$$\left. + \$3,000(\overset{A/G\ 10,\ 4}{1.3812})(\overset{P/A\ 10,\ 4}{3.170}) \right]$$

$$= (0.1)(\$492) + (0.3)(\$4,870) + (0.6)(-\$4,185)$$

$$= -\$998.$$

The expected value for the present-worth of this proposal is $-\$998$ and therefore the proposal is rejected. If the expected value had been positive this proposal would be considered acceptable.

Sometimes it is the life of an asset that is the unknown factor that affects the ultimate earnings produced by the asset. Suppose an investment proposal requiring a \$5,000 outlay will return an annual profit of \$3,000 per year for as long as the asset is productive. The probabilities in Table 15.3 describe the likelihood that the asset will be productive for a specific number of years.

Table 15.3. THE PROBABILITY AN ASSET WILL FUNCTION FOR EXACTLY *N* YEARS

	Number of Years Asset Functions (*N*)					
	1	2	3	4	5	6
Probability Asset is Productive Exactly *N* Years	0.1	0.2	0.2	0.3	0.1	0.1

The expected present-worth for the asset just described is found as follows for an interst rate of 12%.

$$E[PW(12)] = -\$5,000 + (0.1)\left[\$3,000(\overset{P/A\ 12,1}{0.893}) \right] + (0.2)\left[\$3,000(\overset{P/A\ 12,2}{1.690}) \right]$$

$$+ (0.2)\left[\$3,000(\overset{P/A\ 12,3}{2.402}) \right] + (0.3)\left[\$3,000(\overset{P/A\ 12,4}{3.037}) \right]$$

$$+ (0.1)\left[\$3,000(\overset{P/A\ 12,5}{3.605}) \right] + (0.1)\left[\$3,000(\overset{P/A\ 12,6}{4.111}) \right]$$

$$= \$2,771.$$

The positive expected value of present-worth for this investment situation indicates that implementation of this project is desirable.

A company has a large water treatment facility located in the flood plain of a river. The construction of a levee to protect the facility during periods of flooding is under consideration. Data concerning the costs of construction and expected flood damages are shown in Table 15.4.

Table 15.4 PROBABILITY AND COST INFORMATION FOR DETERMINING OPTIMUM LEVEE SIZE

Feet (x) A	Number of years river maximum level was x feet above normal B	Probability of river being x feet above normal C	Loss if river level is x feet above levee D	Initial Cost building levee x feet high E
0	24	0.48	$ 0	$ 0
5	12	0.24	100,000	100,000
10	8	0.16	150,000	210,000
15	3	0.06	200,000	330,000
20	2	0.04	300,000	450,000
25	1	0.02	400,000	550,000
	50̄	1.00		

Using historical records which describe the maximum height reach by the river during each of the last fifty years, the frequencies shown in Column *B* of Table 15.4 were ascertained. From these frequencies are calculated the probabilities that the river will reach a particular level in any one year. The probability for each height is determined by dividing the number of years for which each particular height was the maximum by fifty, the total number of years.

The damages that are expected if the river exceeds the height of the levee are related to the amount by which the river height exceeds the levee. These costs are shown in Column *D*. It is observed that they increase in relation to the amount the flood crest exceeds the levee height. If the flood crest is 15 feet and the levee is 10 feet the anticipated damages will be $100,000, whereas a flood crest of 20 feet for a levee 10 feet high would create damages of $150,000.

The costs of constructing levees of various heights are shown in Column *E*. The company considers 12% to be their minimum attractive rate of return and it is felt that after 15 years the treatment plant will be relocated away from the flood plain. The company wants to select the alternative that minimizes its total expected costs. Since the probabilities are defined as the likelihood of a particular flood level in any one year, the expected equivalent annual costs is an appropriate choice for the basis for comparison.

An example of the calculations required for each levee height is demonstrated for two of the alternatives.

5 Foot Levee

$$A/P \ 12,5$$
Annual Investment Cost $= \$100,000(\ 0.1468 \ )$ $= \$14,682$
Expected Annual Damage $= (0.16)(\$100,000) + (0.06)(\$150,000)$
$\qquad\qquad\qquad\qquad + (0.04)(\$200,000) + (0.02)(\$300,000)$ $= \ \underline{\ 39,000}$

Total expected annual cost $\qquad\qquad\qquad\qquad\qquad\qquad\qquad \$53,682$

10 Foot Levee

$$A/P \ 12,15$$
Annual Investment Cost $= \$210,000(\ 0.1468 \ )$ $= \$30,828$
Expected Annual Damage $= (0.06)(\$100,000) + (0.04)(\$150,000)$
$\qquad\qquad\qquad\qquad + (0.02)(\$200,000)$ $= \ \underline{\ 16,000}$

Total Expected Annual cost $\qquad\qquad\qquad\qquad\qquad\qquad\qquad \$46,828$

The cost associated with the alternative levee heights are summarized in Table 15.5. The levee height that minimizes the total expected annual costs

Table 15.5. SUMMARY OF ANNUAL CONSTRUCTION AND FLOOD DAMAGE COSTS

Levee Height (Feet)	Annual Investment Cost	Expected Annual Damage	Total Expected Annual Costs
0	$ 0	$80,000	$80,000
5	14,682	39,000	53,682
10	30,828	16,000	46,828
15	48,450	7,000	55,450
20	66,069	2,000	68,069
25	80,751	0	80,751

is the levee which is 10 feet in height. The selection of a smaller levee would not provide enough protection to offset the reduced construction costs while a levee larger than 10 feet requires more investment without providing proportionate savings from expected flood damage. The use of expected value in determining the cost of flood damage is reasonable in this case since the 15-year period under consideration allows time for long-term effects to appear.

Use of Probability Distributions for Comparison of Alternatives. Up to now the techniques discussed have relied on the expected value as a basis for considering investments with risk. In many decision situations it is desirable not only to know the expected value of the basis for comparison but to also have a measure of the dispersion of its probability distribution. The variance of a probability distribution provides such a measure and its value in decision making will become evident in the following example.

Suppose a firm has a set of four mutually exclusive alternatives from which one is to be selected. The probability mass functions describing the likelihood of occurrence of the present-worth amounts for each alternative are described in Table 15.6. For example, the probability that Alternative $A2$ will realize a cash flow that has a present-worth equal to $60,000 is 0.4 while the probability that it will have a $110,000 present-worth is 0.2. The expected present-worth and the variance of each probability distribution is shown in Table 15.6. These values are calculated by using the definitions in Section 15.2. For example, the expected present-worth for Alternative $A2$ is

$$E(PW_{A2}) = (0.1)(-\$40,000) + (0.2)(\$10,000) + (0.4)(\$60,000)$$
$$+ (0.2)(\$110,000) + (0.1)(\$160,000) = \$60,000$$

while the variance is

$$\text{Var}\,(PW_{A2}) = E(PW_{A2}^2) - [E(PW_{A2})]^2$$
$$= (0.1)(-\$40,000)^2 + (0.2)(\$10,000)^2 + (0.4)(\$60,000)^2$$
$$+ (0.2)(\$110,000)^2 + (0.1)(\$160,000)^2$$
$$- (\$60,000)^2$$
$$= \$3,000 \times 10^6.$$

Table 15.6. PROBABILITY DISTRIBUTIONS OF PRESENT-WORTH AMOUNTS FOR FOUR ALTERNATIVES

	Present-Worth					Expected Present-Worth	Variance (000,000)
Alternatives	-$40,000 A	$10,000 B	$60,000 C	$110,000 D	$160,000 E	F	G
Alternative $A1$	0.2	0.2	0.2	0.2	0.2	$60,000	$5,000
Alternative $A2$	0.1	0.2	0.4	0.2	0.1	$60,000	$3,000
Alternative $A3$	0.0	0.4	0.3	0.2	0.1	$60,000	$2,500
Alternative $A4$	0.1	0.2	0.3	0.3	0.1	$65,000	$3,850

An examination of the expected present-worth values in Figure 15.6 indicates that there is relatively little difference among the alternatives. However, an examination of the distribution of possible present-worths for each alternative gives additional insight into the desirability of each alternative. First it is important to determine the probability that the present-worth of each alternative will be less than zero. This probability represents the likelihood of the investment yielding a rate of return less than the minimum attractive rate of return. From Table 15.6, Column A

$$P(PW_{A1} \leq 0) = 0.2$$
$$P(PW_{A2} \leq 0) = 0.1$$
$$P(PW_{A3} \leq 0) = 0.0$$
$$P(PW_{A4} \leq 0) = 0.1$$

Since it is desirable to minimize these probabilities, it appears that on this basis Alternative $A3$ is most desirable.

The second important consideration is the variance of these probability distributions. Because the variance indicates the dispersion of the distribution it is usually desirable to try to minimize the variance since the smaller the variance the less the variability or uncertainty associated with the random variable. Alternative $A3$ has the minimum variance as shown in Table 15.6.

It is clear that Alternative $A3$ is more desirable than $A1$ or $A2$ since their expected present-worths are the same and $A3$ is preferred on the basis of the two criteria just discussed. However, the decision is not so obvious when comparing Alternatives $A3$ and $A4$. Alternative $A4$ has a larger expected present-worth but it is not as desirable as $A3$ on the basis of minimum variance and minimum chance of the present-worth being less than zero. In cases such as these the decision maker must weight the importance of each factor and decide if he would prefer more variability in the possible outcomes in order to achieve a higher expected value or less chance of the present-worth being negative. It is possible the relative importance of these three factors could be quantified and then a single basis for comparison could be developed for each alternative.

By having the additional information that can be derived from probability distributions it is likely that a more intelligent decision can be made. Of course the astute decision maker must balance the economic trade-off between the cost of developing better information for decision making and the saving that he hopes to realize from better selection of alternative. Thus it may not be economic to use elaborate techniques to consider small projects while on the other hand the use of more sophisticated analyses may provide substantial payoffs when very large expenditures are being considered.

15.4. DECISION TREES IN THE EVALUATION OF ALTERNATIVES

In many decision-making problems it is desirable to recognize that future decisions are affected by actions that are taken at the present. Too often decisions are made without consideration of their long-term effects. As a result decisions which initially appeared sound may place the decision maker in an unfavorable position with respect to future decisions. For decision problems where consideration of sequences of decisions is important and probabilities of future events are known, the use of *decision-flow diagrams* or *decision trees* for analysis is usually a very effective technique.[1]

The application of decision trees to investment problems will be illustrated by the analysis of the following problem.[2] Suppose a firm is planning to pro-

---

[1] For more detailed information concerning decision trees see "Decision Trees for Decision Making" by J. F. Magee, *Harvard Business Review*, July-August, 1964.

"How to Use Decision Trees in Capital Investment" by J. F. Magee, *Harvard Business Review*, September-October, 1964.

Decision Analysis, by Howard Raiffa, Addision-Wesley Publishing Co., Inc., 1968.

[2] This problem suggested by an example in "Decision Trees for Decision Making" by J. F. Magee, *Harvard Business Review*, July-August, 1964.

duce a product that has never been marketed previously. Because this product is somewhat different from the firm's existing products it will be necessary to construct a separate production facility to manufacture this new product.

Based on information supplied by the firm's marketing group it is believed that the demand for this new product will be significant over the next 10 years. If the product is to be a good seller it is believed that over the next ten years there will be a high demand for the product. If the product becomes a fad then it is anticipated that the demand will be high for the first two years followed by a low demand for the remaining eight years. If the product is a poor seller then the demand is expected to be low for the next ten years. Thus, the demand for this new product is expected to follow one of three demand patterns.

	Demand	Period*	Demand	Period	Designation
Good Seller:	High	1st;	High	2nd	(H_1, H_2)
Fad:	High	1st;	Low	2nd	(H_1, L_2)
Poor Seller:	Low	1st;	Low	2nd	(L_1, L_2)

*In this example the first two years are designated as the first period while the remaining eight years are considered to be the second period.

The first alternative is to build a large plant that would suffice for the full ten years of the product's life. The second alternative is to build a small plant and after two years of observing the product sales make a decision whether to expand the small plant. Which alternative to select is the decision problem confronting management.

The use of a decision tree to display these alternatives along with the possible chance events (demand) that can occur is demonstrated in Figure 15.3. Beginning at the left and moving to the right the decision tree spans 10 years, the life of the product. The nodes of the tree from which the tree's branches emanate are either decision nodes, ☐, or chance nodes, ○. The branches that emanate from a decision node represent alternative courses of actions about which the decision maker must make a choice. On the other hand the branches leaving the chance nodes represent chance events which represent outcomes of Nature. The occurrence of a chance event can be considered to be a random variable over which the decision maker has no control. Chance events are usually controlled by exogenous forces such as weather, sun spots, the market place, etc. In our example the chance events represent the demand for the product over the next ten years.

Starting at the left, the first node represents the choice of building a big plant or a small plant. If a big plant is built, then the next possible events are the chance events which represent the possibility of experiencing a high demand or a low demand for the first two years. If a big plant is built and a high demand is experienced for the first two years, the next chance node represents the possibility of a high demand occurring or low demand

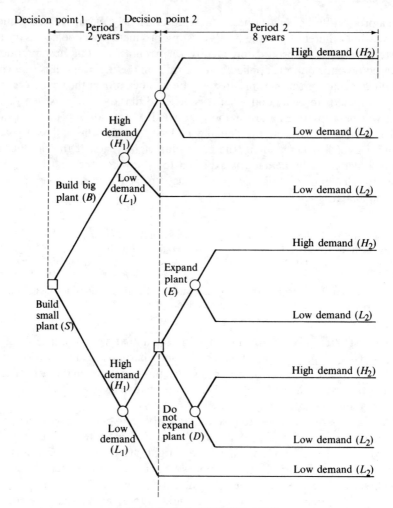

FIGURE 15.3. DECISION TREE.

occurring for the remaining eight years. Thus one sequence of branches beginning at the left and ending at the right represents one of the possible sequence of events that can result from actions by the decision maker and Nature.

Notice that if the small plant is built the decision about whether to expand or not to expand is made after two years only if there is a high demand. If demand is low in the first two years it is known that demand will be low for the remaining eight years. Therefore the decision not to expand is obvious and no decision node is required following low demand in the first two years.

Once the structure of the decision tree is determined the next task is to ascertain the costs and revenues that are associated with each of the decision alternatives and the possible chance outcomes. These amounts should then be written on the appropriate branches of the tree. For our example the costs of plant construction and the net profits expected from sale of the product for the various market conditions and plant sizes are shown in Figure 15.4. Thus the investment required to build the big plant is seen to be $4,000,000 while the net profit for the first two years is $861,000 per year if the big plant is built. If the big plant is built and there is high demand in the first two year period, the net profit is $1,650,000 per year for the remaining eight years, if there is high demand in Period 2.

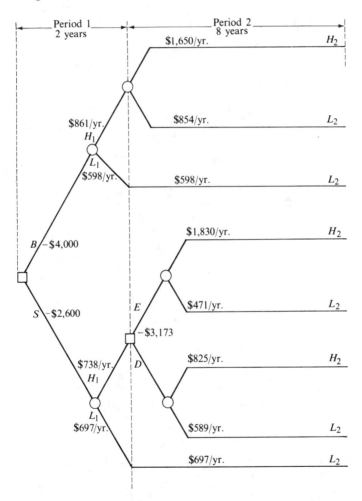

FIGURE 15.4. FORECASTS OF RECEIPTS AND DISBURSEMENTS
PLACED ON DECISION TREE (000'S).

The cost to build the small plant is $2,600,000 and its expansion cost is $3,173,000 if the expansion is undertaken. All the costs are shown as negative values on the decision tree while the revenues are positive amounts. The amounts that are shown on a per year basis represent the annual net profit received during the two years of Period 1 or the 8 years of Period 2.

Since the costs and revenues occur at different points in time over the 10-year study period it is appropriate to convert the various amounts on the tree's branches to their equivalent amounts. For the problem being considered the MARR is 15% and Figure 15.5 shows the receipts and disbursements on the branches transformed to their present-worth equivalents.

To illustrate, if the big plant is built, the present-worth of the net revenues that result from a high demand for the first two years is

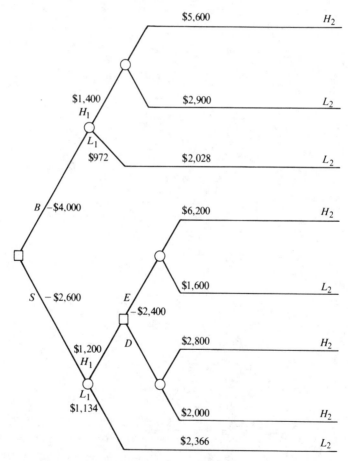

FIGURE 15.5. PRESENT-WORTH AMOUNTS OF RECEIPTS AND DISBURSEMENTS (000'S).

$$PW(15) = \$861,000(\overset{P/A\ 15,\ 2}{1.626}) = \$1,400,000.$$

If the big plant is built and high demand is experienced for the remaining eight years the present-worth of those revenues is

$$PW(15) = \$1,650,000(\overset{P/A\ 15,\ 8}{4.487})(\overset{P/F\ 15,\ 2}{0.7561}) = \$5,600,000.$$

Now that the present-worths of costs and revenues have been determined it is possible to sum these figures along each possible sequence of branches, starting each time at the left-most node. Thus if the big plant is built, the demand is high in Period 1 and Period 2, the total present-worth of that sequence of events is $-\$4,000,000 + \$1,400,000 + \$5,600,000 = \$3,000,000$. This amount is placed at the tip of the right-most branch representing such an outcome. This procedure is repeated for each possible sequence of branches and the resulting amounts are shown in Figure 15.6. If the small plant is built, demand is high for Period 1, the plant is expanded and demand is low in Period 2 the net present-worth of such a sequence of events is $-\$2,600,000 + \$1,200,000 - \$2,400,000 + \$1,600,000 = -\$2,200,000$.

Nature's Tree. Once the cost and revenue information is in the form shown in Figure 15.6, it is necessary to place the probabilities of the chance events occurring at the chance nodes on the decision tree. Suppose that the probabilities of the product being a good seller, a fad, or a poor seller are estimated by the marketing department. These estimated probabilities are

$$P[\text{Good Seller}] = P[H_1 H_2] = \tfrac{2}{5}$$
$$P[\text{Fad}] \qquad\quad = P[H_1 L_2] = \tfrac{1}{5}$$
$$P[\text{Poor Seller}] \ = P[L_1 L_2] = \tfrac{2}{5}.$$

Whenever it is necessary to compute the probabilities for a decision tree it is usually very helpful to first construct Nature's tree; that is, construct a tree that indicates Nature's options as shown in Figure 15.7. In Nature's tree all the nodes are chance nodes. At the beginning of each branch is placed the probability of following that branch given that you are at the node preceding that branch. The probability that branch H_2 is selected given that there has been high demand in Period 1 (H_1) is the conditional probability $P[H_2 \mid H_1]$ and it is placed at the beginning of the H_2 branch that radiates from the H_1 branch. At each tip of Nature's tree is placed the probability that the sequence of events represented by that tip will occur. These probabilities are calculated by multiplying the probabilities on all the branches that lead from the initial node on Nature's tree to each of the tips. Thus to find the probability that H_1 and L_2 occurs, the probability on branch H_1 is multiplied times the probability on branch L_2 following H_1. This calculation gives $P(H_1 L_2) = P(H_1)P(L_2 \mid H_1)$.

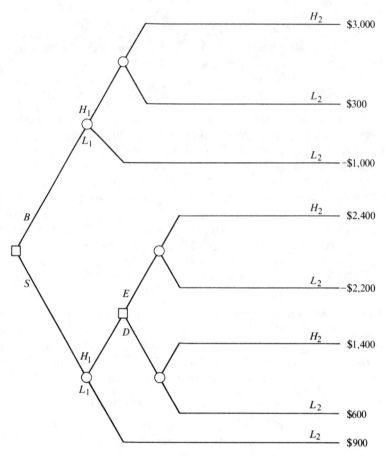

FIGURE 15.6. TOTAL PRESENT-WORTH AMOUNTS FOR EACH POSSIBLE OUTCOME (000'S).

The probabilities that are required for the decision tree are $P(H_1)$, $P(L_1)$, $P(H_2 | H_1)$, and $P(L_2 | H_1)$. By having Nature's tree and the probabilities for the tips of the tree, all that is required to find $P(H_1)$ is to *add* each of the probabilities at the tips that contains an event H_1. Thus,

$$P(H_1) = P(H_1 H_2) + P(H_1 L_2) = \tfrac{2}{5} + \tfrac{1}{5} = \tfrac{3}{5}.$$

Similarly

$$P(L_1) = P(L_1 H_2) + P(L_1 L_2) = 0 + \tfrac{2}{5} = \tfrac{2}{5}.$$

With the $P(H_1)$ now known the conditional probabilities are calculated in the following manner.

$$P(H_2 | H_1) = \frac{P(H_1 H_2)}{P(H_1)} = \left(\frac{\frac{2}{5}}{\frac{3}{5}}\right) = \frac{2}{3}$$

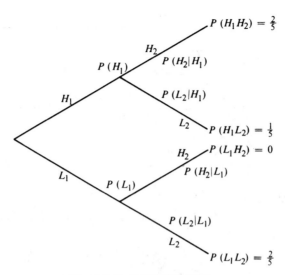

$P(H_1 H_2) = \frac{2}{5}$

H_2

$P(H_1)$ $P(H_2|H_1)$

H_1 $P(L_2|H_1)$

L_2
$P(H_1 L_2) = \frac{1}{3}$

H_2 $P(L_1 H_2) = 0$

$P(H_2|L_1)$

L_1

$P(L_1)$

$P(L_2|L_1)$

L_2
$P(L_1 L_2) = \frac{2}{5}$

FIGURE 15.7. NATURE'S TREE.

$$P(L_2|H_1) = \frac{P(H_1 L_2)}{P(H_1)} = \left(\frac{\frac{1}{5}}{\frac{3}{5}}\right) = \frac{1}{3}$$

These probabilities are then placed at the appropriate chance nodes on the decision tree as is shown in Figure 15.8. For example, the chance node following the decision to expand the small plant has two branches H_2 and L_2. The probability of taking the H_2 branch is the probability that H_2 occurs *given* that H_1 has already occurred. Therefore $P(H_2|H_1) = \frac{2}{3}$ is placed on the H_2 branch and $P(L_2|H_1) = \frac{1}{3}$ is placed on the L_2 branch. For the case where the demand is low in Period 1 the probability $P(L_1) = P(L_1 L_2) = \frac{2}{5}$ since $P(L_2|L_1) = 1$ and $P(H_2|L_1) = 0$.

The "Rollback" Procedure. At this point it is now possible to solve the decision tree in order to see which alternative should be undertaken. The solution technique is relatively simple and it is referred to as the "rollback" procedure. Starting at the tips of the decision tree's branches and working back toward the initial node of the tree the following two rules are used.

1. If the node is a chance node calculate the *expected* value of that node based on the "rolled backed" values on the adjacent nodes to the right of the node being considered.
2. If the node is a decision node *select* the maximum profit or minimum cost from the adjacent nodes to the right of the node being considered.

As each node is considered starting at the right of the decision tree the values calculated for rules 1 and 2 should be placed just to the right of the

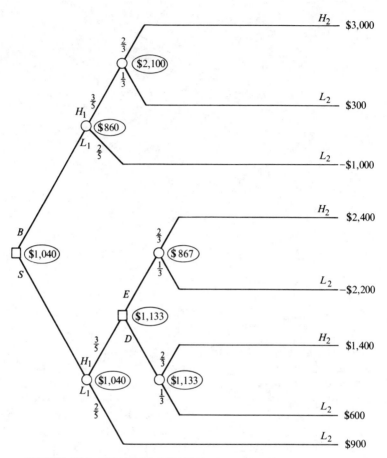

FIGURE 15.8. ROLLBACK SOLUTION OF DECISION TREE (000'S).

node and circled. By working backwards through the decision tree certain alternatives can be eliminated from further consideration and thus the "rollback" procedure is efficient for large decision trees.

To illustrate the rollback technique let's return to Figure 15.8. Starting at the top of the decision tree it is seen that the first node to the left is a chance node and the expected value is $\frac{2}{3}(\$3,000,000) + \frac{1}{3}(\$300,000) = \$2,100,000$. The next chance node that is right-most is the node following the plant expansion branch. The expected value for this node is $\frac{2}{3}(2,400,000) + \frac{1}{3}(-\$2,200,000) = \$867,000$. The other right-most node is a chance node and its expected value is $\frac{2}{3}(\$1,400,000) + \frac{1}{3}(\$600,000) = \$1,133,000$.

The right-most node that has yet to be evaluated is now the decision node concerning the expansion of the small plant. Based on the previous calculations it is seen that if the plant is expanded the expected future profits will be $867,000 while if the plant is not expanded the expected future profits will be $1,133,000. Therefore the optimum policy at this node is to not expand the small plant. The $1,133,000 figure is written after the decision node indicating the expected profit that will be realized if the optimum policy is followed at that node.

Now the next nodes to the left that must be evaluated by the rollback procedure are chance nodes. The expected values at those nodes are calculated from the values on the adjacent nodes to the right that were previously calculated. These two expected values are

$$\tfrac{3}{5}(\$2,100,000) + \tfrac{2}{5}(-\$1,000,000) = \$860,000$$

and $$\tfrac{3}{5}(\$1,133,000) + \tfrac{2}{5}(\$900,000) = \$1,040,000.$$

The last node to be evaluated is the initial decision node. If the big plant is built the expected profits will be $860,000 while if the small plant is built the expected profit is $1,040,000. (This expected profit figure assumes the optimum policy is followed in the future). Therefore the maximum figure is selected for the decision node and the optimum policy is to build the small plant. If the demand is high in Period 1, then the decision not to expand the plant should be followed. If demand is low in Period 1, no expansion of the small plant is the only course of action. If this policy is followed then the expected profit from this venture is $1,040,000. This present-worth amount is positive so this assures the firm an expected return of better than 15% in addition to being the best of the alternatives under consideration.

The Expected Value of Perfect Information. With the information that is available to the firm in our example it is seen that by following an optimal policy the expected present-worth of the profit is $1,040,000. If the firm could obtain additional information about the occurrence of future demand levels it may be that the profit figure could be improved. Usually such additional information costs money since the firm must allocate their resources for further research or they must purchase the information from sources external to the firm (i.e. market research consultants).

It is foolish to pay for additional information if it will not provide additional profits that exceed the cost of the information. In order to evaluate the possible benefits that can be derived from additional information it is necessary to calculate the *expected value of perfect information* (EVPI).

Suppose there is a market research group that does have perfect information about the future demand levels for the product being considered in this example. What would be the most money that the firm would be willing to pay for this additional information? Assume the firm knew with certainty that there would be high demand in both periods (H_1, H_2). Then the optimum policy to follow is to build the big plant since that payoff ($3,000,000) exceeds the payoffs possible from building the small plant and expanding ($2,400,000) or building the small plant and not expanding ($1,400,000). If the demand were known to be (H_1, L_2) then the best strategy is to build the small plant and not expand since that payoff ($600,000) exceeds the returns received if the big plant is built ($300,000) or if the small plant is constructed and then expanded ($-$2,200,000). Thus for each possible state of nature it is necessary to determine the strategy that will maximize the payoff. For our example the three states of nature are (H_1, H_2) (H_1, L_2) and (L_1, L_2) and the maximum payoffs for each state of nature are indicated on Figure 15.9 by the boxes at the tips of the decision tree.

Before receiving the perfect information from the market research group the firm can calculate the *expected profit with perfect information* (EPPI). This is accomplished by summing for each possible state of nature the probability that a particular state will occur multiplied by the maximum

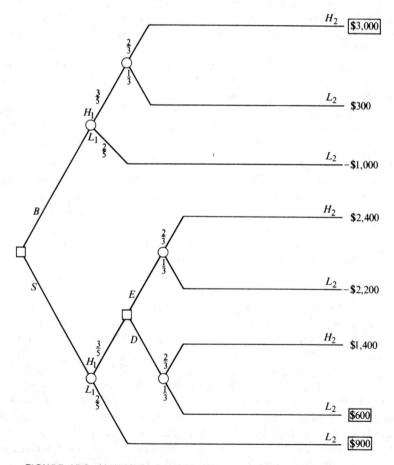

FIGURE 15.9. MAXIMUM PAYOFF FOR THE POSSIBLE OUTCOMES.

payoff achievable for that state of nature. These calculations are shown in Table 15.7 and it is seen that the expected profit for perfect information is

Table 15.7. CALCULATING THE EXPECTED PROFIT FOR PERFECT INFORMATION

State of Nature	Best strategy	Maximum payoff	Probability the state of nature occurs	Expected payoff for each state
A	B	C	D	$C \times D$
H_1, H_2	Big plant	\$3,000,000	$(3/5)(2/3) = 2/5$	\$1,200,000
H_1, L_2	Small plant; no expansion	600,000	$(3/5)(1/3) = 1/5$	120,000
L_1, L_2	Small plant with expansion	900,000	$(2/5)(1)\ \ = 2/5$	180,000
Expected profit with perfect information (EPPI) = \$1,500,000				

\$1,500,000. Since the firm can achieve expected profits of \$1,040,000 without the additional information, the expected value of perfect information is \$1,500,000 − \$1,040,000 = \$460,000. The firm should never pay over \$460,000 for additional information even if that information predicts the future with certainty. Thus the EVPI gives an upper limit reflecting the firms expected improvement in profits if a source of perfect information were available. Information that is less than perfect can only provide a profit improvement that is less than the expected value of perfect information.

15.5. DECISION MAKING WHEN PROBABILITIES ARE UNKNOWN

For many decision problems a particular decision can provide a variety of payoffs because of the possible outcomes of future events. A decision to hold a lawn party can produce a high degree of satisfaction for the guests if the day is sunny while this level of satisfaction would be considerably lower if it rained. These levels of satisfaction would probably be reversed if the decision was made to hold the party inside the house. Thus for the two states of nature, sun or rain, there are differing payoffs depending on the decision that is made concerning the alternative locations for the party.

Decision theory consists of a number of decision rules that evaluate decision alternatives when their payoffs are a function of possible future outcomes or states. These decision rules are applied to a *payoff* matrix that describes for each alternative under consideration the payoff expected for each possible future state that may occur. The following payoff matrix presents the various payoffs (expected present-worths) for three possible future states and four decision alternatives. The likelihood of occurrence of a particular state is considered to be unknown.

Future Possible States

		S_1	S_2	S_3	
	$A1$	$ 8	$ 7	$ 4	
Alternatives	$A2$	10	0	4	(Payoff in thousands of
	$A3$	1	9	5	dollars)
	$A4$	5	6	7	

If Alternative $A1$ is undertaken and the state of nature that occurs is S_2, then the net present-worth of the profit is expected to be $7,000. Similarly, if Alternative $A4$ is implemented and Nature decrees state S_1, the expected return is seen to be $5,000. Each row of payoffs represents the outcomes expected for each state of nature (columns) for a particular alternative (rows). The above payoff matrix will be used to demonstrate the various decision rules that may be helpful in such a decision making situation.

The Maximin or Minimax Rule. The *maximin* decision rule is based on a pessimistic view of the outcomes of nature. That is, the use of this rule implies that it is believed that Nature is going to do her worst. Therefore, the maximin rule chooses the alternative that assures the best of the worst possible outcomes. For the payoff matrix given the maximin rule requires that the minimum value in each row be selected. Then the maximum value of the minimum profit values is chosen and thus $A4$ is the alternative selected. This alternative assures the decision maker of at least $5,000 regardless of Nature's actions.

Alternative	Minimum Profit
$A1$	$4,000
$A2$	0
$A3$	$1,000
$A4$	$5,000* (Maximum)

The Maximax Rule. The *maximax* decision rule is based on the optimistic view that the best state of nature occurs for the alternative accepted. Nature is considered to be benevolent rather than malevolent as it was viewed for the maximin rule. The maximax decision rule requires that the maximum values for each row of the payoff matrix be selected. From these values the maximum value is selected. Using this rule Alternative $A2$ is selected.

Alternative	Maximum Profit
$A1$	$8,000
$A2$	10,000* (Maximum)
$A3$	9,000
$A4$	7,000

The Hurwicz Rule. Because the maximin rule and the maximax rules are extremely pessimistic and optimistic, there appears to be a need for a decision rule that recognizes some middle ground. A compromise between these two extremes is accomplished by allowing the decision maker to select an "index of optimism", α, such that $0 \le \alpha \le 1$. When $\alpha = 0$ the decision maker is pessimistic about Nature while an $\alpha = 1$ indicates optimism about Nature.

Once α is selected the Hurwicz rule requires that for each alternative the expression

$$\alpha[\max_j (V_{ij})] + (1 - \alpha)[\min_j (V_{ij})]$$

be computed where V_{ij} is the payoff for the ith alternative and the jth state of nature. The alternative that possesses the *maximum* value for this expression is the one considered to be most favorable.

For the payoff matrix being considered the Hurwicz rule requires the selection from the following expressions the one that yields a maximum value. Assume $\alpha = 0.15$.

Alternative	$\alpha[\max (V_{ij})] + (1 - \alpha)[\min (V_{ij})]$
A1	(0.15)($8,000) + (0.85)($4,000) = $4,600
A2	(0.15)(10,000) + (0.85)(0) = $1,500
A3	(0.15) (9,000) + (0.85)(1,000) = $2,200
A4	(0.15) (7,000) + (0.85)(5,000) = $5,400

Thus, Alternative A4 is the alternative that is selected by the Hurwicz decision rule for a rather pessimistic outlook about future states of nature. It should be observed that the Hurwicz rule is the same as the *maximin* rule and the *maximax* rule when $\alpha = 0$ and $\alpha = 1$, respectively.

An additional insight into the Hurwicz rule can be obtained by graphing the expressions for each alternative for all the values of α between zero and one. Then it is possible to see the values of α for which a particular alternative is favored. Such a graph is presented in Figure 15.10. From this graph it is seen that Alternative A4 is maximum for $\alpha \le \frac{1}{2}$ with Alternative A1 being maximum for $\frac{1}{2} \le \alpha \le \frac{2}{3}$ and Alternative A2 being maximum for $\frac{2}{3} \le \alpha \le 1$. Notice that there is no value of α for which Alternative A3 is the most favorable alternative.

The Savage rule or minimax regret rule. The *minimax regret* rule is based on the idea that a decision maker wishes to avoid any regret about his decision. If he selects an alternative and the state that occurs is such that he could have done better having selected another alternative he "regrets" his original decision. These regrets are measured as the difference between the best he could achieve with perfect knowledge of the future states and the payoff received for a particular alternative and a future state. Knowing the regret for each alternative and future state the decision maker then finds the maximum possible regret that could occur for each alternative. He then selects the alternative that minimizes his maximum regret.

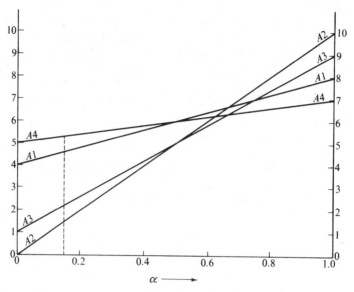

FIGURE 15.10. VALUES FOR THE HURWICZ RULE REPRESENTING FOUR ALTERNATIVES.

To apply the minimax regret rule it is first necessary to compute a *regret matrix*. For each state (column) find the maximum payoff (minimum payoff if the payoffs are costs). From the payoff matrix these values are 10, 9, 7 for S_1, S_2, and S_3 respectively. For each column in the payoff matrix subtract each payoff from the maximum payoff for that column. Thus the regrets for S_1 are $10 - 8 = 2$, $10 - 10 = 0$, $10 - 2 = 8$ and $10 - 5 = 5$ for alternatives $A1$, $A2$, $A3$, $A4$ respectively. Repeating this procedure for each state produces the regret matrix shown below. Now find the maximum

	Future Possible States			
	S_1	S_2	S_3	*Maximum Regret*
$A1$	2	2	3	$3000 (Minimum)
$A2$	0	9	3	9000
$A3$	9	0	2	9000
$A4$	5	3	0	5000

Alternatives

regret for each alternative by selecting the maximum value for each row and placing these values to the right of the regret matrix. Select the alternative which minimizes the maximum regret.

The Laplace Rule. The use of the Laplace rule indicates that the decision maker feels that each future state has the same chance of occurring. Thus for this decision rule the probability of the occurrence of each future state is $1/n$ where n is the number of possible future states.

To determine the alternative to select the expected value or the average of the payoffs for each alternative is computed from the payoff matrix.

Alternative	Average Payoff
A1	$6333 (Maximum)
A2	4667
A3	5000
A4	6000

The maximum value of $6,333 indicates that Alternative A1 is to be selected by the Laplace rule.

Summary. These decision rules are presented to give insights into the types of techniques that may be helpful when probabilistic data are not available. An examination of the outcomes of these five decision rules for the payoff matrix presented indicates that each of the alternatives meets various objectives with varying degrees of success. Thus the choice of the appropriate decision rule for a particular decision problem is dependent on the outlook and objectives of the decision maker. In fact for some problems it may be necessary to develop decision rules that meet objectives other than those embodied in the five decision rules just discussed.

PROBLEMS

1. Because of uncertainties in construction costs, the cost of an automobile tunnel is a random variable described by

Cost (x)	Probability that cost equals x
$10,000,000	0.2
14,000,000	0.4
16,000,000	0.3
20,000,000	0.1

What is the *expected value* of the tunnel cost?

2. A company is considering the purchase of a concrete plant for $4,000,000. The success of the plant depends on the amount of highway construction undertaken over the next 5 years. It is known that there are three possible levels (A, B, and C) of federal support for the construction of new highways. Shown below are the receipts and disbursements (in millions of dollars) the company expects from the concrete plant for each level of government support.

		End of Year			
	0	1	2	3	4
Support Level:					
A	− $4.0	$2.0	$2.0	$2.0	$2.0
B	− 4.0	1.0	1.0	1.0	1.0
C	− 4.0	0.5	0.5	1.0	2.5

The probabilities that support levels A, B, and C are realized over the next 4 years are 0.3, 0.1, 0.6, respectively. If the minimum attractive rate of return is 12%, what is the expected present-worth of this investment opportunity?

3. Uncertainty as to the rate of technological innovation means that a proposed computer system will have a service life which is unknown. If the initial cost of developing a computer system is $10 million and each year of service life produces net revenues of $4.5 million, what is the expected present-worth for an interest rate of 10% if the lifetime of the system is described by the following probabilities:

Lifetime (Years)	Probability
1	0.1
2	0.2
3	0.4
4	0.2
5	0.1

4. A company is considering the introduction of a "new" umbrella although the future success of this product is uncertain. The product will be sold for only 5 years, and if the 5 years are "wet" years, the following cash flows are anticipated.

		Year					
Probability	Cash Flow	0	1	2	3	4	5
0.6	W1	−$4	$2	$3	$2	$1	$0
0.4	W2	− 4	1	2	4	3	2

On the other hand, if the next 5 years are "dry" years, the cash flows are likely to be

		Year					
Probability	Cash Flow	0	1	2	3	4	5
0.5	D1	−$4	$0	$1	$1	$1	$1
0.5	D2	− 4	1	1	2	1	1

The interest rate is 10%.

(a) What is the expected present-worth if the next 5 years are going to be "wet"?

(b) What is the expected present-worth if the next 5 years are going to be "dry"?

(c) If the best available information indicates that the next 5 years will be "wet" with probability 0.7,

    (i) What is the probability that the next 5 years will be "dry"?

    (ii) What is the expected present worth of the proposal?

5. A retail firm has experienced 3 basic responses to its previous advertising campaigns. Presently, the firm has 2 advertising programs under consideration. Because each advertising program has a different emphasis it is expected that the percentage of people responding in a particular manner will vary according to the program undertaken. Shown below are the cash flows that are expected for each of the three possible customer responses. Each cash flow shown assumes 100% response.

PROGRAM A

Response	Percentage Responding	End of Year			
		0	1	2	3
1	20%	−$500,000	$300,000	$200,000	$100,000
2	60%	− 500,000	200,000	200,000	300,000
3	20%	− 500,000	200,000	200,000	200,000

PROGRAM B

Response	Percentage Responding	End of Year			
		0	1	2	3
1	30%	−$600,000	300,000	200,000	200,000
2	40%	− 600,000	200,000	300,000	300,000
3	30%	− 600,000	300,000	300,000	300,000

For an interest rate of 10% which advertising campaign has the largest expected present-worth ?

6. A toy manufacturer must decide whether to market a new doll or to update a doll that is currently being marketed. He has the following information to use in making his decision :

	Initial Cost	Net Return Year	Probability of Duration of Sales (in years)				
			1	2	3	4	5
New doll	$50,000	$25,000	0.1	0.2	0.3	0.2	0.2
Updated old doll	20,000	12,000	0.4	0.3	0.2	0.1	0.0

Using the expected present-worth, which alternative should the manufacturer choose if the interest rate is 15% ?

7. A plant is to be built to produce blasting devices for construction jobs, and the decision must be made as to the extent of automation in the plant. Additional automatic equipment increases the investment costs, but lowers the probability of shipping a defective device to the field which must then be shipped back to the factory and dismantled at a cost of $10. The operating costs are identical for the different levels of automation. It is estimated that the plant will operate ten years, the interest rate is 20%, and the rate of production is 100,000 devices per year for all levels of automation.

Find the level of automation that will minimize the expected annual cost for the investment costs and probabilities given below.

Level of Automation	Probability of Producing a Defective	Cost of Investment
1	0.10	$100,000
2	0.05	150,000
3	0.02	200,000

Level of Automation	*Probability of Producing a Defective*	*Cost of Investment*
4	0.01	$275,000
5	0.005	325,000
6	0.002	350,000
7	0.001	400,000

8. A dam is being planned for a certain river of erratic flow. It has been determined by past experience that a dam of sufficient capacity to withstand various flow rates where the probability of these rates being exceeded in any one year is 0.10, 0.05, 0.025, 0.0125, and 0.00625 will cost $142,000, $154,000, $170,000, $196,000, and 220,000, respectively; will require annual maintenance amounting to $4,600, $4,900, $5,400, $6,500, and $7,200, respectively; and will suffer damage of $122,000, $133,000, $145,000, $170,000, and $190,000, respectively if subjected to flows exceeding its capacity. The life of the dam will be 40 years with no salvage value. For an interest rate of 4% calculate the annual cost of the dam including probable damage for each of the five proposed plans and determine the dam size that will result in a minimum cost.

9. It has been proposed to build a drive-in car wash and the decision must be made as to the number of individual facilities to be provided. A greater number of facilities means fewer customers turned away (and thus more customers serviced) but each additional facility requires additional investment. Assume the interest rate is 10%, an investment life of 10 years, and that each car serviced produces an operating surplus of $1. Using the following data, what number of facilities will result in the highest expected profit?

No. of facilities	Total required Invest- ment	Probability of Averaging n cars per year								
		2,000	4,000	6,000	8,000	10,000	12,000	14,000	16,000	18,000
1	$10,000	0.6	0.4	—	—	—	—	—	—	—
2	18,000	0.2	0.6	0.2	—	—	—	—	—	—
3	25,000	0.1	0.3	0.4	0.2	—	—	—	—	—
4	32,000	0.1	0.1	0.2	0.3	0.2	0.1	—	—	—
5	38,000	0.05	0.05	0.1	0.2	0.4	0.1	0.1	—	—
6	43,500	0.05	0.05	0.1	0.1	0.2	0.3	0.1	0.1	—
7	50,000	0.05	0.05	0.1	0.1	0.2	0.3	0.1	0.05	0.05
8	55,000	0.05	0.05	0.1	0.1	0.2	0.3	0.1	0.05	0.05

10. A company has developed probability distributions representing the probabilities that various annual equivalent amounts will be realized from the three mutually exclusive projects that are under consideration. These distributions are given below.

Annual equivalent profit:	A1	Alternative A2	A3
$ 5,000	0.10	0.00	0.00
10,000	0.10	0.20	0.00
15,000	0.20	0.20	0.20
20,000	0.30	0.20	0.50
25,000	0.30	0.20	0.20
30,000	0.00	0.20	0.10

Calculate the expected annual equivalent profit and the variance for each probability distribution. Which alternative would you consider to be the most attractive?

11. An advertising agency has developed four alternative advertising compaigns for one of its clients. The client has studied the alternatives and has developed probability distributions describing the present-worth of the net profits expected if they invest in a particular advertising program. The probability distributions for the four ad programs are as follows:

Net present-worth of profits

Ad program	− $50,000	− $10,000	$10,000	$50,000	$100,000
A	0.10	0.20	0.30	0.30	0.10
B	0.05	0.15	0.40	0.40	0.00
C	0.40	0.00	0.00	0.00	0.60
D	0.00	0.10	0.40	0.50	0.00

Calculate the mean, the variance, and the probability that the net present-worth of profits is less than zero. Which advertising program would you undertake? Explain your choice.

12. To buy a numerically controlled machine a manufacturer must pay $200,000. If the machine is purchased, there are five different manufacturing processes that can utilize this machine. By using the machine exclusively in process A, B, C, D or E, respectively, the annual income that will be realized is $120,000, $130,000, $150,000, $160,000 or $200,000. The life of the machine is expected to vary according to where the machine is used. The probability that the machine will provide service for exactly 1, 2, 3, or 4 years is shown below for each process.

Life of Machine
(years)

Process	1	2	3	4
A	0.25	0.25	0.25	0.25
B	0.30	0.30	0.30	0.10
C	0.30	0.40	0.25	0.05
D	0.20	0.60	0.20	0.00
E	0.50	0.50	0.00	0.00

Calculate the mean and variance of the present-worth amounts for each of these five processes. The interest rate is 10%. Which process could use this machine most effectively?

13. An inventor has developed an electronic instrument to monitor the impurities in the metal produced by a smelting process. A company that markets this type of equipment is considering the purchase of the patent rights to this instrument for $40,000. The company feels that there is one chance in 5 of the device becoming a successful seller. It is estimated that if the device is successful it will produce net revenues of $150,000 a year for the next 5 years. If the product is not a success no revenues will be received. The company's interest rate is 15%.

(a) Draw a decision tree describing the decision options and determine the best decision policy.

(b) If there is a market research group that can provide perfect information about the success of this product, what would be the most the company should be willing to pay for their service?

(c) Suppose the market research group can make a market survey that with probability 0.7 will give a favorable result if the device will be a success and with probability 0.9 will give an unfavorable result if the device will not sell. How much would this survey be worth to the company?

14. An electronics firm is trying to decide whether or not to manufacture a new communications device. The decision to produce the device means an investment of $5 million, and the demand for such a device is not known. If demand is *high*, the company expects a return of $2.0 million each year for 5 years. If the demand is *moderate*, the return will be $1.6 million each year for 4 years, and a *light* demand means a return of $.8 million each year for 4 years.

It is estimated that the probability of a light demand is 0.1 and the probability of a high demand is 0.5. Interest is 10%.

(a) On the basis of expected present-worth, should the company make the investment?

(b) Someone has proposed that a survey be taken to establish the actual demand to be experienced. What is the maximum value of such a survey?

(c) A survey is available at a cost of $75,000 which has the following characteristics:

| | | Survey results say favorable or unfavorable with following probabilities: | |
		Favorable	Unfavorable
If demand is:	High	0.9	0.1
	Moderate	0.5	0.5
	Low	0.0	1.0

Should the company make the survey? If so, should the company produce the device if the survey says unfavorable? What is the expected profit in this case?

15. A wholesaler is studying his warehouse needs for the next 8 years. At present three alternatives are under consideration: A new warehouse can be built to replace the existing facility, the existing facility can be expanded or the decision to expand can be postponed. If the decision is postponed the wholesaler will wait four years and then decide whether to expand the existing facility or to leave it as is. To expand the present facility now will cost $400,000 while to build a new facility will cost $700,000. Consumer demand for the wholesaler's products is expected to be either high for all eight years (H_1, H_2), high for four years and low for the remaining four years (H_1, L_2) low for the first four years and high for the last four years (L_1, H_2) or low for all eight years (L_1, L_2). Depending on these demand levels the following annual receipts are expected to be received from the two alternatives that required an investment now.

	Demand			
Alternatives:	$H_1 H_2$	$H_1 L_2$	$L_1 H_2$	$L_1 L_2$
Build new warehouse now	$320,000	$160,000	$110,000	$80,000
Expand existing warehouse now	200,000	150,000	100,000	50,000

If the decision to construct additional warehouse capacity is postponed, the cost of expanding the warehouse four years from now is expected to be $600,000. In this case it is anticipated that the annual revenues for high and low demand during the first four years will be $50,000 and $20,000, respectively. The annual revenues for the remaining four years are shown below.

	Demand	
	H_2	L_2
Decision:		
Expand after 4 years	$400,000	$100,000
Do not expand after 4 years	80,000	40,000

The interest rate is 12% and the probabilities that demand will be at a particular level through the 8 year period are

	Demand			
	$H_1 H_2$	$H_1 L_2$	$L_1 H_2$	$L_1 L_2$
Probability	0.3	0.2	0.1	0.4

Using decision tree analysis determine the decision policy that should be followed so that the wholesaler's expected profits are maximized.

16. Urn 1 contains 6 white balls and 4 black balls while Urn 2 contains 7 white balls and 3 black balls. A sample of two balls is going to be taken and on this basis you must decide from which urn the sample was selected. Although you do not see the drawing of the sample it is known that the sample is equally likely to come from either urn. If you make the correct decision you will win $10 and if you make an incorrect decision you win nothing. Using decision tree analysis find the decision policy that will maximize your expected winnings and indicate what the expected winnings will be. (a) Assume that you may choose whether the sample is drawn with or without replacement. (With replacement, the first ball drawn is returned to the urn before the second ball is selected.)

(b) Assume that you may choose whether the second ball shall be drawn with or without replacement after you have seen the result of drawing the first ball.

17. The following pay-off matrix represents the profits expected by a firm for five alternative investments and four different levels of sales. Which alternatives would the firm select if their decisions are based on the (a) maximin rule, (b) maximax rule, (c) Hurwicz rule for $\alpha = .7$, (d) minimax regret rule, and (e) the Laplace rule? What alternatives would be selected using the Hurwicz rule for the range of values, $0 \leq \alpha \leq 1.0$?

	Levels of Sales			
	1	2	3	4
Alternatives:				
A	$15	$11	$12	$ 9
B	7	9	12	20
C	8	8	14	17
D	17	5	5	5
E	6	14	8	19

18. Work Problem 17 for the following payoff matrix.

	Levels of Sales				
Alternative:	1	2	3	4	5
A	$10	$20	$30	$40	$50
B	20	25	25	30	35
C	50	40	5	15	20
D	35	20	20	40	50
E	40	35	30	25	25
F	10	20	45	30	20

19. A construction firm is considering the purchase of a number of different pieces of equipment. The firm knows that, depending on future projects won by bidding, certain types of equipment will have varying costs. Shown below is a cost matrix indicating the equivalent annual costs associated with a particular piece of equipment and its use on a particular project. The equipment alternatives are mutually exclusive and this firm anticipates that they will win the contract on only one of the projects. What piece of equipment would they purchase if they based their decision on the (a) minimax rule, (b) minimin rule, (c) Hurwicz rule for $\alpha = 0.3$, (d) minimax regret rule and (e) the Laplace rule? Graph the Hurwicz rule for each alternative for all values of α.

	Projects		
Equipment:	A	B	C
1	$100	$90	$ 60
2	70	80	90
3	30	30	140
4	100	20	120

20. Work Problem 19 for the following cost matrix.

	Projects					
Equipment:	A	B	C	D	E	F
1	$ 5	$10	$15	$20	$25	$30
2	15	15	15	20	20	20
3	40	30	30	10	15	10
4	10	15	30	40	30	10
5	30	10	10	10	10	30

16

A Plan for Engineering
Economy Studies

In engineering economic analysis, engineering and economic consider-ations are joined. Such analysis can be made haphazardly or it can be made on the basis of a logical plan. The purpose of a plan of analysis is to outline a procedure that will aid in arriving at sound conclusions. A good plan of attack points out the nature and sequence of analysis that will be required in most situations. If followed, a sound plan is effective in eliminating errors of omission for it focuses attention on the steps that should be taken. Where the analysis has to be examined by many persons, a standardized plan of approach will greatly facilitate interpretation.

In the discussion of the engineering process presented in Chapter 1, it was noted that the employment of engineering economy contributed to the crea-tive nature of engineering application. Such employment leads to a plan useful in engineering economy studies involving the creative step, the definitive step, the conversion step, and the decision step. These steps will be discussed in the sections which follow.

16.1. THE CREATIVE STEP

Both individuals and enterprises possess limited resources. This gives rise to the necessity to produce the greatest output for a given input; that is, to operate at high efficiency. Thus, the search is not merely for a fair, plaus-able, or good opportunity for the employment of limited resources, but for the best opportunity.

Engineers, whether engaged in research, design, or in general administra-tive activities, are concerned with the profitable employment of resources. When known opportunities fail to hold sufficient promise for the profitable employment of resources, more promising opportunities are sought. People with vision are those who accept the premise that better opportunities exist than are known to them. This view accompanied by initiative leads to exploratory activities aimed at finding the better opportunities. Exploration, research, investigation, and similar activites are creative. In such activities steps are taken into the unknown to find new possibilities, which may then be evaluated to determine if they are superior to those that are known. Such activities consist essentially of prospecting for new opportunities.

Opportunities are not made; they are discovered. The person who concludes that there is no better way makes a self-fulfilling prophecy. When the belief is held that there is no better way, a search for one will not be made, and a better way will not be discovered.

Search for Facts and New Combinations of Facts. Any situation embraces groups of facts of which some may be known and others unknown. The material out of which new opportunities for profit are to be fashioned are the facts as they exist.

Many successful ideas are merely new combinations of commonly known facts. The highly successful confection widely sold under many names and best described as a piece of chocolate-covered ice cream on a stick is the result of combining several simple known facts or ideas. The principal constituents are a stick, ice cream, and chocolate. It was known that people liked ice cream in conjunction with chocolate. Chocolate coverings for candies had been known for years. The stick had been previously used with candy in the "all-day sucker." Nevertheless, the exploiters of the resulting new combination of these ideas are reported to have made a fortune.

Some successful ideas are dependent upon the discovery of new facts. New facts may become known through research effort or by accident. Research is effort consciously directed to the learning of new facts. In pure research, facts are sought without regard for their specific usefulness, on the premise that a stockpile of knowledge will in some way contribute to man's welfare. Much of man's progress rests without doubt upon facts discovered from efforts to satisfy curiosity.

Both new facts and new combinations of facts may be consciously sought. The creative aspect of engineering economy consists in finding new facts and new combinations of facts out of which may be fashioned opportunities to provide profitable service though the application of engineering.

Aside from the often quoted statement that "inspiration is ninety per cent perspiration," there are few guides to creativeness. It appears that both conscious application and inspiration may contribute to creativeness. Some people seem to be endowed with marked aptitudes for conceiving new and unusual ideas.

It may be presumed that a knowledge of facts in a field is a necessity for creativeness in that field. Thus, for example, it appears that a person who is proficient in the science of combustion and machine design is more likely to contrive a superior internal-combustion engine than a person who has little or no such knowledge. It also appears that knowledge of costs and people's desires as well as of engineering is necessary to conceive of opportunities for profit that involve engineering.

The Economic Opening. The creative step in economy studies consists essentially of finding an opening through a barrier of economic and physical limitations. When aluminum was discovered, uses had to be found for it that would enable it to be marketed, and means had to be found whereby its physical characteristics could be improved and its cost of manufacture reduced. Exploitation of uranium deposits rests upon the phenomenon of fission. The legality of collecting fees for regulations of parking as contrasted to making a charge for the use of parking space on streets was the factor on which exploitation of the parking meter hinged. Titanium has become economically important for its high temperature characteristics. Exploitation of iron ore deposits in Canada and Australia follows exhaustion of ore bodies closer to steel centers.

Economic limitations are continually changing with the needs and wants of people. Physical limitations are continually being pushed back through science. In consequence, new openings revealing new opportunities are continually developing. For each successful venture someone has found an opening through the barrier of economic and physical limitation.

Circumventing Factors Limiting Success. So far, discussion of the creative step has been confined to consideration of seeking new opportunities for profitable employment of resources. Attention will now be turned to a search for means for circumventing factors that limit success of present activities. The aim of circumventing factors limiting success is related to the search for better means for achieving objectives rather than a search for better objectives which was considered in a previous section. Though this is not particularly different in character from the activity first discussed in the creative step, separate consideration is justified because of the more objective approach that is possible.

Where the aim is to improve the success of a given activity, the search may be directed to means for circumventing factors that limit success. A strategic factor is a limiting factor which, if appropriately altered, will remove limitations restricting the success of the entire undertaking. Known strategic factors and those that seem most likely to be strategic factors should be selected for consideration.

The understanding that results from the delineation of limiting factors and their further consideration to arrive at the strategic factors often stimulates the conception of improvements. There is obviously no point in operating upon some factors. Consider for example a situation in which an operator is hampered because he had difficulty in loading his machine with a heavy piece. Three factors are involved: the pull of gravity, the mass of the piece, and the strength of the man. Not much success would be expected from an attempt to lessen the pull of gravity. Nor is it likely that it is feasible to reduce the mass of the piece. A stronger man might be secured, but it seems more logical to consider overcoming the need for strength by devices to supplement the strength of the man. A consideration of the strength factors would thus lead to consideration of devices that might circumvent the limiting factor of strength.

It is possible that consideration of the factor of the piece's weight could have resulted in a reduction in weight by substitution of material, lighter sections, or a similar change. In this connection it may be noted that weight of pipe was considered a strategic factor in the cost of petroleum pipelines. This factor was circumvented by the use of high-strength steels made possible by the development of new welding processes.

Solutions Result from Search for Limiting Factors. In the drilling of oil wells with a rotary drill, a drill bit is rotated at the bottom of the hole by heavy members known as drill pipe joined end to end with screw joints. When it becomes necessary to replace the drill bit, successive joints of drill pipe are raised and unscrewed until the drill bit is reached. One alert person considered the joining method a strategic factor in drilling cost, for each length of drill pipe had to be turned a number of times in making and unmaking joints. After some consideration he devised the taper thread joint, which has saved millions of dollars in drilling oil-wells. This thread becomes disengaged in a few turns, has all the advantages of straight threads and some in addition, and costs about the same as straight thread.

The creative step has been given much emphasis in the discussion above for it is believed to be of first importance in economy studies. It is directly related to the delineation and selection of objectives that are without doubt the most important functions of engineering economy and certainly the first steps toward success in any field of endeavor. Since the mental processes involved are in large measure illogical, this step must be approached with considerable alertness and curiosity and a willingness to consider new ideas and unconventional patterns of thought.

The Creative Conference. An interesting development has been the *creative conference* in which a group of persons seek new opportunities or ways to circumvent limitations. In such a conference a problem is posed for solution. Those in attendance are encouraged to let their imaginations have free reign and suggest solutions to the problem, no matter how fantastic. No criticisms of ideas suggested are permitted on the thought that criticism will inhibit imagination. Ideas produced by the conference are evaluated elsewhere.

16.2. THE DEFINITIVE STEP

The definitive step consists of defining the alternatives that have originated in the creative step or which have been selected for comparison in some other way. In the first stage of the definitive step, the engineer's aim should be to delineate each alternative on the basis of its major and subordinate physical units and activities. The purpose of this stage is to insure that all factors of each alternative and no others will be considered in evaluating it.

The second stage of the definitive step consists of enumerating the prospective items of output and input of each alternative, in quantitative physical terms as far as possible and then in qualitative terms. Though qualitative items cannot be expressed numerically they may often be of major importance. They should be listed carefully so that they may be considered in the final evaluation.

On completion of the definitive step there should be an enumeration giving all items of input and output and the time of their occurrence for each alternative.

Choice Is Between Alternatives. Conception of alternatives is a creative process. A complete, all-inclusive alternative rarely emerges in its final state. It begins as a hazy, but interesting idea. The attention of the individual or group is then directed to analysis and synthesis and the result is a definite proposal. In its final form, an alternative should consist of a complete description of its objectives and its requirements in terms of inputs and outputs.

Except in extreme situations, there are usually several if not many things that are within the powers of an individual or an organization to do. But each choice imposes limitations of resources, time, and place. Thus, though it might be within the power of an individual to go to Place A and Place B, he cannot be at both places simultaneously. Also, he may have the resources, time, talent, and desire to carry Activity A or Activity B to successful conclusions. But he may find that he cannot do both and that he is forced to make a choice between them.

Courses of action between which choice is contemplated are conveniently called alternatives. Both different ends and different methods are embraced by the term alternative. All proposed alternatives are not necessarily attainable. Alternatives are frequently proposed for analysis even though there seems to be little likelihood that they will prove feasible. This is done on the thought that it is better to consider many unprofitable alternatives than to overlook one that is profitable. Alternatives that are not considered cannot be adopted no matter how desirable they might prove to be. Consider a contractor who is trying to evaluate the comparative merit of Method A and Method B when Method C, of which he is not aware, is superior to either of the other methods. The criterion for judging the desirability of an alternative is its result in comparison with the result of other alternatives that may be undertaken.

It may also be noted that there are costs associated with seeking out and deciding upon undertakings. The cost of seeking out desirable undertakings is a charge that must be deducted from the income potentialities of the undertakings that are decided upon. This limits the outlay that can be justified for search on the basis of economy. The measure of the net success of a venture may be thought of as being the difference between its potentialities for income and the sum of the outlay incurred in finding and deciding to undertake it and the outlay incurred in carrying it to completion.

Not All Alternatives Can Be Considered. An objective of economic analysis is to find the best opportunity for the employment of limited resources. However, this objective can rarely be realized; to realize it completely would require that all possible alternatives of a situation be delineated for comparison with each other. It is essential to remember that alternatives are not outlined and evaluated without cost and the passage of time.

As an example, consider the following situation. The attention of a superintendent was directed to a loss of heat from a bare pipe in his plant. The superintendent calculated the heat loss and found it to amount to $260 per month. A covering that would reduce the heat loss by $240 per month was offered by a salesman for $680. The superintendent though he could secure the needed covering for less and actually accepted a bid for $632 three weeks later. Thus the lapse of time in seeking a better alternative had resulted in a saving of $48 on the covering and a heat loss of $180 that could have been prevented by accepting the first offer for the covering.

Similarly, the cost of considering alternatives will ordinarily force a choice before all possibilities are considered. Suppose that a computing service company is seeking a location for another branch office. Such an office might be located in many cities. Obviously, the cost of investigating all the possibilities would be prohibitive. Further suppose that the annual net income would be $10,000 at the first location considered and that an annual net income of $1,000 per year for 10 years is sufficient to justify a present investment of $5,000. If it costs $1,600 to evaluate a location, each investigation of a location after the first would result in an increase in income of

$$\$1,000 \times \frac{\$1,600}{\$5,000} = \$320$$

on the average for each site investigated. If the potential annual income of the first site considered was $10,000, the annual income of the site selected to justify the cost of investigation of a number of sites would be as follows:

Number of Sites Investigated	1	2	4	10	20
Annual Income to Justify Investigation	$10,000	$10,320	$10,960	$12,880	$16,080

Ordinarily too few alternatives are probably considered, but the above analysis shows that there is a limit beyond which considerations cannot be justified. The number of alternatives to consider is in itself a matter of economy worthy of study.

Recognizing that all alternatives cannot be considered, the number of alternatives to be considered is limited on the basis of general assumptions. For instance, a fireman works on the assumption that it is his duty to minimize property loss by fire. When an alarm is sounded, his alternatives are limited to the methods by which he may minimize property loss by extinguishing the fire. The assumption under which he works precludes the necessity for him, for instance, to consider the alternative of letting a particular area be destroyed because mankind might be better served thereby.

Alternatives may be limited in a progressive series of assumptions. If the limiting assumptions are sound, no desirable alternatives will have been excluded. Skill and knowledge that will result in judgment that will enable consideration of none but desirable alternatives without excluding any that are desirable is an important function of intellect.

Consider the problem of selecting a dam site for a reservoir. The cost of constructing a dam with the required storage capacity can be estimated for each point along a river. To make each estimate a detailed study of the foundation requirements, available and required access roads, distance from construction materials and other items would have to be made for each site. Then this large amount of cost data would be compiled and the dam site with the least cost selected.

Actually, the engineer would proceed with detailed cost estimates for only a few desirable alternatives. By utilizing a topographic map, he may pick out a few promising sites and disregard the rest. The engineer may do this with confidence if he is sufficiently familiar with dam construction to know that the excluded sites would have a higher cost. Next, he may make approximate estimates of the cost of these sites assuming normal foundation conditions. Finally, the two or three sites which promise to result in the lowest cost are studied in detail through test borings to obtain accurate foundation cost information. The selection of the site for the dam is then based on detailed accurate cost information for these sites.

At each step in this process there is a chance that the dam site which is most desirable will be eliminated due to lack of complete analysis. The degree of completeness that is allowed at each step depends upon skill and judgment. Thus it is usually best to trust the initial phases of a process such as that described to an experienced person. The detailed analysis which follows may be done by an engineer with less experience.

16.3. THE CONVERSION STEP

In order that alternatives may be compared effectively, it is necessary that they be converted to a common measure. The common denominator usually chosen for economic comparison is value expressed in terms of money.

The first phase of the conversion step is to convert the prospective output and input items enumerated in the definitive step into receipts and disbursements at specified dates. This phase consists essentially of appraising the unit value of each item of output or input and determining their total amounts by computation. On completion, each alternative should be expressed in terms of definite cash flows occurring at specified dates in the future, plus an enumeration of qualitative considerations that it has been impossible to reduce to money terms. For such items the term "irreducibles" is often employed. This term seems to be superior to the term "intangibles" in this connection, for many items described in qualitative terms are clearly known even though not expressible in numbers.

The second phase of the conversion step consists of placing the estimated future cash flows for all alternatives on a comparable basis, considering the time value of money. This involves employment of the techniques presented in previous chapters. Selection of the particular technique depends upon the situation and its appropriateness is a matter of judgment. Consideration of inherent inaccuracies in estimates of the future outputs and inputs may be considered part of the conversion step and should not be overlooked.

The final phase of the conversion step is to communicate the essential aspects of the economy study, together with an enumeration of irreducibles, so that they may be considered by those responsible for making the decision. Responsibility for the acceptance or rejection of an engineering proposal is exercised more often than not by persons who have not been concerned with the technical phases of the proposal. Also, the persons who control acceptance are likely to lack understanding of technical matters.

A proposal should be explained in terms that will best interpret its significance to those who will control its acceptance. The aim of a presentation should be to take persons concerned with a proposal on an excursion into the future to experience what will happen if the proposal is accepted or rejected.

Suppose that a proposal for a new pollution control system is to be presented. Since those who must decide if it should be adopted rarely have the time and background to go into and appreciate all the technical details involved, the significance of these details in terms of economic results must be made clear. Of interest to those in a position to decide will be such things as the present outlay required, capital-recovery period, flexibility of the system in event of production volume changes, effect upon product price and quality, and difficulties of financing. Cost and other data should be broken down and presented so that attention may be easily focused upon pertinent aspects of the proposal. Diagrams, graphs, pictures, and even solid models should be used where these devices will contribute to understanding. The effort expended in developing a sound proposal is often lost though its rejection because of poor presentation.

The aim of economy studies is sound decision. Thus a first consideration should be to present a proposal in terms that will help those who must decide to understand the implications of the proposal to the fullest extent possible.

16.4. THE DECISION STEP

In addition to the alternatives formally set up for evaluation, another alternative is always present. This is the alternative of making no decision on the formal alternative being considered. The decision not to decide may be a result of either active consideration or passive failure to act; it is usually motivated by the thought that there will be opportunities in the future which will prove more profitable than any known at present.

For example, if a venture to build a chemical processing plant is under consideration, there frequently is only one formal alternative to consider. Those concerned will usually decide the issue on how favorably the prospective venture measures up to those generally open for the employment of resources. Thus the decision not to decide is clearly a decision based on a comparison with future, though perhaps unknown, alternatives.

Differences Are Bases for Decision. On completion of the conversion step, quantitative and qualitative outputs and inputs of each alternative, so far as these are known, will form the basis for comparison and decision. Quantitative input may be deducted from quantitative output to obtain quantitative profit, or the ratio of quantitative profit to quantitative input may be found. Each of these measures is then supplemented by what qualitative consideration may have been enumerated.

Decisions between alternatives should be made on the basis of their differences. Thus, all identical factors can be canceled out for the comparison of any two or more alternatives at any step in an economy study. In this process great care must be exercised that factors canceled as being identical are actually of the same significance. Unless it is very clear that factors considered for cancellation are identical, it is best to carry them through the first stage of the decision step. This may entail a greater amount of computation and other paper work, but the slight added complexity and loss of time is ordinarily insignificant in comparison with the value of greater accuracy in decision.

It frequently occurs that alternatives to provide an identical output or service are under consideration. Then output need not be considered and decision is made on the basis of input.

Where all facts about alternatives are known in accurate quantitative terms, the relative merit of each alternative may be expressed in terms of a single number. Decision making in this case is simple. Where all factors are known, evaluation may be made on the basis of reason.

Decisions must always be made in spite of the fact that quantitative considerations are based on estimates that are subject to error and where qualitative knowledge must be used to fill gaps in knowledge. Therefore, it must be remembered that the final calculated amounts embody the errors of the estimated quantities.

When Facts Are Missing Judgment Must Be Used. Where alternatives cannot be completely delineated in accurate quantitative terms, the choice is made on the basis of the judgment of one or more individuals. Judgment seeks to predict the outcome of a course of action. Few believe that any person is endowed with a sixth sense or intuition to enable him to peer into the future yet it appears that some persons perform better than others in this endeavor.

If judgment is not clairvoyant, the ability to predict the future rests entirely on a cause-and-effect relationship. The only bases for the prediction of the outcome of a course of action are the facts in existence at the time the prediction is made. A person's judgment rests, therefore, upon his knowledge of the facts involved and his ability to use these facts.

When complete knowledge of all facts concerned and their relationships exists, reason can supplant judgment and predictions become a certainty. Judgment tends to be qualitative. Reason is both qualitative and quantitative. Judgment is at best an informal consideration and weighing of facts; at its worst it is merely wishful thinking. Judgment appears to be an informal process for considering information, past experience, and feeling in relation to a problem. No matter how sketchy factual knowledge of a situation may be, some sort of a conclusion can always be drawn in regard to it by judgment.

Figure as Far as You Can, then Add Judgment. An important aim of engineering economic analysis is to marshal the facts so that reason may be used to the fullest extent in arriving at a decision. In this way judgment can be reserved for parts of situations where factual knowledge is absent. This idea is embraced in the statement "Figure as far as you can, then add judgment."

Where a diligent search uncovers insufficient information to reason the outcome of a course of action, the problem is to render as accurate a decision as the lack of facts permits. In such situations there is a decided tendency, on the part of many, to make little logical use of the data that are available in coming to a conclusion, on the thought that since some pretty rough estimating has to be done on some elements of the situation, the estimate might as well embrace the entire situation. But an alternative may usually be subdivided into parts, and the available data are often adequate for a complete or nearly complete evaluation of several of the parts. The segregation of the known and unknown parts is in itself additional knowledge. Also, the unknown parts, when subdivided, frequently are recognized as being similar to parts previously encountered and thus become known.

Making the Decision. After a situation has been carefully analyzed and the possible outcomes have been evaluated as accurately as possible, a decision must be made. Even after all the data that can be brought to bear on a situation have been considered, some areas of uncertainty may be expected to remain. If a decision is to be made, these areas of uncertainty must be bridged by consideration of nonquantitative data or, in other words, by the evaluation of intangibles. Some call the type of evaluation involved in the consideration of intangibles *intuition;* others call it *hunch* or *judgment.*

Whatever it be called, it is inescapable that this type of thinking or, perhaps better, this type of feeling, must always be the final part in coming to a decision about the future. There is no other way if action is to be taken. There appears to be a marked difference in people's abilities to come to sound conclusions when some facts relative to a situation are missing. Perhaps much more attention should be devoted to developing sound judgment, for those who possess it are richly rewarded. But as effective as intuition, hunch, or judgment may sometimes be, this type of thinking should be reserved for those areas where facts on which to base a decision are missing.

Where an activity embodies elements of uncertainty it is often wise to hold capital equipment investment to a minimum until outcomes become clearer, even though such a decision may result in higher production costs. Such action amounts to a decision to incur higher costs temporarily in order to reserve the privilege of making a second decision when the situation becomes clearer.

The accuracy of estimates with respect to events in the future is at least to some extent inversely proportional to the span of time between the estimate and the event. It is often appropriate to incur expense for the privilege of deferring decision.

Improving Decision Making Through Feedback. Feedback for the purpose of improving future action is an essential of the decision-making process. It is unlikely that future action can be improved without knowledge of the results of action just taken.

A great deal of time and energy is spent in acquiring information about an undertaking that has been completed. Thus motion pictures of last week's football game may be examined with great care. Corporations make quarterly or annual reports purporting to summarize their activities and to show their status at the end of these periods. The economic justification for acquiring knowledge of the past is that it will enable better decisions to be made in the future. It is a recognized psychological fact that a person cannot improve his skill by practice unless he learns the outcome of his actions. Thus a person could not improve his ability to throw darts if the lights on the target are extinguished from the time a dart leaves his hand until after it is removed from the target.

In a similar way, it is reasonable to believe that skill in making decisions must be dependent upon knowing the result of previous decisions. Since the chief purpose of information of past activity is to aid in performing activities more efficiently in the future, this purpose should be considered in relation to economy. Investigation often reveals that huge sums are spent in business for reports, records, audits, and analyses. It often happens that much of such information is of little use. At the same time some items that would be of great value are not available. The benefits to be gained by the availability of information should always be considered in relation to the cost of collecting it.

QUESTIONS

1. Discuss the potential benefits of the plan for engineering economy studies.
2. Apply the plan for engineering economy studies to an engineering problem of your choice.
3. How is the creative step related to the satisfaction of human wants?
4. List the methods for discovering means to more profitably employ resources.
5. Discuss the nature of the definitive step.
6. Why is it not possible to consider all possible alternatives?
7. What procedure is involved in reducing possible alternatives to a number that is practical to consider?
8. Describe the three phases of the conversion step.
9. What is the nature of the decision step?
10. Discuss the nature of judgment and explain why it is applicable to more situations than reason.
11. Explain why decision must be based on differences occurring in the future.
12. Discuss what is meant by the statement "Figure as far as you can, then add judgement."
13. Why are decisions relative to the future based upon estimates instead of upon the facts that will apply?
14. Why is judgment always necessary to come to a decision relative to an outcome in the future?
15. Give an example of an instance in which feedback is utilized to improve future action.

VI

ECONOMIC ANALYSIS
OF OPERATIONS

17

Organization and Utilization
of Personnel

Man has accumulated an extensive body of physical laws to aid him in dealing with the physical aspect of his environment. With every advance in the physical sciences new relationships are discovered which provide further understanding of physical phenomena. Although much less is known about organized activity and human factors, this important area is now receiving considerable attention.

Most of the analysis presented in previous chapters has been concerned with the economy of physical resources. However, physical resources are of no value in the satisfaction of human wants until they are transformed by human action. A critical element in the development of the forces and materials of nature is human effort acting through organization. In fact, organization is considered by many to be mankind's most important innovation.

17.1. THE ECONOMY OF ORGANIZATION

Organizations consist of the coordinated effort of individuals and have existed since the beginning of time. Through organizations, man can either attain ends that no one can achieve by individual effort or he can attain certain ends more economically.

Labor Saving through Organization. Organized effort is often a means to economy in accomplishment through labor saving. Suppose, for example, that two men adjacent to each other are each confronted with the task of lifting a box onto a loading platform, and that each of the two boxes is too heavy for one man to lift but not too heavy for two men to lift. Assume that the only practical way for one man to accomplish his task is to obtain a hand winch with which the task can be accomplished in 30 minutes' time. If there is no coordination of effort the cost of getting the two boxes onto the platform will be 60 man-minutes.

Suppose that the two men had coordinated their efforts to lift the two boxes in turn and that the time consumed was one minute per box. The two tasks would have been accomplished at the expense of four man-minutes, or about 7% as much as if there had been no coordination of effort.

Coordination of human effort is so effective a means of labor saving that it may be economical to pay for effort to bring about coordination of effort. In the example above, effort directed to bring about coordination would result in a net labor saving of 56 man-minutes of effort.

For further illustration of the creativeness of coordination of human effort, suppose that a water well would have a value of $100 to each of 100 families in a village of a certain undeveloped country. The head of each of the families recognizes this and on inquiry finds that a well will cost $1,000. Each family head, being oblivious of the opportunities for coordination, abandons the well drilling project as unprofitable. If an entrepreneur could bring about a coordination of effort in the village, the net benefit might be as follows:

$$(100 \text{ families} \times \$100) - \$1,000 = \$9,000.$$

Entrepreneurship is a worthwhile and necessary activity in most situations involving organized activity.

The Efficiency of Organization. Organizations are a means for creating and exchanging utilities. The fact that they are creative may be inferred from the following example. A person who is employed by an industrial organization may be presumed to value the wages and other benefits he gets more highly than the efforts he contributes to gain them. The person who sells material to the organization must value them less than the money he receives for them or he would not sell. The same may be said of the seller of equipment. Similarly, a person who loans money to an organization will in the long run receive more in return than he advances or he will cease to loan money. The customer who comes with money in hand to exchange for the products of the organization may be expected to part with his money only if he values it less than he values the products he can get for it. This situation is illustrated in Figure 17.1.

In order that the organization illustrated be successful, not only must the total of the satisfactions exceed the total of contributions, but also each contributor's satisfactions must exceed his contribution as he evaluates them. In other words, each contributor must realize his aspirations to a satisfactory degree or he ceases to contribute. Organizations are essentially devices to which people contribute what they desire less to gain what they desire more. Unless people receive more than they put into an organization they withdraw from it. For an organization to endure, its efficiency—output divided by input—must exceed unity.

FIGURE 17.1. ILLUSTRATION OF THE
ECONOMY OF ORGANIZATION.

The function of a manager is to maintain a system for pooling the activities of people so that each person, including the manager himself, gains more than he contributes and so that individual contributors do not believe that superior alternatives are open to them elsewhere. Those who can successfully manage possess a valuable talent which is usually richly rewarded.

Risks Confronting a New Organization. Experience leads to the conclusion that, despite the most careful planning, a new organization will be more subject to unforeseen hazards or contingencies than one that is well established. Furthermore, the new enterprise usually has very limited reserves to meet even temporary reversals.

Demands for new products are subject to both seasonal and longtime fluctuations. A new enterprise that begins on a long-time decline of demand for its product may fail from this cause. A new enterprise entering a seasonal market may lose much of its first year's income through any cause that may delay completion of its plant in time to take advantage of the active period for its product.

A new enterprise usually needs considerable credit during the period of rapid expansion before it has time to establish itself favorably with sources of credit. It is very vulnerable to any tightening of credit in this period. Even if its need for credit may arise from a more rapid expansion of sales than expected, it may fail from inability to secure funds to bridge the period between expenditures for the manufacture of its product and receipts from its sales.

Established firms do not ordinarily welcome a new competitor into the field with enthusiasm. Thus a new firm often finds active and concerted efforts directed toward discrediting it and its products. This can be very damaging and difficult to offset.

A most important factor in the success of a new venture is the enthusiasm and capabilities of its managers. These men are subject to the hazards of sickness and accident. Impairment of health or death of a capable leader can easily cause a new enterprise to fail.

17.2. SPAN OF CONTROL IN ORGANIZATION

Most of the concrete aspects of products, structures, or services produced by an organization are the result of the activities of those on its lowest level. For example, it is not the orchestra leader, the sales manager, or the foundry superintendent who makes the music, the sale, or the castings. These are made by the men on the lowest level of organization.

This fact in no way minimizes the importance of the functions performed by those of the upper levels or organization. The function of much if not most of the activities of the upper hierarchy of organization is to facilitate the work of those on the lower levels. This is particularly clear in regard to the work of the methods engineer. His work has no significance except that it enables those on the lowest levels to make a product with less effort. Although not as apparent, this seems equally true of the functions of design, production control, stores control, industrial relations, and cost accounting.

Any enterprise has a major objective and subordinate objectives. For example, the ultimate aim of a contractor may be to earn a profit through activities associated with earth moving, such as digging pipeline trenches, excavating for buildings, or building earthen dams. Let earning a profit by digging pipeline trenches be considered to be a major objective designated by A. Objective A is an abstract aim that cannot be realized except by digging pipeline trenches in definite ways in definite places. An objective, a_1, subordinate to A, might be to earn a profit by digging a pipeline trench from Welltown to Pipetown via a certain route and in accordance with certain specifications. Realization of objective A depends upon the realization of objectives a_1, a_2, a_3, and so on. Also each subordinate objective, as for instance a_1, has objectives subordinate to it. And each of these subordinate objectives has its subordinate objectives and so on until the final subordinate objective involves activities of very narrow scope.

Persons in organizational levels above the lowest are for the most part concerned with the determination of objectives and the means for attaining them. The higher the level, the greater the emphasis on general objectives, and the less the emphasis on means. Thus, top levels are concerned predominantly with objectives and the lower levels with the means for accomplishing objectives.

Span of Control and Organizational Levels. When considering organizational structure the first aspect to note is the relationship of span of control and the number of organizational levels. When all levels are complete and the span of control is constant throughout the organization, the relationship between the number of levels, the span of control, and the total number of persons in an organization may be established. Let

N = the total number of persons in an organization;
S = the span of control;
L = the number of levels of organization.

Then

$$N = 1 + S + S^2 + \ldots + S^{L-1}.$$

This result is shown graphically for a number of spans of control and levels of organization in Figure 17.2. From the graph it may be found, for

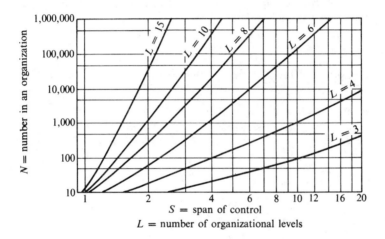

FIGURE 17.2. SPAN OF CONTROL, NUMBER OF ORGANIZATIONAL LEVELS, AND THE NUMBER IN AN ORGANIZATION.

example, that 10,000 people can be organized with a span of control of 6 or 20 and that the number of levels would be 6 and 4, respectively.

It is recognized that the difficulty of communication in organizations is increased by the number of levels. In an organization of a given size the number of levels can be decreased by increasing the span of control. But there are economical limits to the span of control. Thus, one problem in the design of an organization's structure is to determine the most economical compromise between number of organization levels and span of control.

Because of the difficulty of measuring the effect of either the number of organizational levels or the span of control, the thinking about them has been predominantly in qualitative terms. The fact that effective span of control varies widely for different persons and classes of work further complicates the problem. But quantitative analyses can be made of some aspects that will be helpful in improving judgments in regard to the number of organizational levels and the span of control.

It has been pointed out that the primary function of leadership is to facilitate the accomplishment of subordinates. Assuming that this is true, one question that may be considered is the ratio of input of leadership to that of those on the lowest level of an organization through whose efforts the concrete aspects of production are accomplished.

The relationship between the total number of persons in an organization, the span of control, and the number of levels was given by

$$N = 1 + S + S^2 + \ldots + S^{L-1}.$$

Of N employees, $1 + S + S^2 + \ldots + S^{L-2}$ are on levels of organization above the lowest. The number on the lowest level equals S^{L-1}. The ratio r_L of these two groups is expressed by

$$r_L = \frac{1 + S + S^2 + \ldots + S^{L-2}}{S^{L-1}}.$$

As the number of levels increases, the ratio r_L approaches $1/(S - 1)$. For example, for a span of control of six, the following values are obtained.

Number of Levels L	Number in Upper Levels $1 + S + \ldots + S^{L-2}$	Number in Lowest Level S^{L-1}	Ratio r_L
1	0	1	0
2	1	6	0.167
3	7	36	0.194
4	43	216	0.199
5	259	1,296	$\frac{1}{5}$ nearly

Leadership Attention and Performance. The value of a person to an organization may be considered to be the difference between the worth of his output and the cost of the input that he requires in terms of leadership attention and material incentives. The output of a person in an organization consists of his performance that contributes (1) to the attainment of the organization's objectives, and (2) to the establishment and maintenance of a system of cooperative effort.

Examples of the former contributions are such outputs as negotiating a loan on which to base plant expansion, closing a sale for 600 units of product with XYZ Company, designing a new water pump, planning a transformer production layout, turning out 800 motor frame castings, typing a letter to John Doe and Company, helping an electrician install a motor. The value of such contributions depends upon their volume, quality, and timeliness. They are usually concrete and easily observable and are generally quite well understood.

Most things are accomplished in cooperation with others. This necessitates the coordination of several persons' activities in such a way that their joint effort is specialized in a system of cooperation in a way that will accomplish the desired result. An individual may contribute to the establishment and the maintenance of effective cooperative effort; such persons are integrating influences. Those who tend to destroy effective cooperative effort are disruptive influences. Maintenance of effective cooperation is so important in many situations that a member of an organization may be valued primarily for his integrating influence rather than for his concrete output. On the other hand, some star performers, whether on athletic teams or in industrial organizations, are dropped because their disruptive influence outweighs their concrete performance.

A variety of inputs are associated with the utilization of personal service. Among the more tangible are wages and allied compensation, cost of recruitment, and the maintenance of comfort facilities such as heat, light, furniture, and rest rooms. An important additional input is what may conveniently be called leadership attention. Leadership attention to a subordinate takes the form of directions, instructions, encouragement, discipline, transfer to more suitable work, and compensation for a shortcoming in ability or performance.

It is observable that there is a great difference in the amount of leadership attention required to secure satisfactory performance from individuals. Some individuals require so little leadership that they need only be given general directions. Some have shortcomings requiring so much leadership attention that the cost of this input exceeds the value of their output.

Ratio of Leadership to Subordinate Input. The number of subordinates assigned to a leader for the accomplishment of a given task may be varied. Such variation may be thought of as a variation in the ratio of leader to subordinate input on a man-to-man basis. Compensation may be considered and the ratio of leader of subordinate input may be obtained in terms of cost.

A function of leadership activities is to utilize the effort of subordinates in such a manner that their output will be greater than it would be without leadership. Thus, the amount by which the output of unsupervised workers must be increased by leadership to justify an expenditure for leadership can be calculated. Let

S = number of workers per leader, that is, span of control;
W = average daily wage per worker;
W_1 = average daily wage per leader;
P = ratio of leader's to subordinate's wage rate = $W_1 \div W$;
U = average daily output per unsupervised worker;
R = ratio of daily output of supervised subordinate to daily output of unsupervised worker;

RU = average daily output per supervised subordinate;
T = leader's time per worker per 480-minute day.

The daily wage cost per supervised group of S subordinates and one leader is equal to $SW + W_1 = SW + PW$. The daily output per supervised group of S subordinates and one supervisor is equal to SRU, and the cost per unit output for a supervised group of S subordinates and one supervisor is equal to $(SW + PW)/SRU$.

Let cost per unit output for a supervised group of S subordinates and one leader equal the cost per unit output of unsupervised workers; then

$$\frac{SW + PW}{SRU} = \frac{W}{U}$$

Solving for R,

$$R = 1 + \frac{P}{S}$$

Also

$$T = \frac{480}{S}$$

Values for R and T corresponding to values of S from 1 to 20 and for values of P from 1 to 3 are given in Figure 17.3.

Suppose that the question under consideration is whether 6 or 8 is the better span of control on the basis of which to organize a function, and that the ratio of leader's to subordinate's wage considered desirable is 1.2. From Figure 17.3 the value of R for a span of 6 is 1.20 and for a span of 8 is 1.15

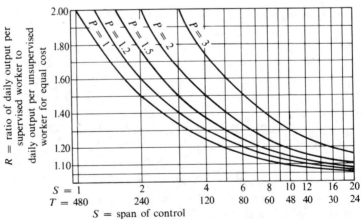

S = span of control
T = leaders time per subordinate in minutes per 480 min. day
P = ratio of leaders to subordinates wage rate

FIGURE 17.3. SUPERVISION, SPAN OF CONTROL, AND WORKER OUTPUT.

when $P = 1.2$. To justify a span of 6 as against a span of 8, one would have to be satisfied that the output per subordinate would be increased to the extent that R would be increased from 1.15 to 1.20 or by

$$\frac{1.20 - 1.15}{1.15} = 0.04, \quad \text{or } 4.3\%.$$

Supervision per worker can be increased from 60 to 80 minutes per subordinate per day or

$$\frac{80 - 60}{60} = 0.33, \quad \text{or } 33\%.$$

17.3. THE SELECTION OF PERSONNEL

Viewed objectively, subordinates are merely their leader's means for reaching ends. Of the several productive factors used in industry, none has such a variety of characteristics as personnel. Characteristics of both body and mind are of concern. One individual may accomplish several times as much of a task requiring physical strength, dexterity, or acuteness of vision as another. The range of mental proficiencies for given tasks is even wider.

The physical and mental endowments of peoples appear to be subject only to limited change by training and leadership. For this reason it is important that personnel be selected that have inherent characteristics as nearly compatible as possible with the work they are to perform and the position they are to occupy. Contrasted with the meticulous care with which materials are selected, the selection of personnel in most concerns is given but superficial attention.

The Range of Human Capacities. The ratio of the poorest to the best performance time is often as much as 1 to 3 in common operations. It is apparent that the cost of labor input per unit of output might be materially reduced by selecting only employees capable of better than average performance. In many cases superior employees receive little or no greater compensation than less capable employees. This is particularly true where performance is measured with difficulty. Superior managerial ability, for example, may go unrecognized and therefore unrewarded for long periods of time because it cannot be measured. But even where superior ability is proportionally compensated for, savings may still result.

Superior ability that remains unused is not an advantage and may often be a disadvantage because of the resulting frustration and dissatisfaction of the employee. Generally speaking, the work assignment should require the workman's highest skill.

For many tasks, tests can be devised for selection of prospective employees with superior ability. Representative of these are a battery of three tests for the selection of mail distributors for the Postal Service. It was found, from extensive experimental trials, that of those making the highest 25% of scores on these tests, over 93% would be above average in proficiency. The economic desirability of having most employees above average in ability for the task under consideration (even through compensation be in proportion to ability, which it rarely is) can hardly be overestimated.

In this relatively new field, results of the installation of a program of selection practices are most difficult to estimate. Since the input cost of a program can usually be estimated with reasonable accuracy, it may be helpful in arriving at a decision to calculate the benefits necessary to justify a contemplated selection program.

Suppose that a certain plant employs 40 operators. The average output is 28 pieces per hour, the average spoilage is 1.7%, and each rejected piece results in a loss of $0.60. Records reveal that average spoilage for the best 20 operators is 1.1%, whereas that for the poorest 20 operators is 2.3%. A consultant agrees to prepare a set of tests and administer them for a year for $1,800. Administration subsequent to the first year is expected to cost $3 per operator hired or $140 per year based upon the estimated turnover.

Let P equal the percentage reduction in spoilage during first year to justify expenditure of $1,800. Then $P \times 40$ operators $\times$ 28 pieces per hr. $\times$ 2,000 hr. per year $\times$ $0.60 per reject $=$ $1,800 form which $P = 0.00134$ or 0.134%.

This percentage reduction in spoilage seems attainable in view of the fact that a program that will reject applicants of less than average ability in relation to spoilage will result in an average reduction of $1.7 - 1.1$ or 0.6%. Moreover, it is reasonable to expect other benefits from the program.

By tending to place people in work for which they are best fitted, sound selection is generally beneficial. Even those who are rejected will generally be benefited; for, if continually rejected for work for which they are unfitted, they must eventually find a job that will permit them to work at their highest skill.

The Economy of Proficiency. Value of proficiency is not necessarily directly proportional to degree of proficiency. Thus, a baseball player whose batting average is 0.346 will ordinarily command more than twice as much salary as one whose average is 0.173, if they are equal in other respects. In some activities the compensation of those with unusually high proficiencies is extremely high, but those with ordinary abilities can find no market for their services. Consider acting; the motion picture industries pay fabulous salaries for those whose box office appeal is high when they undoubtedly could cast all their productions with volunteer actors of fair ability without cost.

The cost of human effort is a considerable portion of the total cost of carrying on nearly all production and construction activities. People are employed with the idea of earning a profit on the skills they possess. In general, the higher a person's proficiency in any skill, be it manual dexterity, creativeness, leadership, inventiveness, or physical ability, the greater his value to his employer. Consider a machine operation for which a workman is paid $4 per hour for producing 10 pieces per hour on a machine whose rate is $6 per hour. In this example the cost per piece is equal to

$$\frac{\$4 + \$6}{10} = \$1.00.$$

If a second worker of less proficiency completes 9 pieces per hour, his relative worth to his employer (as compared with that of the first workman) is calculated as follows: Let W equal the hourly pay of the second workman to result in a cost per piece of $1.00. Thus,

$$\frac{W + \$6}{9} = \$1.00$$

and

$$W = \$3.00$$

It will be noted in this example that a 10% reduction in the ability of the workman to turn out pieces results in his services being worth 75% as much per hour to his employer as the services of the first workman, all other things being equal.

If the second workman also received $4 per hour, the cost per piece of the 9 pieces he turns out in an hour will be as follows:

$$\text{cost per piece} = \frac{\$4 + \$6}{9} = \$1.11.$$

It is conceivable that the resulting cost per piece may be so high that the employer cannot profit if he must pay the second workman $4 per hour.

The comparative worth of a third workman, who produces only 5 pieces per hour, is calculated as follows: Let W equal the hourly pay of the third workman to result in a cost per piece of $1.00.

$$\frac{W + \$6}{5} = \$1.00$$

and

$$W = -\$1.00.$$

The negative result is illustrative of the fact that there are levels below which proficiencies have little or no economic value. Thus, the fixing of wage rates by minimum wage laws or other artificial limitations may result in unemployment for persons of low output and increased demand for persons of high proficiency.

The Economy of Specialization. To specialize is to restrict a person or thing to a particular activity, place, time, or situation. Specialization in this sense results when there is a division of labor, as for example, when operations in a process performed by a single person are divided among a number of persons. It is clear that machines can also be specialized in this sense, that is, specialized as to activity.

Persons or things can also be specialized as to use, place, or time. Thus a person whose assignments require that he report to work in Atlanta is specialized as to place. A passenger train that operates on a schedule is specialized as to time and place; so is a person who works in one place from nine to five o'clock, five days a week.

Specialization is of interest in relation to economy studies because it is often a means whereby the cost of accomplishing a given result can be reduced. In any design or planning effort the desirability of specialization should be considered as a routine matter. It is widely recognized that specialization, particularly as it relates to people, may result in improved performance.

Specialization usually requires preplanning and specialized arrangements, the cost and maintenance of which may more than offset the advantage gained. The gains of a specialization of one kind may require specialization of another that is too costly. For instance, it might be desirable to have two men perform one each of two activities that must be performed one after the other, unless the new arrangement would also require the men to be specialized as to time. The latter specialization might easily result in one man's being idle while the other works.

Specialization is subject to the law of diminishing returns. It is apparent that it would not be economical to specialize to such an extent, for example, that one truck driver made only right turns and another made only left turns.

Generalization, the reverse of specialization, may also be a factor in economy. Thus in machine design it is often desirable to have one part serve several purposes; the crankcase of an engine, for instance, serves as a base for the engine, a container for oil, and a support for the crankshaft and cylinder. The term "all-around man" is evidence of generalization with respect to personnel.

The Economy of Dependability. There is an economy of each of the innumerable personal qualities, but few transcend dependability in importance. Dependability in relation to organized action means behaving as agreed upon or as expected. For example, the chairman of an industrial committee has called a meeting of twelve persons but arrives fifteen minutes after the specified time. If the average salary rate of committee members is $8 per hour the lack of dependability of the chairman may have resulted in a loss of time of $12 \times 0.25 \times \$8$, or $24. This direct loss points out the value of dependability, but it probably is not as great as the frustration and loss of confidence that the lack of dependability engenders in the persons affected by it.

Dependability rests upon ability, consideration for others, willingness to cooperate, and honesty. Lack of dependability in one person, regardless of the reason for it, invariably results in the need for increased vigilance and effort on the part of others. People in supervisory positions are keenly aware of the costs of deviation from dependability, and they thus prize dependability highly.

Willful undependability and dishonesty are particularly difficult to cope with because even infrequent dishonesty leads to the necessity for a supervisor to be continually prepared against the entire range of defections that a dishonest person may cause. Thus, for many positions, even extremely limited dishonesty may outweigh the honest services a person may render.

17.4. INCENTIVES IN ORGANIZED ACTIVITY

A person will accept and continue to work in a situation he believes to hold a net advantage for him in comparison with other opportunities that he knows to be open to him. The advantages of a situation may be considered to be positive incentives and the disadvantages may be considered to be negative incentives. People are induced to act on the basis of net incentives. Incentives are personal. A thing prized highly by one person may be regarded with indifference, or even with disgust, by another.

Material Incentives. Material incentives offered to induce people to work in free enterprise are wages; wage supplements such as contributions to insurance, savings, and housing programs; and the physical condition of the plant and its locale. Up to the point where it is sufficient to provide the basic necessities of life, material compensation is one of the strongest incentives. But even wages must be paid out in certain ways for greatest effectiveness. Recognition of this fact at one time led to some very complex incentive wage plans. But, today with our more advanced techniques and greater experience, the trend is to keep the plan itself as simple as possible. Merit rating and job evaluation are other techniques that are being used to secure greater output for an input of wages.

Although material incentives as represented by compensation are emphasized, they have limited effectiveness, particularly after the individual receiving them is provided with physical necessities. This may be inferred from the fact that it is generally recognized that such valuable qualities as devotion, dependability, loyalty, initiative, industry, and honesty can be purchased for money in only a small degree. High rates of absenteeism often accompany high rates of compensation. Productivity per man does not necessarily increase with an increase in wages, and it has been known to fall. Ordinarily, doubling of wages cannot be expected to incite people to work twice as hard, if for no other reason than their physical limitations.

The overemphasis on material incentives stems, no doubt, in some measure from the sales effort, most readily discernible in modern advertising, directed at inculcating a desire for material things. The latter is a means for increasing output so that advantage may be taken of the economies of mass production. It has also resulted in demands for higher wages for a given service, or in other words, has resulted in a decrease in the value of wages as an incentive.

There are a number of material incentives that do not result in enlarged employee earnings. These are embodied in the objective plant conditions and involve such things as cleanliness, quality of buildings and equipment, maintenance, lighting, heating, air conditioning, provision of parking space, and recreational facilities. These incentives seem to have more or less universal appeal and are evaluated in comparison with the conditions existing in other plants. They can be applied on an impersonal basis. Good physical conditions are often an important factor in inducing a favorable response, particularly in conjunction with other incentives of more personal nature, but alone their effectiveness appears to be very limited.

Nonmaterial Incentives. There seems to be a wide individual divergence in the effectiveness of nonmaterial incentives. Some people are greatly influenced by the companionship afforded by fellow workers. Others can be induced to perform superlatively by a knowledge that such performance will result in praise or prestige. An important incentive for ambitious people is the opportunity a situation affords for advancement. The opportunity for enlarged participation in the affairs of the organization with which they are associated is a very strong incentive for many. For some this takes the form of being able to have a hand in directing the activities of the group; others will consider being "in the know" a reward for faithful service.

One of the strongest incentives is the opportunity for self-expression through creativeness in work that is in accord with natural aptitudes of the person concerned. This incentive is not feasible in work requiring strict obedience to specifications, policies, and superiors.

Leaders should strive to organize their subordinates in relation to the work to be done, so that each worker may receive the greatest satisfaction in doing that which contributes most toward reaching the objectives to be accomplished. When the work itself provides an incentive for doing it in whole or in part, there is true economy.

Nonmaterial incentives are not provided without cost, and they can easily be more costly for the beneficial motivation they provide than material incentives. They must, to a large extent, be applied upon a personal basis, for their effectiveness depends upon the characteristics of the persons to whom they are offered. For example, the prestige of a private office might be a great incentive to some employees, but not to others.

A peculiarity of nonmaterial incentives is that they often affect others beside the person to whom they are directed. Thus, giving one employee a private office might prove to be a negative incentive to several others and result in a net loss. On the other hand, if all employees of a group were given private offices, the beneficial motivation might be very slight.

Because of their personal nature, the successful and extensive offering of nonmaterial incentives depends upon the selection of personnel who are easily pleased and upon leaders who have an aptitude for distributing incentives to their subordinates in such a way that each feels he has received favorable individual attention. The selection of employees who are easily and favorably motivated is an economy, since such employees require less input of leadership attention.

The cost and effectiveness of nonmaterial incentives rests almost entirely upon the personal attributes of leaders. Nonmaterial incentives that are offered to a subordinate are controlled in large measure by leaders. Thus, the selection of leaders is an important factor in economy, although it appears not to receive the attention it warrants.

17.5. THE ECONOMY OF RESOURCE INPUT

In order for organizations to remain operational, they must receive both tangible and intangible inputs. Incentives, as discussed in the previous section, are inputs of a particular type directed to the person for his individual benefit. Other resource inputs are directed to the person for his use as a contributor to the organization. To reach a decision relative to the economic desirability of a proposed resource input, a relationship between the cost of the input and the value of the expected output should be established in terms of a comparable measure such as money.

Evaluating a Tangible Resource Input. The cost of a tangible resource input such as equipment for use by an individual in his work can usually be obtained without great difficulty. But the value of output is often very difficult, if not impossible, to obtain. In such case, an analysis is necessary to determine the increment of output that must be obtained for an increment of resource input.

As an example suppose that the desirability of providing an electronic technician with automated test equipment is being considered. Under the present set-up the annual cost of wages and fringe benefits is $12,200. Under the proposed arrangement, the annual cost would be increased by the equivalent annual cost of the test equipment. Suppose that this increase is $800 and that the technician must increase his productivity in direct proportion to the increase in cost of his wages, fringe benefits, and equipment.

With this assumption, the increase in productivity must be at least

$$\frac{(\$12,200 + \$800) - \$12,200}{\$12,200} = 0.0656 \text{ or } 6.56\%.$$

The decision is now reduced to consideration of the estimated increase in productivity with the minimum increase needed to pay for the automated test equipment.

Evaluating an Intangible Resource Input. All discernible human activity is of a physical nature. Man achieves his ends by manipulation of his physical environment. This is as true of the administrator as it is of a machinist, for whatever thoughts the former may have, he cannot transmit them except as they are manifested in physical ways which can be perceived by others.

The manipulation of the physical environment necessary to produce a desired result can be learned only as the result of experience or of instruction. Experience is often very costly, but the knowledge that a single person has gained may often be transmitted through the processes of instruction at relatively little cost. For example, the ingredients and the method of heat-treating a useful alloy may be learned only after years of costly experimentation, but the knowledge gained may be taught and become useful to other persons at relatively little cost.

Instruction is an economical method of transmitting useful knowledge and therefore an economical means of increasing the abilities of persons to achieve ends not otherwise attainable or at less cost. Through instruction, the best knowledge of the most capable persons can be made available to all. Thus, instruction is a resource input which can be of great value to an organization.

Industrial and business organizations are making increasing use of instructional programs to improve the competence of employees. The worth of such programs is often difficult to evaluate. However, the cost of an instructional program can usually be estimated with a fair degree of accuracy. Based on the cost, the increase in the outputs of participants necessary to justify a program of instruction can be calculated.

Consider the following example involving instruction as an intangible resource input. The program of instruction is to be 40 hours in duration for a class of 10 engineers at a cost of $40 per hour plus salary costs of participants. Assume that the average salary of participants is $11,600 for a 40 hour, 50-week year and that the value of the instruction will be applicable for one year only. Under these assumptions, the average increase in productivity per engineer during the year to justify the instruction is

$$\frac{\left[10(\$11,600) + \frac{10(\$11,600)}{50} + 40(\$40) \right] - 10(\$11,600)}{10(\$11,600)}$$

$$\frac{\$119,920 - \$116,000}{\$116,000} = 0.0338 \text{ or } 3.38\%.$$

The decision regarding the intangible input now hinges on the estimated output increase compared with the minimum needed to break even. It should be recognized in the decision that the required productivity increase of 3.38% is based on a payout in one year when it is likely that the value of instruction will exist for several years.

17.6. ECONOMICAL SIZE OF REPAIR CREWS

When nonrepetitive work that precludes preplanning is performed, the number of men in a crew is an important consideration from the standpoint of the crew's effectiveness. Loss in effectiveness may arise from one or more characterisics inherent in the size of the crew. For example, as the size of the crew increases, loss of effectiveness will result from the time of going to and from the job.

As an absurd illustration, assume that a maintenance job involving 60 man-minutes of work is to be performed at such a distance that it will require 10 minutes to go to the job and 10 minutes to return from the job. Assume further that the work to be performed can be done effectively by either a one-man or a 60-man crew. Labor cost is taken at $4.80 per man-hour, or $0.08 per man-minute. A comparison of the over-all effectiveness of each crew is revealed by the following analyses:

ONE-MAN CREW

Activity	Minutes Required in Performance	Man-Minutes
Walk to job	10	10
Perform job	60	60
Walk from job	10	10
Total man-minutes		80
Total cost, 80 man-min. @ $0.08 = $6.40		

SIXTY-MAN CREW

Activity	Minutes Required in Performance	Man-Minutes
Walk to job	10	600
Perform job	1	60
Walk from job	10	600
Total man-minutes		1260
Total cost, 1260 man-min. @ $0.08 = $100.80		

The difference between $100.80 and $6.40, or $94.40, is a measure of the loss of effectiveness in travel time because of an excessive crew size.

There are many situations in which economy depends upon a balance between losses associated with machine down time and the cost associated with maintaining a crew to undertake preventative maintenance and repair to minimize such losses. Suppose that the failure of a machine will result in a loss of $8 per hour during the period that it is inoperative or "down" for repairs. Past experience shows that a typical repair for this machine can be made by one man in 9 hours, by two men in 5 hours, by three men in 4 hours, by four men in 3.50 hours, and by five men in 4 hours. The wage rate of repair men is taken to be $4 per hour. The costs incident to the failure of the machine are given in Table 17.1.

Table 17.1. COSTS INCIDENT TO REPAIR AND DOWN TIME

Number of men in repair crew	1	2	3	4	5
Hours required to make repair	9	5	4	$3\frac{1}{2}$	4
Man-hours required to make repair	9	10	12	14	18
Labor cost of making repair at $4 per hour	$36	$40	$48	$56	$72
Down-time cost at $8 per hour for machine	$72	$48	$32	$28	$32
Resultant of labor cost of making repair and down-time loss of machine	$108	$88	$80	$84	$104

The economy associated with the loss of use of the machine suggests that the repair crew should be the size that will result in minimum down time. A person concerned only with the down-time loss of a machine would select a 4 man crew, which would hold the down-time loss to $28 per repair. The economy of the repair activity points to a crew of one man so that the repair may be made most efficiently and at least cost. A person primarily interested in the efficient utilization of repair men and low repair costs would have his men work singly. Usually it is the overall economy that is desired. In this case the greatest overall economy occurs when crews of 3 men are used resulting in a total cost of $80.

PROBLEMS

1. Suppose that there are 100 boulders in a path to a spring used regularly by 50 cave men. Each cave man considers the inconvenience and concludes that the present worth of the benefit to him would be $20 if the boulders were removed. Since the cost of removing each boulder is $2, each abandons the thought of clearing the path. An entrepreneur, perceiving the potential benefit of organization, convinces the group that they should coordinate their efforts. If each cave man removes his share of the 100 boulders how much can the entrepreneur charge for his service if the benefit-cost ratio must be at least 2?

2. A group of 60 employees, whose average salary is $8,800 per year, is directed by 5 foremen, whose average salary is $12,000 per year. It is suggested that the number of foremen be increased to 6.

(a) What % increase in average output per man must result for the two plans to incur equal cost?

(b) If the suggested plan is adopted, what will be the % increase in average time that can be spent in supervision per worker?

3. A foreman receiving a salary of $14,500 per year supervises the work of 18 men whose average annual salary is $9,000 per year. It is proposed that the employment of a second foreman will improve the total situation.

(a) Assuming that the two foremen would have to spend 2 hours each day coordinating their work, what will be the average increase in the time spent in supervision per day per man?

(b) What average % increase in output is necessary to justify the employment of the second foreman?

4. Foreman A directs the work of 12 men. He plans his work "as he goes" during the eight-hour work period. Foreman B also directs the work of 12 men, but spends two hours per day in planning the work to be done, thus reducing the time available for active supervision. What percentage more effective must his supervision be per unit of time so that his supervisory effectiveness equals that of Foreman A?

5. An engineer can do certain required computations in 3 hours or he can delegate the work to an engineering aid. If the work is delegated, it will take 0.75 hour to explain the computational procedure and 0.50 hour to check the results. The actual calculations will take 4 hours to do if done by the aid. If the engineer receives a salary of $18,600 per year and the aid receives $7,200 per year, what are the comparative costs for each of the methods for a working year of 2,080 hours?

6. A foreman is in charge of a construction crew of eight men and takes great pride in the amount of work he personally performs on the job. Observation shows that the accomplishment of the men is impaired by lack of direction as a result of the foreman's active participation in the work to be done. The foreman receives $5.10 and his men $3.80 per hour. If the foreman does one-third as much work as the average of his men would do if they were properly directed, what loss in the effectiveness of the crew will just be compensated for by the actual work performed by the foreman?

7. A foreman supervises the work of Mr. B, and eight other men. The foreman states that "Mr. A requires twice as much and Mr. B requires half as much of my time as the average of my men." Mr. A's output is 8 units per day and Mr. B's output is 7 units per day. On the basis of equal cost per unit, what monthly salary is justified for Mr. B if the foreman receives $920 per month and Mr. A receives $680 per month?

8. A manufacturer has been awarded a contract to make 6,500 mechanisms for a 50-cap blasting machine. Assembly of the mechanism can be performed by highly skilled operators working individually or by workmen of lesser skill working as a team, if they are provided with specialized assembly equipment and given special supervision. It is estimated that the highly skilled workmen, whose wage rate is $5.80 per hour, will require an average of three hours per assembly.

An assembly line that will permit specialization of labor can be set up for an initial cost of $9,500. The assembly can then be performed by four men whose individual wage rate is $4.10 per hour supervised by a foreman whose wage rate per hour is $6.40. It is estimated that the average time for this crew to complete an assembly will be 30 minutes per unit. If the assembly line equipment will be worthless after completion of the contract which method should be used?

9. In a certain manufacturing activity, 40 employees are engaged in identical activities. The average output of the group as a whole is 46.4 units per hour. The average output of the less productive half is 40.2 satisfactory and 1.4 unsatisfactory units per hour and the average for the more productive half is 52.6 satisfactory and 0.8 unsatisfactory units per hour. The employees work on a straight piecework plan and receive $3.30 per hundred satisfactory units. The firm sustains a loss of $0.07 for each unsatisfactory unit. One machine is required for each employee. Each machine has an annual fixed cost of $320 and a variable cost of $0.18 per hour. Supervision and other overhead costs are estimated at $720 per employee per year. The average employee works 1,900 hours per year. How much could be paid annually for a selection, training, and transfer program which would result in raising the average productivity of the entire group to 52.6 satisfactory and 0.8 unsatisfactory units per hour?

10. A small amount of collating is done in an office. A collating device which permits assembly of papers to be accomplished in about one-half the time required without it can be purchased for $125. On the basis of a service life of five years, an employee wage rate of $2.90, and a 50% time saving, calculate the number of hours of use per year to justify the purchase of the device if the interest rate is 8%.

11. Assume that N units of product can be made manually by one man in a year of 2,000 working hours, that the man's wage rate is W dollars per hour, and that a labor saving machine whose annual capital and operating cost is equivalent to $R \times 2,000$ hours of labor, will, if used throughout the year, reduce the amount of labor needed to produce the N units by $S \times 2,000$ hours. Write equations for the unit cost of product when manually made and when made with the aid of the machine. Determine the ratio of R to S when the costs are equal.

12. A pneumatically operated wrench costing $62 will permit the installation of a certain bracket 75% faster than with a manual wrench. On the basis of a service life of 5 years, an employee wage rate of $4.10 per hour, and an interest rate of 8% compounded continuously, calculate the number of hours of use per year that will justify the purchase of the automatic wrench.

13. Two typewriters are under consideration costing $340 and $460, respectively. A secretary will use a typewriter 70% of her working time. The annual direct and indirect cost of the secretary is $6,100. What % increase in effectiveness will be necessary to justify the price of the higher priced typewriter if each machine has a service life of 4 years and the interest rate is 9%?

14. A manufacturing concern has sales offices in five states in addition to the main plant and offices. Each branch sales manager is paid $16,000 per year and the vice-president for sales receives $24,000 per annum. At the present time, average annual sales amounts to $2,300,000. An annual sales conference is being considered which will meet at the main plant for 2 days each year.

(a) If each branch sales manager is paid $280 for travel and other expenses, what is the total cost of the conference on the basis of a 240-day work year?

(b) What % increase in sales attributable to the conference is necessary to justify the annual meeting if the gross profits are 12% of sales?

15. A manufacturer of a special purpose precision machine conducts a 2-week course for machine operators employed by its customer companies. It is generally accepted that the output of operators attending such a course is increased by 5%. A company employing operators at a salary of $190 per week is considering the 2-week course.

    (a) What is the total cost of training per operator if the fee is $250 and the operator is paid 15% extra for attending the course?

    (b) What would be the rate of return on the investment in the training program?

16. A machine tool operator produces 46 units per hour of which an average of 6 are defective. He is paid on a straight piecework basis at the rate of $5 per 100 satisfactory units. The firm sustains a loss of $0.02 per defective unit produced. The machine used in the process has a total operating cost of $3 per hour which includes depreciation, return on investment, and overhead. The operator works 1,800 hours per year. If a training program is initiated, it is anticipated that the production rate will increase to 52 units per hour only 2 of which will be defective.

    (a) What is the maximum amount that the firm can spend on the training program per year?

    (b) What would be the effect of the training program on the hourly wage of the operator?

18

Operations Involving
Production Facilities

Most functions of an enterprise may be classified under the heading of operations. Included here are the routine activities associated with engineering, finance, manufacturing, marketing, industrial relations, and so forth. Such operations are the result of organized activity directed to the satisfaction of human wants through production or construction. The purpose of evaluating operations is for the attainment of objectives through the selection of operational alternatives that are most favorable from an economic standpoint.

In this chapter techniques of economic analysis are presented for plant and equipment operations. Operations of this type concern the physical facilities required by organizations to produce goods and service.

18.1. LINEAR BREAK-EVEN ANALYSIS

There are two aspects of an industrial enterprise. One consists of assembling labor, facilities, and material for the production of goods or services. The other consists of the distribution of the goods or services that have been produced. The success of an enterprise depends upon its ability to carry on these activities to the end that there may be a net difference between receipts for goods and services sold and the input necessary to produce and distribute them.

If receipts and costs are assumed to be linear functions of the quantity of product to be made and sold, analysis of their relationships to profit is greatly simplified. If this assumption is made, the patterns of income and costs will appear as in Figure 18.1.

In Figure 18.1 fixed production costs are represented by the Line *HL*.

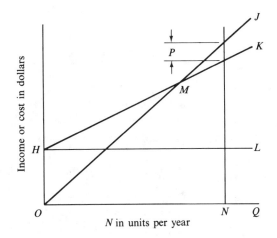

FIGURE 18.1. GENERAL GRAPHICAL REPRESENTA-
TION FOR INCOME, COST, UNITS OF OUTPUT, AND
PROFIT.

The sum of variable production costs and distribution costs is represented by the Line HK. Income from sales is represented by the Line OJ. It must be remembered that this linear representation is only an approximation of actual operating conditions.

Analysis of existing or proposed operations that are represented with a break-even model can be made mathematically or graphically. For an illustration of the mathematical methods, let

$N =$ number of units of product made and sold per year.

$R =$ the amount received per unit of product in dollars. R is equal to the slope of OJ.

$I = RN$, the annual income from sales in dollars. $I = RN$ is the equation of Line OJ.

$F =$ fixed cost in dollars per year, represented by OH and HL.

$V =$ variable cost per unit of product. V is equal to the slope of HK.

$TC =$ the sum of fixed and variable cost of N units of product, $F + VN$. $TC = F + VN$ is the equation of Line HK.

$P =$ annual profit in dollars per year. $P = I - TC$. Negative values of P represent loss.

$M =$ break-even point. At this point $P = O$.

$Q =$ capacity of plant in units per year.

The break-even point occurs when income is equal to cost. In Figure 18.1 this occurs where Lines OJ and HK intersect. At this point $I = C$ and $RN = F + VN$. Solving for N,

$$N = \frac{F}{R - V}$$

which is the abscissa value of the break-even point.

If $F/(R - V)$ is substituted for N is $I = RN$ or $TC = F + VN$, the ordinate of the break-even point may be found. The value of the ordinate in terms of dollars of income or cost will be

$$I = R\left(\frac{F}{R - V}\right) \text{ and } TC = F + \frac{VF}{R - V}.$$

As an example, suppose it is desired to find the break-even point when $R = \$11$, $F = \$4,000$, and $V = \$5$.

$$N = \frac{F}{R - V} = \frac{\$4,000}{\$11 - \$5} = 667 \text{ units per year}$$

and

$$I = C = RN = \$11 \times 667 = \$7,337.$$

Since P is the annual profit it is often desirable to have a relationship that expresses P as a function of the number of units made and sold, N. This relationship may be derived as follows:

$$P = I - C$$
$$= RN - (F + VN)$$
$$= (R - V)N - F.$$

For example, if the annual profit is required when $R = \$11$, $F = \$4,000$, $V = \$5$, and $N = 800$, the following calculations may be made:

$$P = (R - V)N - F$$
$$= (\$11 - \$5)800 - \$4,000$$
$$= \$6(800) - \$4,000 = \$800.$$

Even though the analysis presented was based upon the assumption that fixed cost, variable cost, and income are linear functions of the quantity made and sold, it is very useful in evaluating the effect on profit of proposals for new operations not yet implemented and for which no data exists. Consider a proposed activity consisting of the manufacturing and marketing of a certain plastic drafting template for which the sale price per unit is estimated to be $0.30. The machine required in the operation will cost $1,400 and will have an estimated life of 8 years. It is estimated that the cost of production including power, labor, space, and selling expense will be $0.11 per unit sold. Material will cost $0.095 per unit. An interest rate of 8% is considered necessary to justify the required investment. The costs associated with this activity are:

A/P 8,8	Fixed Costs (*Annual*)	Variable Costs (*Per Unit*)
Capital recovery and return $1,400(0.17401)	$244	
Insurance and taxes	34	
Repairs and maintenance	22	$0.005
Material		0.095
Labor, electricity, and space.....................		0.110
Total.......................................	$300	$0.210

The difficulty of making a clear-cut separation between fixed and variable costs becomes apparent when attention is focused on the item for repair and maintenance in the above classification. In practice it is very difficult to distinguish between repairs that are a result of deterioration that takes place with the passage of time and those that result from the wear and tear of use. However, in theory, the separation can be made as shown in this example and is in accord with fact, with the exception, perhaps, of the assumption that repairs from wear and tear will be in direct proportion to the number of units manufactured. To be in accord with actualities, depreciation also undoubtedly should have been separated so that a part would appear as variable cost.

In this example, $F = \$300$, $TC = \$300 + \$0.21\ N$, and $I = \$0.30\ N$. If F, TC, and I are plotted as N varies from 0 to 6,000 units, the result will be as shown in Figure 18.2.

The cost of producing templates will vary with the number made per year. The production cost per unit is given by $F/N + V$. If production cost per

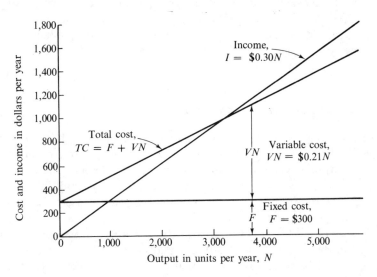

FIGURE 18.2. FIXED COST, VARIABLE COST, AND INCOME PER YEAR.

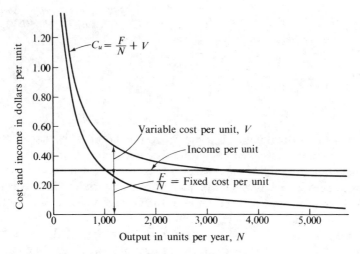

FIGURE 18.3. FIXED COST, VARIABLE COST, AND INCOME
PER UNIT.

unit, variable cost per unit, and income per unit are plotted as N varies
from 0 to 6,000 the results will be as shown in Figure 18.3. It will be noted
that the fixed cost per unit may be infinite. Thus, in determining unit costs,
fixed cost has little meaning unless the number of units to which it applies
are known.

Income for most enterprises is directly proportional to the number of units
sold. However, the income per unit may easily be exceeded by the sum of
the fixed and the variable cost per unit for low production volumes. This is
shown by comparing the total cost per unit curve and the income per unit
curve shown in Figure 18.3.

18.2. NON-LINEAR BREAK-EVEN ANALYSIS

Table 18.1 gives cost data for a hypothetical plant which has a maximum
capacity of 10 units of product per year and rates of production ranging
from 0 to 10 units per year. These data and those presented in Tables 18.2
and 18.3 will be used as a framework for an analysis of relationships between
production costs, distribution cost, income, and profit where the functions
are nonlinear.

The annual total production cost of the hypothetical plant as its rate of
production varies from 0 to 10 units per year is given in Column (2) to Table
18.1. It should be specifically noted that no distribution costs are embraced
in these data.

Annual fixed cost is given as $200 per year in Column (3). The annual
fixed cost in this example should be considered to be the annual cost of

maintaining the plant whose capacity is 10 units per year in operating condition at a production rate of zero units per year. The fixed cost of a plant is analogous to the standby costs of a steam boiler whose fire is banked but which is ready to furnish steam on short notice.

The difference between the annual total production cost in Column (2) and the fixed production cost in Column (3) constitutes the annual variable cost and appears in Column (4). Average unit production, fixed and variable costs appear in Columns (5), (6), and (7) respectively.

The increment production costs per unit given in Column (8) were obtained by dividing the differences between successive values of annual total production cost as given in Column (2) by the corresponding difference between successive values of annual output given in Column (1). In this case the divisor will be equal to unity but it should be understood that an increment of production of several units may be convenient in some analyses.

The values in all columns of Table 18.2 except those in Column (6) and the values of all columns of Tables 18.2 and 18.3 have been plotted with respect to the annual output in number of units in Figures 18.4, 18.5 and 18.6, and have been keyed for identification.

Consider Curve (A) in Figure 18.4. Total production cost curves of actual operations take a great variety of forms. Ordinarily, however, a producing unit will produce at minimum average cost at a rate of production between zero output and its maximum rate of output. Thus, the average unit cost of production will decrease with an increase in the rate of production from zero

Table 18.1. RELATIONSHIP OF PRODUCTION COST, NET INCOME FROM SALES, PROFIT, AND THE NUMBER OF UNITS MADE AND SOLD PER YEAR

(1) Annual output, number of units	(2) Annual total production cost, A, Fig. 18.4	(3) Annual fixed production cost, B, Fig. 18.4	(4) Annual variable production cost, C, Fig. 18.4	(5) Average production cost per unit, D, Fig. 18.4	(6) Average fixed cost per unit	(7) Average variable cost per unit, E, Fig. 18.4	(8) Incremental production cost per unit, F, Fig. 18.4
0	$200	$200	$ 0	$ ∞	$ ∞	$ 0	
1	300	200	100	300.00	200.00	100.00	$100
2	381	200	181	190.50	100.00	90.50	81
3	450	200	250	150.00	66.67	83.30	69
4	511	200	311	127.75	50.00	77.75	61
5	568	200	368	113.60	40.00	73.60	57
6	623	200	423	103.83	33.33	70.50	55
7	679	200	479	97.00	28.57	68.48	56
8	740	200	540	92.50	25.00	67.50	61
9	814	200	614	90.44	22.22	68.22	74
10	924	200	724	92.40	20.00	72.40	110

until a minimum average cost of production is reached and then the average cost per unit will rise.

The average total production cost is given by Curve (D) in Figure 18.4. It will be noted that it reaches a minimum at a rate of nine units per year. The average variable production cost is given in Curve (E); its minimum occurs at eight units per year. Curves (D) and (E) should be considered in relation to Column (6) of Table 18.1. It should be noted that fixed cost per unit is inversely proportional to the number of units per year. Reduction of fixed cost per unit is an important factor tending toward lower average cost with increases in rates of production.

Incremental production costs are given in Curve (F); their minimum value is reached at six units per year. The form of this curve shows that the incremental cost of production per unit declines until 6 units per year are produced and then rises.

An interesting fact to observe is that the incremental cost Curve (F) intersects the average total production cost Curve (D) and the average variable production costs Curve (E) at their minimum points. The incremental cost Curve (F) is a measure of the slope of curve (A). It may be noted that the slope of Curve (A) decreases until six units per year is reached and then increases. The same could have been said of a variable production cost Curve (C) if one had been plotted.

Pattern of Income and Cost of Distribution. Further consideration will be given to the curves of Figure 18.4 after income from sales of product and the cost of distribution have been explained. For simplicity, the cost of dis-

Table 18.2. RELATIONSHIP OF GROSS INCOME FROM SALES, NET INCOME FROM SALES, DISTRIBUTION COST, AND NUMBER OF UNITS SOLD PER YEAR

(1) Annual output, number of units	(2) Annual gross income from sales, J, Fig. 18.5	(3) Annual distribution cost, K, Fig. 18.5	(4) Annual net income from sales, G, Fig. 18.5	(5) Incremental distribution cost per unit, L, Fig. 18.5	(6) Incremental net annual income from sales per unit, H, Fig. 18.5
0	$ 0	$ 0	$ 0	—	—
1	140	38	102	$ 38	$102
2	280	72	208	34	106
3	420	104	316	32	108
4	560	136	424	32	108
5	700	170	530	34	109
6	840	211	629	41	99
7	980	261	719	50	90
8	1,120	325	795	64	76
9	1,260	405	855	80	60
10	1,400	505	895	100	40

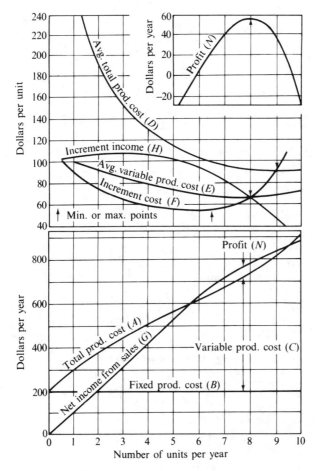

FIGURE 18.4. GRAPHICAL PRESENTATION OF THE
DATA GIVEN IN TABLE 18.1.

tribution will be considered to be a summation of an enterprise's expenditures
to influence the sale of its products and services. Such items as advertising,
sales administration, salesmen's salaries, and expenditures for packaging and
decoration of products done primarily for sales appeal will be included in
the cost of distribution. Gross annual income from sales is the total income
received from customers as payment for products. In Columns (2) and (3)
of Table 18.2, gross annual income from sales and the annual cost of distribution are given for various rates of product sales of the firm for which cost-
of-production data were given in Table 18.1 and Figure 18.4.

The difference between the gross annual income from sales and the annual
cost of distribution is equal to the annual net income from sales given in

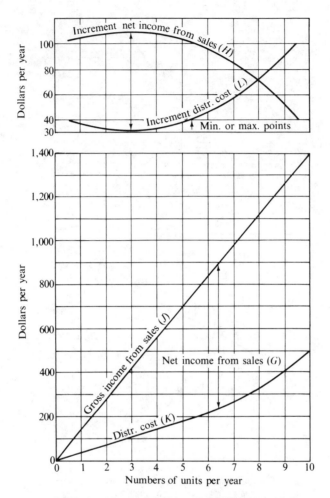

FIGURE 18.5. GRAPHICAL PRESENTATION OF THE
DATA GIVEN IN TABLE 18.2.

Column (4). Incremental distribution costs per unit are given in Column (5) and incremental net income from sales per unit appears in Column (6). The values of each of these columns have been plotted with respect to annual output in Figure 18.5. The gross annual income from sales, Curve (J) in Figure 18.5, is a straight line and is typical of situations in which products are sold at a fixed price. The annual incremental distribution cost, Curve (L), first falls slightly until a minimum is reached and then rises. This is typical of situations in which sales effort is relatively inefficient at low levels and where sales resistance increases with increased number of units sold.

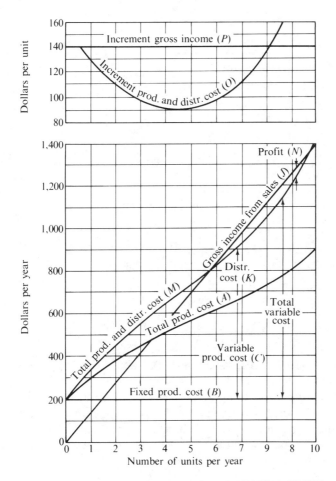

FIGURE 18.6. GRAPHICAL PRESENTATION OF THE
DATA GIVEN IN TABLE 18.3.

A pattern of distribution cost, made up of fixed and variable costs similar to those of production, might have been taken. Ordinarily, fixed distribution costs are a minor consideration; they have been omitted from this analysis in the interest of simplicity.

A representation of the net income from sales is given as (G) in Figure 18.5. The positive difference between net income from sales and the total cost of production, Curve (A) in Figure 18.4, represents profit. Profit is shown as (N) in Figure 18.4, it should be noted that the maximum profit, or minimum loss if no profit is made, occurs at a rate of production at which the incremental income Curve (H) intersects the incremental cost Curve (F),

Table 18.3. RELATIONSHIP OF GROSS INCOME, PRODUCTION COST, DISTRIBUTION COST, PROFIT, AND NUMBER OF UNITS MADE AND SOLD PER YEAR

Annual output, number of units	Annual total production cost, A, Fig. 18.6	Annual distribution cost, K, Fig. 18.6	Annual total of production and distribution cost, M, Fig. 18.6	Annual gross income from sales, J. Fig., 18.6	Annual profit N, Figs. 18.4 and 18.6	Incremental total production and distribution costs, O, Fig. 18.6
0	$200	$ 0	$ 200	$ 0	−$200	
1	300	38	338	140	−198	$138
2	381	72	453	280	−173	115
3	450	104	554	420	−134	101
4	511	136	647	560	−87	93
5	568	170	738	700	−38	91
6	623	211	834	840	6	96
7	679	261	940	980	40	106
8	740	325	1,065	1,120	55	125
9	814	405	1,219	1,260	41	154
10	924	505	1,429	1,400	−29	210

or at eight units per year. From Column (8), Table 18.1, and Column (6), Table 18.2, it may be observed that the incremental product on cost and incremental net income for a ninth unit are $74 and $60 respectively. Thus, a loss of $14 profit would be sustained if the activity were increased to nine units. When the profit motive governs, there is no point in producing beyond the point where the incremental cost of the next unit will exceed the incremental income from it.

If the sales effort in the example above is resulting in sales of nine or ten units per year, a number of steps might be taken. The sales effort could reduced, causing sales to drop. The price could be increased, causing sales to decrease and the income per unit to increase. The plant could be expanded or other changes could be made to alter the pattern of production costs. Consideration of any of the steps above would require a new analysis embracing the altered factors.

Consolidation of Production and Distribution Cost. It is common practice to consolidate production and distribution cost for analysis of operations. To illustrate this practice an analysis of the example above will be made in this manner. The method is illustrated in Table 18.3, whose values have been plotted as several curves in Figure 18.6.

Profit in this case will be equal to the difference between gross income (J)

and total production and distribution cost (M). The point of maximum profit will occur when the incremental gross income curve (P) is intersected by a rising incremental production and distribution cost Curve (O). This occurs at a rate of 8 units per year. It would be noted that if production reaches 10 units, a loss of $29 will be sustained from the year's operation.

18.3. EQUIPMENT SELECTION FOR EXPANDING OPERATIONS

In the early stages of the enterprise, when production volume is low, it will usually prove to be economical to purchase equipment whose fixed costs are low. In the latter stages, when sales are approaching the ultimate level, high fixed cost equipment permitting low variable production costs may be most economical. In this connection consider the following example.

It is estimated that annual sales of a new product will begin at 1,000 units the first year and increase by increments of 1,000 units per year until 4,000 units are sold during the fourth and subsequent years. Two proposals for equipment to manufacture the product are under consideration.

Proposal A involves equipment requiring an investment of approximately $10,000. Annual fixed cost with this equipment is calculated to be $2,000 and the variable cost per unit of product will be $0.90. The life of the equipment is estimated at four years.

Proposal B involves equipment requiring an investment of approximately $20,000. Fixed cost of this equipment is estimated at $3,800 per year and variable cost per unit of product will be $0.30. The life of this equipment is also estimated at four years.

On the basis of the ultimate annual production of 4,000 units, cost per unit will be as follows:

Proposal A

Annual cost for 4,000 units of product,	
$2,000 + (4,000 × $0.90)	$5,600.00
Cost per unit, $5,600 ÷ 4,000	1.40

Proposal B

Annual cost for 4,000 units of product,	
$3,800 + (4,000 × $0.30)	$5,000.00
Cost per unit, $5,000 ÷ 4,000	1.25

On the basis of the ultimate rate of production, Proposal B is superior to Proposal A. On the basis of the total production during the life of the equipment, the following analyses apply:

PROPOSAL A

Year of Life	No. of Units Made	Fixed Cost	Variable Cost
1	1,000	$2,000	1,000 × $0.90 = $ 900
2	2,000	2,000	2,000 × $0.90 = 1,800
3	3,000	2,000	3,000 × $0.90 = 2,700
4	4,000	2,000	4,000 × $0.90 = 3,600
	10,000	$8,000	$9,000
Cost per unit = ($8,000 + $9,000) ÷ 10,000 = $1.70			

PROPOSAL B

Year of Life	No. of Units Made	Fixed Cost	Variable Cost
1	1,000	$ 3,800	1,000 × $0.30 = $ 300
2	2,000	3,800	2,000 × $0.30 = 600
3	3,000	3,800	3,000 × $0.30 = 900
4	4,000	3,800	4,000 × $0.30 = 1,200
	10,000	$15,200	$3,000
Cost per unit = ($15,200 + $3,000) ÷ 10,000 = $1.82			

The calculated advantage of Proposal A over Proposal B would have been increased by considering the time value of money. But perhaps even more important than the difference in cost per unit, particularly for a new enterprise that must conserve its funds or where there is considerable doubt that production schedules will be reached, is the lesser investment required by Proposal A.

At rates of operation approaching capacity, single-purpose machines may be expected to produce more economically than general-purpose machines. Specialized facilities usually result in higher fixed costs and lower variable costs than general-purpose facilities. Consequently specialized equipment is generally advantageous for high volumes of work and general-purpose equipment is advantageous for low volumes of work.

Single-purpose machines particularly are subject to obsolescence owing to change in design or demand for the product on which they are specialized. Soundness of analyses involving specialized facilities rests heavily upon the accuracy of estimates of the volume of work to be performed by them.

18.4. ECONOMIC OPERATION OF EQUIPMENT

Most production facilities may be operated at rates below, equal to, or above their normal capacity. Facilities are usually inefficient and costly to operate when utilized at a rate of output below their normal capacity. For example, if a production department is allotted more space than is needed for its efficient operation, a number of losses may be expected to arise. The fixed costs such as maintenance, taxes, insurance, and interest will be higher than necessary. Heat, light, and janitor service will be wasted in the unused space, and the cost of supervision and material handling may be expected to be higher than necessary.

The load that is considered proper to impose on equipment is usually indicated by its makers. Such indicated normal loads are often not the optimum for economy. Operation of equipment above its normal capacity will result in increased production at the expense of incresed power consumption and shortened life. But, in many cases, the overall result is a reduced cost per unit produced. If such is the case, the equipment in question should be overloaded so that the resulting economy may be realized.

The Economy of Loading to Normal Capacity. A mining company that is expanding its operations is in need of direct current for an electrolytic process. During the next year, an average of 12 kilowatts of direct current energy will be needed. Planned expansion of the existing operations is expected to increase the rate at which direct current is needed by 2 kilowatts per year until 30 kilowatts are needed. At this point, the demand for current will remain constant. Energy will be needed 2,000 hours per year regardless of the rate at which it is used.

Investigation reveals that the need for direct current can best be provided by an a-c to d-c motor-generator set. These sets can be purchased in a variety of capacities. Plan A involves the purchase of a 20-kilowatt set now for five years of use at which time the demand will have reached 20 kilowatts. At this time, a 30-kilowatt set will be purchased.

Efficiency-load curves provided by the vendor show that the 20-kw set has efficiencies of 48, 69, 78, and 76% at $\frac{1}{4}, \frac{1}{2}, \frac{3}{4}$, and full load respectively. The 30-kw set is slightly more efficient at its $\frac{1}{4}, \frac{1}{2}, \frac{3}{4}$, and full load. The purchase rate of a-c current is $0.02 per kilowatt-hour. Interest is taken at 10%.

Table 18.4. PRESENT-WORTH COST OF CONSUMED POWER FOR PLAN A

Year	Output rate d-c current, in kw (A)	Efficiency at d-c output rate (B)	Input rate a-c current, A ÷ B (C)	Annual power bill C × $0.02 × 2,000 hr. (D)	Present worth of annual power bill D × (P/F 10, n) (E)
1	12	74	16.2	$ 648	$589
2	14	77	18.2	728	602
3	16	78	20.5	820	616
4	18	78	23.1	924	631
5	20	76	26.3	1,052	653

Taxes, insurance maintenance, and operating costs will be neglected in the interest of simplicity. The 20-kw set will cost $1,230 installed. It is estimated that $800 will be allowed for the 20-kw set on the installed purchase price of $1,580 of the 30-kw set 5 years hence. An analysis of the present-worth cost of the power consumed for each of the 5 years is given in Table 18.4.

The present-worth of providing 5 years of service under Plan A is calculeted as follows:

Total present worth of power bill for 5 years (ΣE in table) $3,091
Present-worth cost of 20-kw set installed $1,230
$$P/F\,10,\,5$$
Present worth of 20-kw set trade-in receipt, $800(0.6209) .... −497
$$P/F\,10,\,5$$
Present worth of 30-kw set, $1,580(0.6209) 981
                                                                      ─────────
                                                                      $4,805

Plan B involves the purchase of the 30-kilowatt set now so that it will be available to meet the anticipated demand. The present-worth cost of the power consumed for each of the first five years is given in Table 18.5.

Table 18.5. PRESENT-WORTH COST OF CONSUMED POWER FOR PLAN B

Year	Output rate d-c current, in kw (A)	Efficiency at d-c output rate (B)	Input rate a-c current, $A \div B$ (C)	Annual power bill $C \times \$0.32 \times$ 2,000 hr. (D)	Present worth of annual power bill, $D \times (\;\;\; P/F\,10,\,n\;\;\;)$ (E)
1	12	65	18.5	$ 740	$673
2	14	70	20.0	800	661
3	16	74	21.6	864	649
4	18	77	23.4	936	639
5	20	79	25.3	1,012	628

The present-worth cost of providing five years of service under Plan B is calculated as follows:

Total present worth of power bill for 5 years ($\sum E$ in table) $3,250
Present-worth cost of 30-kw set installed $1,580
                                                                      ─────────
                                                                      $4,830

Power costs beyond the first five years have not been taken into account in the analysis above because they will be equal in subsequent years since a 30-kilowatt set will be used regardless of whether Plan A or B is adopted. The assumption that a 30-kilowatt unit five years old is equivalent to a new set was made to simplify the analyses and because electrical equipment ordinarily has a long service life.

The Economy of Loading above Normal Capacity. The life of a piece of equipment is usually inversely proportional to the load imposed upon it. But the output is directly proportional to the load imposed. Where such is the case there exists a least cost load that will determine the level of operation for maximum economy.

In some cases, particularly with short-lived assets, the economic load may be determined by experiment. If the effect of various loadings upon the life, maintenance, and other operating costs of a unit of equipment are known or can be estimated with reasonable accuracy, it is practical to determine the load that will result in the greatest economy.

Consider the following example. A concern has an automatic plastic moulding machine that produces one piece for each revolution it makes. It is now being operated at the rate of 150 r.p.m. and produces 18.6 pounds of product per hour. The machine costs $6,150 and is estimated to have an operating life of 10,000 hours. Direct wages plus labor overhead applicable to the operation amount to $1.37 per hour. Average present maintenance and power are $0.072 and $0.064 per hour, respectively. Output of product and cost of power used per hour are estimated to be directly proportional to the r.p.m. of the machine. Since centrifugal forces on machine parts increase with the square of the speed, it is estimated that maintenance costs will increase in proportion to the square of the speed, and that the useful life of the machine will be inversely proportional to the square of the speed. On the basis of these assumptions Table 18.6 is constructed.

Table 18.6. RELATIONSHIP OF COST TO OPERATING SPEED

Operating speed in r.p.m.	Estimated life in hr., $10{,}000 \times (150)^2/A^2$	Average output in pounds per hr. $18.6 \times A/150$	Labor cost per hour	Average maintenance cost per hr., $\$0.072 \times A^2/(150)^2$	Power cost per hr., $\$0.064 \times A/150$	Average depreciation cost per hr., $\$6{,}150 \div B$	Total cost of operation per hr., $D + E + F + G$	Cost per pound, $H \div C$
A	B	C	D	E	F	G	H	I
150	10,000	18.6	$1.37	$0.072	$0.064	$0.615	$2.121	$0.114
200	5,630	24.8	1.37	0.128	0.085	1.092	2.675	0.108
250	3,600	31.0	1.37	0.200	0.107	1.708	3.385	0.109

On the basis of the estimated use, it would be desirable to increase the speed of the machine to 200 r.p.m. The expected saving per pound of product would be

$$(\$0.114 - \$0.108) \div \$0.114 = 5.3\%.$$

This is a worthwhile saving, particularly since it shortens the capital recovery period and so lessens the possibilities of losses from obsolescence and inadequacy.

Economic Load Distribution between Machines. In many cases two or more machines are available for the same kind of production. Thus two or more boilers may be available to produce steam or two or more turbine generators may be on hand to produce power. If the total load is less than the combined capacity of the machines that are available, the economy of operation will, in some measure, depend upon the portion of the total load that is carried by each machine.

As an example of a simple method of determining the load distribution between two machines suppose that two Diesel-engine-driven d-c generators, one of 1,000 kw capacity and the other of 500 kw capacity, are available. The efficiencies for different outputs and the corresponding inputs for each are given in Table 18.7.

Table 18.7. EFFICIENCIES FOR DIFFERENT OUTPUTS

Output, in kw	Machine A, 1,000 kw Output		Machine B, 500 kw Output	
	Efficiency in per cent	Input, in kw	Efficiency in per cent	Input in kw
100	15.1	663	20.7	483
200	22.0	909	27.9	719
300	26.3	1,141	31.8	943
400	29.5	1,356	32.5	1,231
500	31.2	1,603	32.0	1,563
600	32.3	1,858		
700	32.5	2,154		
800	32.6	2,454		
900	32.4	2,778		
1,000	32.3	3,096		

Suppose that the total load to be met at a certain time is 1,200 kw. This load may be distributed in several ways as is shown in Table 18.8. It is observed that the desired output of 1,200 kw can be produced with a minimum input when the larger unit carries 800 kw and the smaller unit 400 kw.

In the example above input can readily be converted to units other than kw. Suppose that it is desired to express input in terms of dollars. Let it be assumed that fuel oil having an energy content of 18,800 Btu weighs 7.48 pounds per gallon and costs $0.084 per gallon delivered. One kw-hr. is equivalent to 3,410 Btu. From these data the cost per kw-hr. is calculated to be

Table 18.8. DISTRIBUTION OF LOAD BETWEEN MACHINES

Output in kw			Input in kw		
Machine A	Machine B	Total	Machine A	Machine B	Total
1,000	200	1,200	3,096	719	3,815
900	300	1,200	2,778	943	3,721
800	400	1,200	2,454	1,231	3,685
700	500	1,200	2,154	1,563	3,717

$$\frac{\$0.084 \times 3,410}{7.48 \times 18,800} = \$0.002037.$$

At this rate of cost for fuel, 1,200 kw-hr. output of energy can be produced for $3,685 \times \$0.002037 = \$7.51$ if the loads on the two machines are 800 kw and 400 kw, respectively. If the load distribution had been 1,000 kw and 200 kw on the machines, the cost would have been \$7.78, an increase over the better method of operation of \$0.27, or $\$0.27/\$7.51 = 3.7\%$. This is a rather high rate of saving considering the ease with which it is made. The saving is a result of a correct decision based on knowledge of load distribution.

PROBLEMS

1. The total cost of manufacturing 260 units of a product is \$3,200. If 340 units are manufactured, the total cost will be \$3,800.
   (a) What is the average manufacturing cost for the first 260 units?
   (b) What is the variable cost per unit?
   (c) What is the total fixed cost?
   (d) What is the average fixed cost per unit for the first 260 units?

2. The fixed cost of a machine (depreciation, interest, space charges, maintenance, indirect labor, supervision, insurance, and taxes) is F dollars per year. The variable cost of operating the machine (power, supplies, and similar items, excluding direct labor) is V dollars per hour of operation. N is the number of hours the machine is operated per year, TC the annual total cost of operating the machine, TC_h the hourly cost of operating the machine, t the time in hours to process one unit of product, M the machine cost of processing a unit of product, and n the number of units of product processed per year. In terms of these symbols, write expressions for (a) TC, (b) TC_h, (c) M.

3. In Problem 2, $F = \$600$ per year, $t = 0.2$ hour, $V = \$0.50$ per hour, and n varies from 0 to 10,000 in increments of 1,000.
   (a) Plot values of M as a function of n.
   (b) Write the expression for TC_u the total cost of direct labor and machine cost per unit using the symbols given in Problem 2 and letting W equal the hourly cost of direct labor.

4. A semiautomatic arc welding machine that is used for a certain joining process costs \$10,000. The machine has a life of 5 years, and a salvage value of \$1,000. Maintenance, taxes, interest, and other fixed costs amount to \$500 per year. The cost of

power and supplies is $3.20 per hour, and the operator receives $4.60 per hour. The cycle time per unit of product is 60 minutes. If interest is 7% calculate the cost per unit of processing the product if (a) 200, (b) 600, (c) 1,200, (d) 2,500 units of product are made per year.

5. An analysis of the current operations of a firm results in the conclusion that the production cost is $10,900 + 80x + 1500x^{1/2}$ where x is the number of units pro-duced. The selling cost is $500y^{1/2}$ where y is the number of units sold during the year. The selling price is $180 per unit. Plot the production cost, the selling cost, the total production and selling cost, the gross income from sales, and profit for production levels ranging from 0 to 200 units per year. Determine the level of produc-tion for maximum profit.

6. A manufacturing concern estimates that its expenses per year for different levels of operation would be as follows:

Output in units of product	0	10	20	30	40	50
Administrative and sales	$4,900	$ 5,700	$ 6,200	$ 6,700	$ 7,100	$ 7,500
Direct labor and materials	0	2,500	4,600	6,400	8,100	9,800
Overhead expense ..	4,120	4,190	4,270	4,350	4,440	4,550
Total	$9,020	$12,390	$15,070	$17,450	$19,640	$21,850

(a) What is the incremental cost of maintaining the plant ready to operate (the incremental cost of making zero units of product)?

(b) What is the average incremental cost per unit of manufacturing the first incre-ment of 10 units of product per year?

(c) What is the average incremental cost per unit of manufacturing the increment of 31 to 40 units per year?

(d) What is the average total cost per unit when manufacturing at the rate of 20 units per year?

(e) At a time when the rate of manufacture is 20 units per year, a salesman reports that he can sell 10 additional units at $260 per unit without disturbing the market in which the company sells. Would it be profitable for the company to undertake the production of the 10 additional units?

7. An analysis of an enterprise and the market in which its product is sold results in the following data:

Level of Operation, Units of Product	Production Cost, Dollars	Net Income from Sales, Dollars	Level of Operation, Units of Product	Production Cost, Dollars	Net Income from Sales, Dollars
0	$10,900	$ 0	700	$41,700	$43,000
100	17,600	8,000	800	44,800	46,800
200	23,000	15,400	900	48,000	50,000
300	27,500	22,000	1,000	51,300	52,700
400	31,500	28,000	1,100	54,800	54,800
500	35,100	33,500	1,200	58,500	56,300
600	38,500	38,500			

(a) Determine the profit for each level of operation.

(b) Plot production cost, net income from sales, profit, average incremental production cost per unit, and average incremental net income from sales per unit for each
level of operation.

8. The product of an enterprise has a fixed selling price of $62. An analysis of production
and sales costs and the market in which the product is sold has produced the following results:

Level of Operation, Units of Product	Total Production and Selling Cost, Dollars	Level of Operation, Units of Product	Total Production and Selling Cost, Dollars
0	$13,200	600	$35,000
100	17,900	700	40,400
200	21,400	800	47,100
300	24,600	900	55,600
400	27,200	1,000	65,400
500	30,600		

(a) Determine the profit for each level of operation.

(b) Plot production and selling cost, income from sales, profit, average incremental
production and selling cost per unit, and average incremental income per unit, for
each level of operation.

9. The annual fixed production and selling cost of a company producing a certain product is estimated at $40,000. The variable cost is estimated at $65 per unit. Dissatisfied
with the small margin of profit, the company is studying a plan to improve their
product and increase selling effort. The plan of improvement requires an additional
annual fixed expenditure of $10,000 on production and selling. Also, an additional
$20 per unit will be spent to improve the product from the standpoints of quality
and customer appeal. On the basis of a market survey, estimated sales, with and
without the improvements, at various selling prices are given below:

Unit Selling Price	Sales Without Improvements	Sales With Improvements
$350	0	55
300	20	180
250	90	360
200	225	610
150	460	920
125	680	—
110	1,010	—

Plot the total production and selling cost of each plan, and plot the total income from
each plan for the various selling prices against output as the abscissa. Determine
whether or not the plan for improvements should be adopted and determine the
production and selling price that will result in the maximum profit.

10. A manufacturer of radio receivers has an annual fixed cost of $300,000 and a variable
cost of $6.60 per unit produced. An engineering proposal under consideration
involves redesign of the present AM unit to include FM. The production engineering
staff estimates that the redesigned product will increase the annual fixed cost by

$200,000 and the variable cost by $2.60 per unit. A market survey produced the following information:

Unit Selling Price	Sales With AM Only	Sales With AM and FM
$35.00	0	6,200
30.00	3,000	21,000
25.00	11,000	39,500
20.00	25,500	64,000
15.00	48,000	100,000
12.50	74,000	—
11.00	98,000	—

(a) Determine the selling price for maximum annual profit if only AM receivers are manufactured; if AM and FM receivers are manufactured.

(b) Should the design change be adopted?

11. A company has priced its product at $1 per pound, and is operating at a loss. Sales at this price total 850,000 pounds per year. The company's fixed cost of manufacture and selling is $480,000 per year and the variable cost is $0.46 per pound. It appears, from information obtained by a market survey, that price reductions of $0.05, $0.10, $0.15, and $0.20 per pound from the present selling price will result in total annual sales of 1,030,000 pounds, 1,190,000 pounds, 1,360,000 pounds, and 1,480,000 pounds per year, respectively.

(a) Calculate the annual profit that will result from each of the selling prices given, assuming variable cost per unit will be the same for all production levels.

(b) Determine graphically the annual profit that will result from each of the selling prices by the use of a break-even type chart.

12. A certain firm has the capacity to produce 650,000 units of product per year. At present, it is operating at 64% of capacity. The firm's annual income is $416,000. Annual fixed costs are $192,000 and the variable costs are equal to $0.356 per unit of product.

(a) What is the firm's annual profit or loss?

(b) At what volume of sales does the firm break even?

(c) What will be the profit or loss at 70, 80, and 90% of capacity on the basis of constant income per unit and constant variable cost per unit?

13. A company is operating at capacity for one shift per day. Annual sales are 3,600 units per year and the income per year and the income per units is $200. Fixed costs are $280,000 and total variable costs are $440,000 per year at the present rate of operation.

If output can be increased, 400 additional units can be sold at $200 per unit during the coming year. These additional units can be produced through overtime operation at the expense of a 20% increase in the unit variable cost of the additional units. The elasticity of demand for the product in question is believed to be such that an increase in selling price to $208 will result in curtailing demand to 3,600 units. For greatest profit in the coming year should output be increased or should selling price be increased?

14. A manufacturing company owns two Plants, A and B, that produce an identical product. The capacity of Plant A is 60,000 units annually while that of Plant B is 80,000 units. The annual fixed cost of Plant A is $260,000 per year and the variable

cost is $3.20 per unit. The corresponding values for Plant B are $280,000 and $3.90 per unit. At present, Plant A is being operated at 35% of capacity and Plant B is being operated at 40% of capacity.

(a) What are the unit costs of production of Plant A and Plant B?

(b) What is the total cost and the average cost of the total output of both plants?

(c) What would be the total cost to the company and the unit cost if all production were transferred to Plant A?

(d) What would be the total cost to the company and the unit cost if all production were transferred to Plant B?

15. A firm is considering the expansion of its operations to include a plating department that will require 5,000,000 killowatt-hours of electrical energy per year. The maximum rate of consumption is estimated to be 1,400 kilowatts. The power can be purchased from a power company for $0.013 per kilowatt-hour providing the firm will construct the required 25 miles of transmission line and furnish necessary transformers. The cost of the high tension line will be $9,500 per mile and the transformers required will cost $22,000. The life of these assets will be 30 years with a salvage value of $12,000.

As an alternative, the firm can construct and operate its own power plant. The plant can be located on a river which would furnish adequate cooling water for condensing purposes if a steam plant is selected or for cooling if a diesel plant is installed. A good location is also available for a hydrogenerating plant 36 miles from the firm's plant site. Costs involved in the construction and maintenance of the three types of power plants are as follows:

	Steam	Diesel	Hydro
Total investment per kw of capacity	$210.00	$260.00	$310.00
Transmission equipment cost..........	0	0	$364,000
Fuel required in pounds per kw-hr.	1.85	0.68	0
Cost of fuel per pound	$0.00375	$0.0092	0
Annual cost of maintenance	$27,000	$19,000	$15,000
Taxes and ins. as a % of investment ..	2	2	2
Life of plant and equipment in years ..	20	16	30
Salvage value as % of original cost	10	12	5

Determine the cost of each of the four methods of supplying the required power if the interest rate is 6% compounded continuously.

16. Electric lamps rated at 200 watts and 110, 115, or 120 volts can be purchased for $0.46 each. When the lamps are placed in a 115-volt circuit, the following data apply:

	110 v Lamp	115 v Lamp	120 v Lamp
Average watts input	209.1	195.2	182.8
Average lumens output per watt of input	18.30	16.84	15.48
Average life of lamp in hours	420	750	1,340

Energy costs $0.042 per 1,000 watt-hours. Determine the cost per million lumen hours for each lamp under the three conditions.

17. The normal operating speed of wire weaving looms is 300 r.p.m. At this speed the looms have a life expectancy of 5 years, an annual maintenance cost of $200,

and an annual power cost of $70. The machines cost $4,000 and have negligible salvage value. Annual space charges are estimated at $170 per machine. The operator's rate is $4.45 per hour and each operator runs three machines regardless of their speed. The interest rate is 8%

Experimental work on the operation of these machines has resulted in the following data:

Speed r.p.m.	Life in Years	Annual Maintenance Cost	Annual Power Cost
250	7.2	$120	$ 60
300	5.0	$200	$ 70
350	3.7	$320	$ 90
400	2.8	$480	$120
450	2.2	$680	$160

Summarize the items of cost in a table and determine the most economical r.p.m. at which to run the machines. Assume that a loom operating at a normal operating speed of 300 r.p.m. will produce 300 units of product per year (each year consists of 40 hours a week for 50 weeks) and that production is directly proportional to the speed of the loom.

18. In the operation of a certain type of equipment, repairs average $6 each and the cost of a shutdown for emergency repairs entails an additional loss of $14. In the past, there has been no routine periodic inspection of the equipment. It is believed that periodic inspections would result in a worthwhile saving.

In order to get data on which to determine the frequency of periodic inspections for greatest economy, five groups of an equal number of machines A, B, C, D, and E were inspected 0, 1, 2, 3, and 6 times per week, respectively. The cost per inspection was found to be $1.25. The plant operates 6 days per week, 50 weeks per year. The results of a year's run are shown below:

Machine Group	Inspections Per Week	Emergency Repairs	Non-emergency Repairs
A	0	31	27
B	1	20	36
C	2	13	41
D	3	9	43
E	6	3	47

What number of inspections should be made per week and what will be the total yearly cost for inspection, loss due to emergency repairs, and cost of repairs for this number of inspections per week?

19

Mathematical Models
for Operations

Mathematical models are finding common use in connection with many operational systems. Such models are quantitative abstractions of those aspects of the system important to its description and control. By formulating a mathematical model it is possible to determine values for policy variables which lead to least cost or maximum profit operation.

The delay in developing models for operational systems may be attributed to an early preoccupation with the physical environment. During much of history the limiting factors were predominantly physical. But, with the accumulation of knowledge about physical phenomena, complex operational systems have been developed to produce goods and services. Decision makers are becoming increasingly aware that experience, intuition, and judgment must be augmented with analysis techniques using mathematical models if such systems are to be operated economically.

19.1. MODELS FOR INVENTORY OPERATIONS

When the decision has been made to procure a certain item, it becomes necessary to determine the procurement quantity that will result in a minimum cost. The demand for the item may be met by producing a year's supply at the beginning of each year, or by procuring a day's supply at the beginning of each day. Neither of these extremes may be the most economical in terms of the sum of costs associated with procurement and holding the item in inventory. Models for inventory operations are used to determine the most economical procurement quantity.

The Economic Purchase Quantity. If the demand for an item is met by purchasing once per year, the cost incident to purchasing will occur once, but the large quantity received will result in a relatively high inventory holding cost for the year. Conversely, if orders are placed several times per year, the cost incident to purchasing will be incurred several times per year, but since small quantities will be received, the cost of holding the item in inventory will be relatively small. If the decision is to be based on economy of the total operation the purchase quantity that will result in a minimum annual cost must be determined. Let

TC = total yearly cost of providing the item;
D = yearly demand for the item;
N = number of purchases per year;
t = time between purchases;
Q = purchase quantity;
C_i = item cost per unit (purchase price);
C_p = purchase cost per purchase order;
C_h = holding cost per unit per year made up of such items as interest, insurance, taxes, storage space, and handling.

It it is assumed that the demand for the item is constant through-out the year, the purchase lead time is zero, and no shortages are allowed; the resulting inventory system may be represented graphically as shown in Figure 19.1.

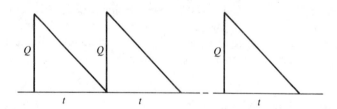

FIGURE 19.1. GRAPHICAL REPRESENTATION OF AN INVENTORY PROCESS FOR PURCHASING.

The total yearly cost will be the sum of the item cost for the year, the purchase cost for the year, and the holding cost for the year. That is,

$$TC = IC + PC + HC.$$

The item cost for the year will be the time cost per unit times the yearly demand in units, or

$$IC = C_i(D).$$

The purchase cost for the year will be the cost per purchase times the number of purchases per year, or

$$PC = C_p(N).$$

But since N is the yearly demand divided by the purchase quantity

$$PC = \frac{C_p(D)}{Q}.$$

Since the interval, t, begins with Q units in stock and ends with none, the average inventory during the cycle will be $Q/2$. Therefore, the holding cost for the year will be the holding cost per unit times the average number of units in stock for the year, or

$$HC = \frac{C_h(Q)}{2}$$

The total yearly cost of providing the required item is the sum of the item cost, purchase cost, and holding cost, or

$$TC = C_i(D) + \frac{C_p(D)}{Q} + \frac{C_h(Q)}{2}.$$

The purchase quantity resulting in a minimum yearly cost may be found by differentiating with respect to Q, setting the result equal to zero, and solving for Q as follows:

$$\frac{dTC}{dQ} = -\frac{C_p(D)}{Q^2} + \frac{C_h}{2} = 0$$

$$Q^2 = \frac{2C_p(D)}{C_h}$$

$$Q = \sqrt{\frac{2C_p(D)}{C_h}}$$

As an example of the use of this model assume that the annual demand for a certain item is 1,000 units. The cost per unit is $6 delivered. Purchasing cost per purchase order is $10 and the cost of holding one unit in inventory for one year is estimated to be $1.32.

The economic purchase quantity may be found by substituting the appropriate values in the derived relationship as follows:

$$Q = \sqrt{\frac{2(\$10)(1,000)}{\$1.32}}$$

$$= 123 \text{ units.}$$

Total cost may be expressed as a function of Q by substituting the costs and various values of Q into the total cost equation. The result is shown in

Table 19.1. The tabulated total cost value for $Q = 123$ is the minimum cost purchase quantity for the conditions specified. Total cost as a function of Q is illustrated in Figure 19.2.

Table 19.1. TABULATED VALUES OF TOTAL COST AS A FUNCTION OF PURCHASE QUANTITY

Purchase Quantity	Total Cost
50	$6,233
100	$6,166
123	$6,162
150	$6,165
200	$6,182
300	$6,231
400	$6,289
600	$6,413

The Economic Production Quantity. When the decision has been made to produce a certain item, it becomes necessary to determine the production quantities which are determined in a manner similar to determining economic purchase quantities. The difference in analysis is due to the fact that a purchased lot is received at one time while a production lot accumulates as it is made. Let

$TC =$ total yearly cost of providing the item;
$D =$ yearly demand for the item;
$N =$ number of production runs per year;
$t =$ time between production runs;
$Q =$ production quantity;
$C_i =$ item cost per unit (production cost);

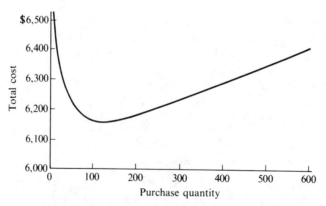

FIGURE 19.2. TOTAL COST AS A FUNCTION OF PURCHASE QUANTITY.

C_s = set-up cost per production run;

C_h = holding cost per unit per year made up of such items as interest, insurance, taxes, storage space, and handling;

R = production rate.

If it is assumed that the demand for the item is constant, the production rate is constant during the production period, the production lead time is zero, and no shortages are allowed; the resulting inventory system may be represented graphically as in Figure 19.3.

The total yearly cost will be the sum of the item cost for the year, the set-up cost for the year, and the holding cost for the year. That is,

$$TC = IC + SC + HC.$$

The item cost for the year will be the item cost per unit times the yearly demand in units, or

$$IC = C_i(D).$$

The set-up cost for the year will be the cost per set-up times the number of set-ups per year, or

$$SC = C_s(N).$$

But since N is the yearly demand divided by the production quantity

$$SC = \frac{C_s(D)}{Q}.$$

When items are added to inventory at the rate of R units per year and are taken from inventory at a rate D units per year, where R is greater than D, the net rate of accumulation is $(R - D)$ units per year. The time required to produce D units at the rate R units per year is D/R years. If D units are made in a single lot, the maximum accumulation in inventory will be $(R - D)$ D/R. Since no units will be in storage at the end of the year, the average number in inventory will be

$$\frac{(R - D)\dfrac{D}{R} + 0}{2} = (R - D)\frac{D}{2R}.$$

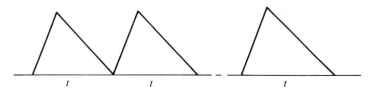

FIGURE 19.3. GRAPHICAL REPRESENTATION OF AN INVENTORY PROCESS FOR PRODUCTION.

If N lots are produced per year, the average number of units in storage will be

$$(R - D)\frac{D}{2RN}.$$

But, since $N = D/Q$, the average number of units in storage may be expressed as

$$(R - D)\frac{Q}{2R}.$$

The holding cost for the year will be the holding cost per unit times the average number of units in storage for the year, or

$$HC = C_h(R - D)\frac{Q}{2R}.$$

The total yearly cost of producing the required item is the sum of the item cost, set-up cost, and holding cost, or

$$TC = C_i(D) + \frac{C_s(D)}{Q} + C_h(R - D)\frac{Q}{2R}.$$

The production quantity resulting in a minimum yearly cost may be found by differentiating with respect to Q, setting the result equal to zero, and solving for Q as follows:

$$\frac{dTC}{dQ} = -\frac{C_s(D)}{Q^2} + \frac{C_h(R - D)}{2R} = 0$$

$$Q^2 = \frac{C_s(D)2R}{C_h(R - D)} = \frac{C_s(D)2}{C_h\left(1 - \frac{D}{R}\right)}$$

$$Q = \sqrt{\frac{2C_s(D)}{C_h\left(1 - \frac{D}{R}\right)}}.$$

As an example of the use of this model assume that the annual demand for a certain item is 1,000 units. The cost of production is $5.90 per unit and includes the usual cost elements of direct labor, direct material, and factory overhead. The set-up cost per lot is $50 and the item can be produced at the rate of 6,000 units per year. The cost of holding one unit in inventory for one year is estimated to be $1.30.

The economic production quantity may be found from the derived relationship by substituting the appropriate values as follows:

$$Q = \sqrt{\frac{2(\$50)(1,000)}{\$1.30\left(1 - \frac{1,000}{6,000}\right)}}$$

$$= 302 \text{ units.}$$

Total cost may be expressed as a function of Q by substituting costs and various values of Q into the total cost equation. The result is shown in Table 19.2. The tabulated total cost value for $Q = 302$ is the minimum cost production quantity for the condition specified. Total cost as a function of Q is illustrated in Figure 19.4.

Table 19.2. TABULATED VALUES OF TOTAL COST AS A FUNCTION OF PRODUCTION QUANTITY

Production Quantity	Total Cost
100	$6,454
150	$6,314
200	$6,258
300	$6,229
302	$6,228
400	$6,241
500	$6,270
600	$6,307

The "Make or Buy" Decision. The question of whether to manufacture or purchase a needed item may be resolved by the application of minimum cost analysis for multiple alternatives. The alternative of producing may be compared with the alternative of purchasing if the minimum cost procurement quantity for each is computed and used to find the respective total cost values. Choice of the total cost value that is a minimum identifies the best of the two alternatives.

For example, suppose that an item will have a yearly demand of 1,000 units and that the costs associated with purchasing and producing are the same costs which were assumed in the previous two sections; that is,

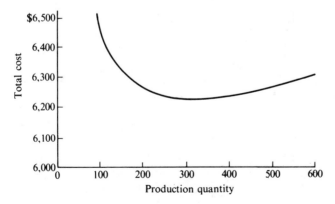

FIGURE 19.4. TOTAL COST AS A FUNCTION OF PRO-
DUCTION QUANTITY.

	Purchase	Produce
Item cost	$ 6.00	$ 5.90
Purchase cost	$10.00	—
Set-up cost	—	$50.00
Holding cost	$ 1.32	$ 1.30

For the conditions assumed, total cost as a function of the purchase quantity was given in Table 19.1 and total cost as a function of production quantity was given in Table 19.2. Total cost as a function of purchase quantity was graphed in Figure 19.2 and total cost as a function of production quantity was graphed in Figure 19.4. If Figure 19.2 and 19.4 are superimposed, the result is as shown in Figure 19.5.

The decision of whether to produce or purchase may be made by examining and comparing the minimum cost for each alternative. In this case, the decision to purchase will be the least cost alternative and will result in a saving of $6,228 less $6,162, or $66 per year. If the decision were made on the basis of item cost alone, the needed item would have been supplied by producing with a resultant loss of $66 per year.

The optimal procurement policy may be formally stated by specifying that the item will be procured when the available stock falls to zero units on hand, for a procurement quantity of 123 units, from the purchasing source. In essence the policy states when, how much, and from what source. The policy, if the decision has been made to purchase, states when and how much, the source being fixed by restriction. Therefore, the decision of "make or buy" is essentially a release of the restriction that the source is fixed. This analysis may be extended to any number of sources; for example, it may be used to compare re-manufacturing with purchasing, to compare alternate manufacturing facilities, or to evaluate alternate vendors.

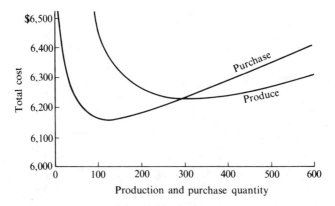

FIGURE 19.5. TOTAL COST AS A FUNCTION OF PRODUCTION AND PURCHASE QUANTITIES.

19.2. ANALYSIS OF WAITING LINE OPERATIONS

A general waiting line or queuing system is shown in Figure 19.6. The system exists because the population shown demands service. To satisfy the demand, a decision maker must specify the level of service capacity at the service facility. This is a problem in operations economy which will involve altering the service rate at existing channels or by adding or deleting channels. One approach to the problem is illustrated by the example of this section.

Failures of equipment do not occur with regularity. In addition, the amount of repair work needed for any given failure may be expected to vary above and below the average amount needed over a long period of time. Each of these events, the occurrence of a failure and the amount of repair needed, are a function of many unpredictable chance causes.

If the number of repair crews provided is just sufficient to take care of the average amount of repair work needed, there will be a considerable backlog of work waiting to be done and the cost of equipment down time will be high. Where down time is costly, it may be wise to maintain repair crew capacity in excess of that needed for the average repair load even though this will result in some idleness on the part of the repair crews.

As an example of the analysis required in finding the economical number of repair crews, consider an illustration from the petroleum industry. In petroleum production, heavy equipment is used to pump oil to the surface. When this equipment fails, it is necessary to remove it from the well for repairs. The required repairs are made by crews of three to five men who are equipped with heavy, portable machinery to pull the pumping equipment from the well.

Between the time the pumping equipment fails and the time it is repaired, the well is idle. This results in a loss, known as *lost production*, equal to the amount of oil that would have been produced if there had been no failure of equipment. In some cases lost production may be only production that is deferred to a later date. In other cases lost production is partially lost because of drainage to competitors' wells. This loss may be quite large. Two

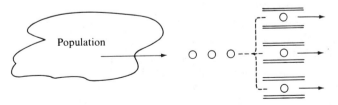

FIGURE 19.6. A MULTIPLE-CHANNEL WAITING LINE SYSTEM.

days' down time of a well producing 200 barrels of oil per day at $3.80 per barrel, for instance, when drainage to a competitor's well is judged to be half of lost production, results in a loss of $380.

The down time of a well is made up of the time that the well is idle before the repair crew gets to it and the time it is idle while repair is in process. Since the rate at which repairs can be made is substantially controlled by the repair equipment, reduction in loss is brought about by reducing the time a well is idle awaiting repair crews. This is done at the expense of having excess repair crews. Thus the problem is to balance the number of repair crews with the lost production associated with delays of repairs to wells. Consider the following data adapted from an actual situation in which the wells are operated 24 hours a day, 7 days per week:

N = number of oil wells in field = 30;
U = average interval at which individual wells fail = 15 days;
T = average actual time for company crew to repair well failure = 12 hours;
C_1 = hourly cost for company-operated repair unit and repair crew = $30 per hour;
C_2 = lost production per well when "down" = $380 per day.

Under the present situation, analysis has revealed that operation of the equivalent of one repair unit 24 hours per day, seven days per week can keep up with the repair of failures in the long run. This means that one channel is sufficient to service the 30 wells. However, serious losses are arising from delays in getting to wells after notification of their failure. Delay is due to the chance bunching of well failures. For instance, one period during which no wells fail may be followed by a period of an above-average number of failures. Since repairs cannot be made before failures occur, it is clear that crews sufficient to take care of the average number of failures will have a backlog of failures awaiting them most of the time.

The first step in an analysis to determine the most economical number of repair crews is to determine the number of wells that may be expected to fail during each and every day of a period in the future. The number of failures to expect in the future may be estimated on the basis of past records or by mathematical analysis. The former method is used in the example because it is more revealing.

In the solution of this example the pattern of well failures of a previous 30-day period selected at random will be considered to be representative of future periods. A 30-day period will be taken in the interest of simplicity, even though experience has shown that longer periods are advantageous.

Since the company's unit can repair two wells per day when operating three eight-hour shifts per day, unrepaired wells will be carried over one day each day that the number of failures plus the carry-over from the previous day exceeds two.

Line *A* of Figure 19.7 shows the number of wells that failed during the 30-day period selected at random and considered to be typical. The number of wells failing during any one day ranged between 0 and 5, and their total was 57.

Line *B* gives the number of wells repaired each day with the company's unit and crews. During the 30-day period, 57 wells failed and 55 were repaired. Thus, in spite of the periodic backlogs, there was idle time of the unit and crews on the days numbered 3, 6, 7, and 8—sufficient to repair 5 wells.

The carry-over of unrepaired wells is given in Line *C*. The total carry-over for the 30-day period is 52 well-days.

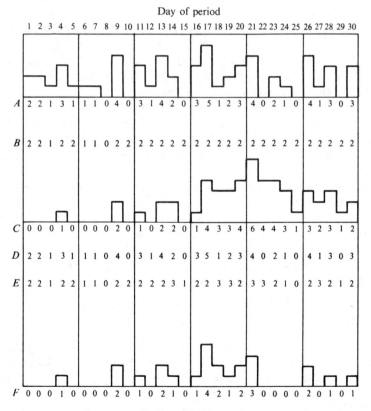

FIGURE 19.7. PATTERN OF OIL-WELL FAILURE AND TWO
REPAIR METHODS.

On the present basis of operation, employing the equivalent of one unit 24 hours per day, the total cost incident to well failures and repair during the 30-day period is calculated as follows:

$$TC = C_1(24)(30) + C_2(52)$$
$$= \$30(24)(30) + \$380(52)$$
$$= \$41,360.$$

Analysis of the pattern of occurrence of unrepaired wells carried over reveals that, once a backlog has accumulated, wells may remain unrepaired and unproductive for a long period. As a remedy the supervisor in charge considers the feasibility of hiring an additional unit and a crew (operating with two channels) whenever a backlog of unrepaired wells has accumulated. He finds that a repair unit and crew can be hired on short notice when needed for $40 per hour. Because of greater travel distance and other reasons, a hired unit and crew has been found to require an average of 16 hours to repair a well.

The supervisor wishes to determine the effect of a policy of hiring an additional unit and crew for a 16-hour period on days when there is a carry-over of two or more unrepaired wells from the previous day. He assumes that the hired unit and crew will be paid for a minimum of 16 hours each time it is asked to report, whether or not it is used.

If this policy had been in effect during the 30-day period under considera-tion, the additional unit and repair crew would have been hired for 16 hours on the days numbered 10, 14, 18, 19, 21, 22, and 27; the total number of wells repaired during each day would have been as given in Line E of Figure 19.7.

Under this plan of operation, the total carry-over would have been 24 well-days as is shown in Line F.

Line E indicates that there was idle time of units and crew during the days numbered 3, 6, 7, 8, 10, 15, 24, 25, and 29—sufficient for the repair of 11 wells.

On the basis of the analysis above, the total cost incident to well failures and repair for the 30-day period is calculated as follows:

$$TC = \$30(24)(30) + \$40(16)(7) + \$380(24)$$
$$= \$35,200.$$

The decision to hire an additional unit and crew on days when there is a carry-over of two or more unrepaired wells from a previous day would result in a reduction of the total cost of operations by $41,360 less $35,200, or $6,160 per month if future months' experience the same pattern of failures. Actually, this is an average value which will be achieved in the long run if the data used exhibits an average pattern.

19.3. LINEAR PROGRAMMING OF OPERATIONS

Operational situations involving many activities which compete for scarce resources can often be optimized by use of the general linear programming model. Usually any operation involving a linear effectiveness function and linear resource constraints can be expressed in the format of the linear programming model formally stated as optimize the effectiveness function

$$E = \sum_{j=1}^{n} e_j x_j$$

subject to the constraints

$$\sum_{j=1}^{n} a_{ij} x_j = b_i \qquad i = 1, 2, \ldots, m$$

$$x_j \geq 0 \qquad j = 1, 2, \ldots, n.$$

Optimization requires either maximization or minimization of E depending upon the measure of effectiveness involved. The decision maker has control of the variables, x_j. Not directly under his control are the effectiveness coefficients, e_j, the constants a_{ij}, and the constants, b_i.

Consider the following production example: Two products compete for scarce machine time during the production process. A single unit of product A requires 2.2 minutes of punch press time and 4.0 minutes of assembly time. The profit for product A is \$0.60 per unit. A single unit of product B requires 3.0 minutes of punch press time and 2.5 minutes of welding time. The profit for product B is \$0.75 per unit. The available capacity of the punch press department is 1,200 minutes per week. The welding department has an available capacity of 500 minutes per week and the assembly department can supply 1,600 minutes of capacity per week. The manufacturing and other data for this production situation are summarized in Table 19.3.

Table 19.3. PRODUCTION DATA FOR TWO PRODUCTS

Department	Product A	Product B	Capacity
Punch Press	2.2	3.0	1,200
Welding	0	2.5	500
Assembly	4.0	0	1,600
Profit	\$0.60	\$0.75	

In this example, two products compete for scarce production time. The objective is to determine the quantity of product A and the quantity of product B to produce so that total profit will be maximized. This will require maximizing the linear effectiveness function

$$TP = \$0.60A + \$0.75B$$

subject to the linear constraints

$$2.2A + 3.0B \leq 1200$$
$$0A + 2.5B \leq 500$$
$$5.0A + 0B \leq 1600$$

where both A and B must be ≥ 0.

The graphical equivalent of the algebraic statement of this two product problem is shown on Figure 19.8. The set of linear restrictions define a region of feasible solutions. This region lies below $2.2A + 3.0B = 1,200$ and is restricted further by the requirements that $B \leq 200$, $A \leq 400$, and that both A and B be nonnegative. Thus, the scarce production time (resources) determine which combinations of these activities are feasible and which are not feasible.

The production quantity combinations of A and B that fall within the region of feasible solutions constitute feasible production programs. That combination of A and B which maximizes profit is sought. The relationship of A and B is $A = 1.250 B$ and is based on the relative profit of each product. The total profit will depend upon the production quantity combination chosen. Thus, there is a family of isoprofit lines, one of which will have at least one point in the region of feasible production quantity combinations and have a maximum distance from the origin. The member that satisfies this condition intersects the region of feasible solutions at the extreme point

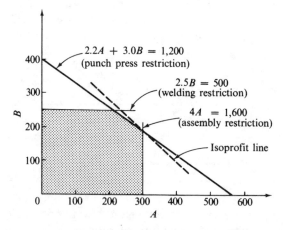

FIGURE 19.8. GRAPHICAL SOLUTION FOR
PROFIT MAXIMIZATION.

$A = 300$, $B = 180$. This is shown as a broken line in Figure 18.14, and represents a total profit of $0.60(300) + \$0.75(180) = \$315$. No other production quantity combination would result in a higher profit.

This illustration dealt with only two activities making the graphical solution method practical. Usually, however, several activities are involved in a linear programming application. In this case an optimization technique known as the simplex method may be used.

19.4. OPERATING TO A VARIABLE DEMAND

The demand for many manufactured goods is seasonal. Such variation in demand reflects changes in human wants caused by seasonal changes such as temperature, rainfall, and hours of sunshine. Some items may vary in demand because of social customs as is the case with the sale of fireworks.

The seasonal-goods manufacturer may make his product at a relatively low rate throughout the year and store it until it is needed. Or he may acquire sufficient facilities to manufacture the product at a rate equal to the demand during the period in which it is sold.

The disadvantage of the first plan is that storage cost is relatively high, and the disadvantage of the second method is that a rather large equipment investment will be required. The most desirable plan may be a compromise of the two plans above.

The method of solution of this and similar situations may be illustrated by an example. Let it be assumed that 36,000 units of a product are sold during a four-month period each year as follows:

Month of Year	Number Sold
9th	2,000
10th	10,000
11th	15,000
12th	9,000

One machine can make 36,000 units of this product during a year. The fixed charges on the required machine for such items as interest, taxes, insurance, space to house machine, and depreciation due to causes exclusive of usage amounts to $2,000 per year. The cost for depreciation due exclusively to usage, power, supplies, maintenance, and so forth amounts to $0.25 per hour or $40 per month of operation on the basis of a 160-hour month.

Although one machine can meet the demand for the product, the accumulation of finished products throughout the year will result in considerable expense for storage. The expense for storage can be reduced by using more machines to shorten the period required to make the year's needs.

The costs associated with plans of production based on using one, two, and three machines will be determined. For an output of 36,000 units per machine, 12, 6, and 4 months respectively will be required to manufacture the annual output of 36,000 units with one, two, and three machines. The number sold each month and the number in storage at the end of each month during the year are given in Table 19.4 and are shown graphically in Figure 19.9.

Table 19.4. STORAGE REQUIREMENTS FOR MANUFACTURING TO A SEASONAL DEMAND

Month	Sales During Month	Number in Storage at End of Month for		
		1 Machine	2 Machines	3 Machines
1		3,000		
2		6,000		
3		9,000		
4		12,000		
5		15,000		
6		18,000		
7		21,000	6,000	
8		24,000	12,000	
9	2,000	25,000	16,000	7,000
10	10,000	18,000	12,000	6,000
11	15,000	6,000	3,000	0
12	9,000	0	0	0

The average number in storage for one, two, and three machines is 13,083, 4,083, and 1,083, respectively. These results are obtained by adding the average of the number of units of product in storage at the beginning and end of each month in the year for each plan and dividing the resulting sum by 12.

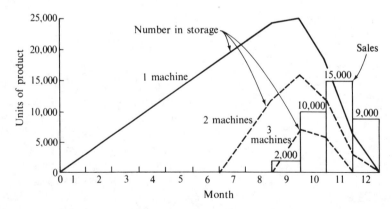

FIGURE 19.9. GRAPHICAL PRESENTATION OF STORAGE REQUIREMENTS.

The product is valued at $3 per unit. The sum of interest, taxes, and insurance is taken as 10% of the unit cost and storage costs as $0.20 per unit per year of storage.

On the basis of the above data the cost with one, two, and three machines will be as follows.

One Machine

Fixed charges on machine, 1 × $2,000	$2,000
Variable charge on machine, 12 × $40	480
Interest, taxes, and insurance, 13,083 × $3 × 0.10	3,925
Storage cost, 13,083 × $0.20	2,617
Total cost with one machine	$9,022

Two Machines

Fixed charges on machines, 2 × $2,000	$4,000
Variable charge on machines, 6 × 2 × $40	480
Interest, taxes, and insurance, 4,083 × $3 × 0.10	1,225
Storage cost, 4,083 × $0.20	817
Total cost with two machines	$6,522

Three Machines

Fixed charges on machines, 3 × $2,000	$6,000
Variable charge on machines, 4 × 3 × $40	480
Interest, taxes, and insurance, 1,083 × $3 × 0.10	325
Storage cost, 1,083 × $0.20	217
Total cost with three machines	$7,022

On the basis of the analysis above, a saving of $500 per year would result from using two machines in place of three and a saving of $2,500 would result from using two machines in place of one. It should be realized that there may be considerable hazard in the use of any of the three plans. If only one machine is used, there is a possibility that the product may become outmoded before the sales period or that sales may not be up to expectations. If three machines are used, there is a large investment that may not be recovered before the machines are rendered obsolete or inadequate.

PROBLEMS

1. A contractor has a requirement for cement in the amount of 45,000 bags per year. Cement costs $1.40 per bag and it costs $25 to process a purchase order. Holding cost is $2.30 per bag per year due to the high rate of spoilage. Find the minimum cost purchase quantity.

2. A foundry uses 3,600 tons of pig iron per year at a constant rate. The cost per ton delivered to the foundry is $70. It costs $42 to place an order and $8 per ton per year for storage. Find the minimum cost purchase quantity.

3. The demand for 1,000 units of a part to be used at a uniform rate throughout the year may be met by manufacturing. The part can be produced at the rate of 3 per

hour in a department which works 1,880 hours per year. The set-up cost per lot is estimated to be $40 and the manufacturing cost has been established at $5.20 per unit. Interest, insurance, taxes, space, and other holding costs are $3.10 per unit per year. Calculate the economic manufacturing quantity.

4. An aircraft manufacturer uses 1,000,000 special rivets per year at a uniform rate. The rivets are made on a single-spindle automatic screw machine at the rate of 3,000 per hour. The manufacturing cost is $0.0052 per rivet and the storage cost is estimated to be $0.0008 per rivet per year. The set-up cost is $40 per set-up. Calculate the economic production quantity and the allowable variation in this quantity for a maximum deviation from the minimum cost of 5%. Assume the machine operates 1,700 hours/year.

5. Rederive the economic purchase quantity model to reflect holding cost for interest, insurance, taxes, and handling based on the average inventory level and the holding cost for space based on the maximum level.

6. Show that the economic production quantity model reduces to the economic purchase quantity model as R approaches infinity.

7. The annual demand for an item that can be either purchased or produced is 3,600 units. Costs associated with each alternative are as follows:

	Purchase	Produce
Item cost	$ 3.90	$ 3.75
Purchase cost per purchase	$15.00	—
Set-up cost per set-up	—	$175.00
Holding cost per item per year	$ 1.25	$ 1.25
Production rate per year	—	15,000

Should the item be manufactured or purchased? What is the economic lot size for the least cost alternative?

8. The annual demand for a certain item is 1,000 units per year. Holding cost is $0.85 per unit per year. Demand can be met by either purchasing or producing the item with each source described by the following data:

	Purchase	Produce
Item cost	$ 8.00	$ 7.40
Procurement cost	$20.00	$80.00
Production rate	—	2,500 per year

Find the minimum cost procurement source and calculate the economic advantage over the alternate source. What is the minimum cost procurement quantity?

9. For a certain group of machines, a repair crew requires one day to repair a machine that breaks down. A loss of $180 is incurred for each day that a machine is "carried over" unrepaired. Each crew costs $75 per day to equip and maintain. The pattern of machine failures on succeeding days of a 30-day period is 7, 5, 8, 0, 3, 4, 5, 0, 2, 3, 8, 4, 5, 6, 1, 3, 4, 0, 2, 8, 6, 5, 5, 3, 4, 6, 7, 1, 3, 2. Assume that all breakdowns occur at the first of each day.

(a) Determine the idle time of crews when 4, 5, or 6 crews are employed.

(b) What is the optimum number of crews to employ?

10. A company has 24 identical machines that are operated 24 hours per day, 360 days per year. On the average, each machine breaks down at intervals of 6 working days.

The time required to repair each machine is 8 man-hours. Repairmen receive $4 per hour and work 8 hours per day. A loss of $100 is sustained for each day that a machine is "carried over" unrepaired.

(a) Determine the number of machines that may be expected by chance to go "down" during each day of a month by tossing a die 24 times to represent the 24 machines. Let the downs for each day be represented by the aces that come up.

(b) Assuming that all downs as found in part (a) occur at 12 midnight, determine the most economical number of regular repairmen to employ, if no extra repairmen or overtime wages are to be allowed and regular repairmen are not used except to repair the machines in question. Assume that the sequence for the 30-day month will be repeated 12 times during the year.

11. Solve graphically for the values of A and B that maximize total profit expressed as

$$TP = \$0.28A + \$0.36B$$

subject to the constraints

$$A \leq 42$$
$$B \leq 30$$
$$A + B \leq 62$$
$$A \geq 0 \quad \text{and} \quad B \geq 0.$$

12. A small machine shop has capability in turning, milling, drilling, and welding. The machine capacity is 16 hours per day in turning, 16 hours per day in milling, 8 hours per day in drilling, and 8 hours per day in welding. Two products, designated A and B, are under consideration. Each will yield a net profit of $0.25 per unit and will require the following amount of machine time:

	Product A	*Product B*
turning	0.064	0.106
milling	0.106	0.053
drilling	0.000	0.080

Solve graphically for the number of units of each product that should be scheduled to maximize profit.

13. A refinery produces three grades of gasoline; premium, regular, and economy. Each grade requires straight gasoline, octane, and additives which are available in the amount of 3,200,000; 2,400,000; and 1,100,000 gal per week, respectively. A gallon of premium requires 0.22 gal of straight gasoline, 0.50 gal of octane, and 0.28 gal of additives. One gal of regular requires 0.55 gal of straight gasoline, 0.32 gal of octane, and 0.13 gal of additives. A gallon of economy requires 0.72 gal of straight gasoline, 0.20 gal of octane, and 0.08 gal of additives. A profit of $0.048, $0.040, and $0.029 per gallon is received for premium, regular, and economy, respectively. Formulate the effectiveness function and constraints which may be used to find the number of gallons of each grade to produce which will lead to a maximum profit.

14. A manufacturer of a seasonal item finds that his sales are at the rate of 10,000 units per month, each month of the year except September and October when the rate of sales is 40,000 units per month. This product is made on single-purpose machines that have an initial cost of $12,000 each, an estimated life of 6 years, and no salvage

value. Fixed charges other than interest and depreciation on each of these machines amount to $400 and the variable cost amounts to $0.42 per hour of operation. Machine operators are paid $2 per hour. Each machine normally produces 25 units of product per hour, and the plant operates 160 hours per month. The material entering the product is received as used and costs $2.80 per unit of product. Charges for interest, taxes, and insurance on the average inventory of finished goods are estimated to be 9% of the variable cost of manufacture and materials. Charges for storage of inventory amount to $0.10 per unit per year of average inventory. How many machines should be employed in the production of the item for minimum cost if the interest rate is 8%?

15. The expected demand for a certain item during the months of January through June is 10,000 per month. It is expected that the demand will be 30,000 per month during the months of July through September and 20,000 per month during the months of October through December. The item can be ordered for a January 1 delivery at $3.95 per unit, for an April 1 delivery at $4 per unit, and for an August 1 delivery at $4.05 per unit. The item cannot be delivered at other times during the year. The cost to initiate and complete a purchase is $150 and the storage cost is $0.10 per unit per year. Interest, insurance, taxes, and other storage costs are estimated to be 9 per cent of the investment in average inventory.

As an alternative, the item can be manufactured on single purpose machines with the following characteristics:

Machine cost	$12,000
Service life in years	6
Salvage value	0
Annual fixed cost excluding depreciation	$ 400
Variable cost per hour of operation	$ 0.42
Hours available per month	160
Production in units per hour	25
Interest rate	8%

The direct labor cost for the operator is $2.80 per hour. The direct material cost per unit is $3. Calculate the number of machines for minimum cost if the manufacturing alternative is chosen and compare this with the least cost purchase policy.

APPENDICES

A

Zero-One
Programming

The statement of an investment decision problem as a zero-one programming problem allows the application of solution algorithms that optimize some objective (i.e. the maximizing of present-worth) while considering constraints on the decision variables. These constraints include budget restrictions and dependence relationships among the investment proposals. Although a complete discussion of the solution techniques used to solve a zero-one programming problem is inappropriate for this book, a basic understanding of a zero-one programming solution can be obtained graphically.

To illustrate suppose one is attempting to determine which alternative to undertake when there are two independent proposals under consideration. Proposal A has a first cost of $6,000 and the $PW_A(\text{MARR}) = \$1,000$ while Proposal B has a first cost of $3,000 and the $PW_B(\text{MARR}) = \$2,000$. If the money available for investment is $7,500, the decision problem formulated as a zero-one programming problem appears as

$$\text{Maximize } Z = \$1,000 X_A + \$2,000 X_B$$

subject to the restriction
(budget constraint)

$$\$6,000 X_A + \$3,000 X_B \le \$7,500$$
$$X_A, X_B \text{ are 0 or 1.}$$

This problem is depicted graphically in Figure $A1$. The four heavy black points represent the four possible alternative solutions to the problem.

	X_A	X_B
Do Nothing	0	0
Accept A only	1	0
Accept B only	0	1
Accept A and B	1	1

429

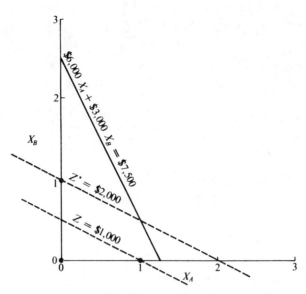

FIGURE A-1 GRAPHICAL SOLUTION OF A ZERO-
ONE PROGRAMMING PROBLEM.

The solid line representing the budget constraint indicates that only those solutions that are located below or on the line are feasible.

The objective of the problem is to find the highest value of $Z = \$1,000X_A + \$2,000X_B$ that passes through any of the feasible solutions. A dotted line in Figure $A1$ represents the value of $Z = \$1,000$. To increase the value of Z a line parallel to the dotted line is moved upward. It is seen in Figure A1 that there is a feasible alternative $(X_A = 0, X_B = 1)$ that can be reached by increasing the value of Z. Thus the maximum is reached when Z has a value Z^* such that $Z^* = \$1,000X_A + \$2,000X_B$ passes through $(X_A = 0, X_B = 1)$. Z^* is equal to \$2,000 under these conditions. The optimum decision for the budget constraint is to select Proposal B and to reject Proposal A.

If Proposal B is contingent on Proposal A then another constraint must be added to the original formulation of the problem. That is, add the constraint

$$X_B - X_A \leq 0.$$

If X_A is one then X_B can be one or zero but if X_A is zero then X_B can only be zero. The original problem with the added contingent constraint is shown in Figure A2. It is seen that the once feasible solution $(X_A = 0, X_B = 1)$ is now infeasible. Thus the optimum solution is $(X_A = 1, X_B = 0)$, select Proposal A only.

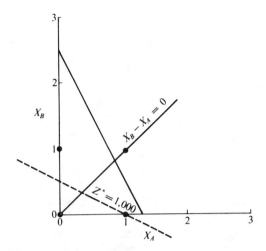

FIGURE A-2 ZERO-ONE PROGRAMMING PROBLEM
WITH CONTINGENT CONSTRAINT.

B

The MAPI Formulas

The MAPI Formulas are replacement formulas, advocated by the Machinery and Allied Products Institute. These formulas are developed in *Dynamic Equipment Policy*[1] by George Terborgh, Director of Research of MAPI, in the *MAPI Replacement Manual*[2], and in *Business Investment Policy*.[3]

Sponsored as they are by an organization influential in the manufacture and use of tools in industry, and because of the attention that has been accorded them in the industrial press, the MAPI formulas merit the attention of persons interested in engineering economy.

The MAPI formulas are used to determine whether equipment should be replaced. They take into account capital costs, interest, operating costs, and costs associated with physical impairment and obsolescence.

Terborgh has introduced a few new terms. These will be explained as needed. The term *defender* is used to designate a present asset whose replacement is under consideration. The term *challenger* is an asset under consideration as a replacement for another asset, usually the present defender. The MAPI formulas can be concisely developed and explained on the basis of the tabulations in Table B.1.

In Column B, the P's are the first costs and Q, $Q - b$, $Q - 2b$, etc. are the first year operating costs of a succession of challengers; these costs are assumed to have become available at one year intervals. The term b is a measure of improvement of an asset and represents a decrease in operating cost or an improvement of income (an income is treated as a negative cost). Column B represents total annual capital and operating costs that would

---

[1]George Terborgh, *Dynamic Equipment Policy* (New York: McGraw-Hill Book Company, Inc., 1949).

[2]*MAPI Replacement Manual* (Washington, D.C.: Machinery and Allied Products Institute, 1950).

[3]George Terborgh, Business Investment Policy (Washington, D.C.: Machinery and Allied Products Institute, 1958).

Table B.1. COSTS OF THREE PLANS OF PROVIDING A SERVICE

End of year A	Costs of available series of challengers, B	Costs of defender, C	Costs of challenger, C
1	$P + Q + 0$	$p + q + 0$	$P + Q + 0$
2	$P + Q - b$	$q + m'_c$	$Q + m'$
3	$P + Q - 2b$	$q + 2m'_c$	$Q + 2m'$
4	$P + Q - 3b$		$Q + 3m'$
5	$P + Q - 4b$		$Q + 4m'$
6	$P + Q - 5b$		$Q + 5m'$
7	$P + Q - 6b$		
8	etc.		

result if a succession of assets were used for one year and then replaced. For simplicity of exposition the salvage value is taken to be zero. The plan of a yearly replacement in Column *B* takes full advantage of asset improvements as they develop but suffers the disadvantage of high capital costs.

Column *C* represents the costs associated with continuing to use a defender for three years. The term *p* represents the salvage value of the defender—the amount that would be received for the defender if it were retired at the beginning of the first year. For simplicity of exposition the defender's future salvage is taken to be zero. The annual costs are represented by $q, q + m'_c$, and $q + 2m'_c$, where m'_c represents costs associated with physical impairment.

Column *D* represent the costs associated with the acquisition and use of a challenger. Its first cost is represented by *P* and annual costs by *Q*, $Q + m', + Q + 2m'$, etc., of which m' represents costs arising from physical impairment. Again, for simplicity of exposition the asset is assumed to have zero salvage value for any life.

To take into account obsolescence and to simplify comparison of continuing with the defender or installing the defender, a second tabulation is given in Table B.2. Table B.2 is the result of subtracting *Q* and adding *b*, 2*b*, 3*b*, etc., to Columns *B*, *C*, and *D* of Table B.1, year-by-year.

From the pattern of comparative costs in Column *D*, the life for minimum comparative cost can be found by the method explained with an example not involving interest following Table 8.7 in Chapter 8. Where interest is involved, $m' + b$ is replaced by *g* and the general expression for equivalent annual cost of the challenger as in Column *D*, Table B.2 is

$$C = (P - F)(\overset{A/P\,i,\,n}{}) + Fi + \frac{g}{i}\left(1 - n(\overset{A/F\,i,\,n}{})\right)$$

Table B.2. COMPARATIVE COSTS OF THREE PLANS OF PROVIDING A SERVICE

Year number, A	Comparative costs of available series of challengers, B	Comparative costs of defender, C	Comparative costs of challenger, D
1	P	$p + q - Q + 0$	$P + 0$
2	P	$+ q - Q + m'_c + b$	$+ m' + b$
3	P	$+ q - Q + 2m'_c + 2b$	$+ 2m' + 2b$
4	P	$- Q + 2b$	$+ 3m' + 3b$
5	P	$- Q + 2b$	$+ 4m' + 4b$
6	P	$- Q + 2b$	$+ 5m' + 5b$

or

$$C = (P - F)\frac{i(1 + i)^n}{(1 + i)^n} - 1 + Fi + \frac{g}{i} - \frac{ng}{i}\frac{i}{(1 + i)^n} - 1$$

where F = salvage value of asset on retirement (zero value for conditions in Table B.1). The minimum value of this equation is the *adverse minimum* in Terborgh's terminology.

At this point it is well to mention assumptions which are basic to the MAPI formulas. These are,

1. The future challengers will have the same adverse minimum as the present one.
2. The present challenger will accumulate operating inferiority at a constant rate over its service life.

The condition of the first of these assumptions will be met if the first costs of successive available replacements are equal to P, if the basic annual operating cost of each successive annually available replacement is equal to Q, $Q - b$, $Q - 2b$, $Q - 3b$, etc., and if the annual opeating costs of successive available replacements increase at a constant rate of m'. The condition of the assumption can be met in other ways that are not simply represented. For example, year to year variations in the first costs of assets could be compensated for appropriate variations in the other quantities involved.

The conditions of the second assumption are met by the increase of the challenger's comparative operating cost at the constant annual rate of $m' + b = g$ as shown in Column D, Table B.2.

For purposes of illustration assume that the minimum value of the equation for the asset in Column D, Table B.2, occurs for a life of six years. If the original challenger in Column D were retired at the end of six years, it would be replaced by a challenger whose first cost would be P and whose first year operating cost would be $Q - 6b$ and whose operating cost in succeeding years would increase at the rate of m' per year. If Columns B and D, Table B.1, were now extended and if Q were subtracted and $6b$ were added to Columns B and D in Table B.1, the result in Columns B and C of Table B.2 would be a repetition of the pattern of costs shown for years 1 to 6. Thus a life of six years would be the life of the second asset for minimum cost. For the conditions given, replacement at six-year intervals forever would result in minimum equivalent cost.

The adverse minimum of the defender may be determined by using the equation for finding the adverse minimum of the challenger by substituting p for P, $f =$ salvage value at retirement for F and $g' = m' + b$ for g. However, this method is not usually used because if the adverse minimum of a defender has not been reached at the time replacement is being considered, it will be reached shortly. For short periods of time it is simpler to calculate the adverse minimum of a defender on the basis of estimated annual operating costs and losses in salvage value in successive years.

For illustration, assume that the minimum cost of the defender in Column C, Table B.2, occurs for a life of 3 years. Then the equivalent annual cost of the three annual costs of $q - Q$ will be equal to $q - Q$. In Terborgh's terminology $q - Q$ is the *defender's next year's operating inferiority*. Thus the defender's adverse minimum will be the defender's next year's operating inferiority plus the cost of capital recovery for the defender. When the defender's adverse minimum is greater than the challenger's adverse minimum, replacement is indicated. If the defender's costs are increasing and its salvage value at the time of comparison is zero the capital recovery will be zero, leaving only the defender's next year's operating inferiority to compare with the challenger's adverse minimum.

Because of the difficulty of differentiating the expression

$$C = (P - F)\frac{i(1 + i)^n}{(1 + i)^n - 1} + Fi + \frac{g}{i} - \frac{ng}{i}\frac{i}{(1 + i)^n - 1}$$

approximations are used by Terborgh.

The expression for the zero salvage value case is

$$C = \frac{P}{n} + \frac{Pi}{2} + g\frac{(n - 1)}{2} \quad \text{and} \quad \frac{dc}{dn} = -\frac{P}{n^2} + \frac{g}{2} = 0$$

$$n = \sqrt{\frac{2P}{g}}.$$

This value of n results in

$$C_{\min} = \sqrt{2Pg} + \frac{Pi - g}{2} = \text{adverse minimum.}$$

This method of arriving at the adverse minimum requires that the terms

P, i, and g be known or estimated. The life, n, need not be known but it may be calculated.

Terborgh advances a second method for determining the adverse minimum which requires a life estimated but eliminates the need for estimating g.

For the zero salvage value case, begin with

$$C = \frac{i(1+i)^n}{(1+i)^n - 1} + \frac{g}{i} - \frac{ng}{i} \frac{i}{(1+i)^n - 1}$$

If C is plotted against n as an abscissa, the expression will result in a curve that is concave upward. Thus there are two points on the curve, one of which is on either side of the curve's minimum point for which the ordinates will be equal. Assume that the first point to the right of the axis occurs for an abscissa equal to n and that the second point occurs for an abscissa equal to $n + 1$.

Now the equivalent annual cost of the first n years of the asset's life will be equal to C when $n = n$. Operating costs for the year, n, equal $(n - 1)g$. To continue with the asset through the year $(n + 1)$ will give rise to additional operating costs but to no additional capital costs. The additional operating costs will be equal to ng. The equivalent annual cost C will be a minimum when the cost of extending the life of the asset one more year equals or exceeds the equivalent annual cost to date. (See the discussion preceding Table 8.3). Thus if ng is the cost of extending the asset's life for year $n + 1$, C will be a minimum for n years when

$$C = ng \quad \text{and} \quad g = \frac{C}{n}.$$

This value of g can now be substituted in the above equation to find the value of C for given values of n, P and i, and

$$C = P\frac{i(1+i)^n}{(1+i)^n - 1} + \frac{C}{ni} - \frac{nC}{ni} \frac{i}{(1+i)^n - 1}$$

solving for C results in

$$C_{\min} = \frac{Pm^2}{ni - 1 + \dfrac{1}{(1+i)^n}}.$$

The method above of arriving at the adverse minimum requires that the terms P, i, and n be known or estimated. The value of g may be calculated if desired. A version of the above expression for taking salvage values into account is given in the *MAPI Replacement Manual*.

Because of the way in which this equation is derived, it does not give a true minimum but only a result that will give equivalent annual costs that are equal for the years, n and $n + 1$, on either side of the minimum. The expression gives values that are a few per cent less than the true minimum, the amount of deviation being most for small values of n. Nearly true minimum values will be obtained if $(n - \frac{1}{2})$ is substituted for all values of n in the above expression.

Persons interested in the replacement problem will find the philosophy, concepts and procedures developed in the three publications which have been quoted both interesting and provocative. Although the publications are several years old the concepts are still valid and useful in replacement studies.

C

Service Life Predictions

In economy problems, patterns of the future depreciation of all manner of assets must be predicted. This is a difficult task. When a machine depreciates through use, a prediction must be made of the extent to which it will be used. If depreciation is caused by the elements, the rate at which deterioration progresses must be established. Even more difficult are the predictions that seek to determine when a machine will become obsolete because of new inventions and new needs, or inadequate owing to unanticipated demand.

Much has been written concerning the service life of equipment. Compilations summarizing the depreciation of many types of equipment in many different situations are available.

Unfortunately such data are only of limited value as a basis for predicting the service life of a particular item of equipment. For the most part, the information that is available consists of tables giving the average life of various types of structures, machines, and so forth. These have been prepared by people of various degrees of competence and ability. In any event, they are largely based on judgments. One difficulty with such tables is that the conditions under which the facilities were used are not sufficiently described to enable application to be made to a particular situation.

Mortality data are very useful for purposes of adjudicating a fair cost for services. Such data may also be very useful in making decisions involving great numbers of units used under similar conditions, such as railroad ties, telephone poles, and electric light bulbs. However, where the future service life of a single unit is a factor in decision, particularly where obsolescence may intervene, mortality data may be of limited use.

Where the service lives of a large number of assets used under similar conditions are known, a number of useful curves can be drawn. Mortality data for 30,000 wooden telegraph poles are given in Table C.1,[1] for ease of

---

[1]Marston, Anson, and Agg, *Engineering Valuation* (New York: McGraw-Hill Book Company, Inc., 1936).

Table C.1. COMPILATION OF THE MORTALITY DATA OF 30,000 WOODEN TELEGRAPH POLES TREATED WITH COAL TAR

Age interval, years (1)	Units retired during age interval, % (2)	Survivors at beginning of age interval, % (3)	Service during age interval, %-years (4)	Remaining service at beginning of age interval, %-years (5)	Expectancy at beginning of age interval, years (6)	Probable life at beginning of age interval, years (7)
$0-\frac{1}{2}$	0.00	100.00	50.00	1,067.88	10.68	10.68
$\frac{1}{2}-1\frac{1}{2}$	0.35	100.00	99.82	1,017.88	10.18	10.68
$1\frac{1}{2}-2\frac{1}{2}$	0.74	99.65	99.28	918.06	9.21	10.71
$2\frac{1}{2}-3\frac{1}{2}$	1.45	98.91	98.19	818.78	8.28	10.78
$3\frac{1}{2}-4\frac{1}{2}$	3.19	97.46	95.86	720.59	7.39	10.89
$4\frac{1}{2}-5\frac{1}{2}$	2.96	94.27	92.79	624.73	6.63	11.13
$5\frac{1}{2}-6\frac{1}{2}$	5.68	91.31	88.47	531.94	5.83	11.33
$6\frac{1}{2}-7\frac{1}{2}$	6.11	85.63	82.58	443.47	5.18	11.68
$7\frac{1}{2}-8\frac{1}{2}$	7.28	79.52	75.88	360.89	4.54	12.04
$8\frac{1}{2}-9\frac{1}{2}$	9.63	72.24	67.42	285.01	3.95	12.45
$9\frac{1}{2}-10\frac{1}{2}$	10.37	62.61	57.43	217.59	3.48	12.98
$10\frac{1}{2}-11\frac{1}{2}$	10.33	52.24	47.07	160.16	3.07	13.57
$11\frac{1}{2}-12\frac{1}{2}$	9.54	41.91	37.14	113.09	2.70	14.20
$12\frac{1}{2}-13\frac{1}{2}$	9.06	32.37	27.84	75.95	2.34	14.84
$13\frac{1}{2}-14\frac{1}{2}$	7.26	23.31	19.68	48.11	2.06	15.56
$14\frac{1}{2}-15\frac{1}{2}$	6.44	16.05	12.83	28.43	1.77	16.27
$15\frac{1}{2}-16\frac{1}{2}$	3.77	9.61	7.73	15.60	1.62	17.12
$16\frac{1}{2}-17\frac{1}{2}$	3.01	5.84	4.33	7.87	1.35	17.85
$17\frac{1}{2}-18\frac{1}{2}$	1.84	2.83	1.91	3.54	1.25	18.75
$18\frac{1}{2}-19\frac{1}{2}$	0.50	0.99	0.74	1.63	1.65	20.15
$19\frac{1}{2}-20\frac{1}{2}$	0.12	0.49	0.43	0.89	1.82	21.32
$20\frac{1}{2}-21\frac{1}{2}$	0.13	0.37	0.31	0.46	1.24	21.74
$21\frac{1}{2}-22\frac{1}{2}$	0.21	0.24	0.13	0.15	0.63	22.13
$22\frac{1}{2}-23\frac{1}{2}$	0.03	0.03	0.02	0.02	0.67	23.17
$23\frac{1}{2}-24\frac{1}{2}$	0.00	0.00	0.00	0.00	0.00	23.50
Total..	100.00	...	1,067.88	Av. life = 1,067.88 ÷ 100 = 10.68 yrs.		

understanding it may be assumed that all the poles were installed at the same time. On the basis of this supposition, the age interval in Column 1 will also be the elapsed time since installation. From this data several useful curves may be drawn as is shown in Figure C.1.

Examination of the mortality frequency curve shows that some poles do not survive as long and some poles survive much longer than the average of the group, but that the greatest rate of retirement centers around the average age. In order to maintain service, poles that are retired are immediately replaced. These replacements are subject to the same mortality frequency curve as the original group. Thus after a few years there are renewals of original poles, renewals of renewals, renewals of renewals of renewals, and

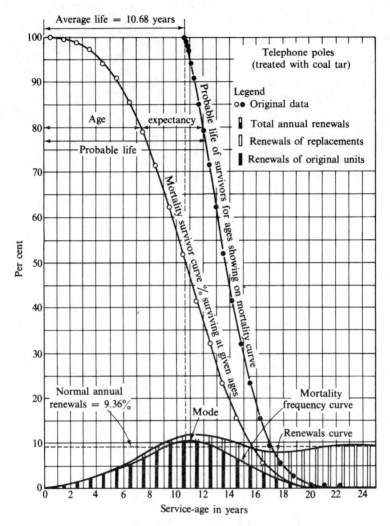

FIGURE C-1 MORTALITY CURVE FOR 30,000 WOOD TELE-
GRAPH POLES.

so forth. A summation of these renewals is given by the renewals curve in
Figure C.1.

The mortality survivor curve shows the number of original poles surviving
at any service age. This curve is very similar to the survivorship curves for
humans which are used in life insurance actuarial work.

The curve showing probable life of survivors is useful in predicting the
life of those of the original groups that remain in use. Note that the probable

life of the group at zero service age is equal to the average life of the group. The curves shown apply only to assets that have the same types of mortality frequency curve and have an average service life of 10.68 years.

To extend the use of this type of data Kurtz[2] has generalized its application by setting up a group of mathematical expressions representative of mortality experience of a large number of items of property. These expressions facilitate mathematical analysis of probems related to mortality. They are particularly valuable for the valuation of property composed of original and renewal items or exclusively of renewal items, such as railroad ties on an old road bed. The Kurtz expressions also greatly facilitate the extension of analyses over periods in the future.

---

[2]Edwin B. Kurtz, *The Science of Valuation and Depreciation* (New York: The Ronald Press, Inc., 1937).

D

Interest Tables

The computational time required in the solution of engineering economy problems can be reduced by the use of tabular values corresponding to the interest formulas developed in the text. Use of the recommended factor designations during the formulation of the problem simplifies the substitution of the appropriate tabular value. Reduction of the resulting expression to a numerical answer may be accomplished by hand, by a slide rule, or by a desk calculator.

Individual tabular values are given to a number of decimal places sufficient for most practical applications. Since the end result of an engineering economy study is based on estimated quantities and will be used in decision making, the analysis does not need to be accurate to a great many decimal places. In order to develop results with a degree of accuracy beyond that which may be obtained by the use of tabular values it will be necessary to use the formulas directly and employ logarithmic operations. Of course, the formulas may be used where their direct application will be more convenient than the use of tabular values.

If the magnitude of the problem requires the use of an electronic digital computer, the continuous compounding formulas lend themselves more readily to programming than do the annual compounding expressions. The burdensome portions of the formulas, involving powers of e, are easily reduced by the use of standard subroutines.

Table D.1. ½% INTEREST FACTORS FOR ANNUAL COMPOUNDING INTEREST

	Single Payment		Equal Payment Series				Uniform gradient-series factor
	Compound-amount factor	Present-worth factor	Compound-amount factor	Sinking-fund factor	Present-worth factor	Capital-recovery factor	
n	To find F Given P F/P i,n	To find P Given F P/F i,n	To find F Given A F/A i,n	To find A Given F A/F i,n	To find P Given A P/A i,n	To find A Given P A/P i,n	To find A Given g A/G i,n
1	1.005	0.9950	1.000	1.0000	0.9950	1.0050	0.0000
2	1.010	0.9901	2.005	0.4988	1.9851	0.5038	0.4988
3	1.015	0.9852	3.015	0.3317	2.9703	0.3367	0.9967
4	1.020	0.9803	4.030	0.2481	3.9505	0.2531	1.4938
5	1.025	0.9754	5.050	0.1980	4.9259	0.2030	1.9900
6	1.030	0.9705	6.076	0.1646	5.8964	0.1696	2.4855
7	1.036	0.9657	7.106	0.1407	6.8621	0.1457	2.9801
8	1.041	0.9609	8.141	0.1228	7.8230	0.1278	3.4738
9	1.046	0.9561	9.182	0.1089	8.7791	0.1139	3.9668
10	1.051	0.9514	10.228	0.0978	9.7304	0.1028	4.4589
11	1.056	0.9466	11.279	0.0887	10.6770	0.0937	4.9501
12	1.062	0.9419	12.336	0.0811	11.6189	0.0861	5.4406
13	1.067	0.9372	13.397	0.0747	12.5562	0.0797	5.9302
14	1.072	0.9326	14.464	0.0691	13.4887	0.0741	6.4190
15	1.078	0.9279	15.537	0.0644	14.4166	0.0694	6.9069
16	1.083	0.9233	16.614	0.0602	15.3399	0.0652	7.3940
17	1.088	0.9187	17.697	0.0565	16.2586	0.0615	7.8803
18	1.094	0.9141	18.786	0.0532	17.1728	0.0582	8.3658
19	1.099	0.9096	19.880	0.0503	18.0824	0.0553	8.8504
20	1.105	0.9051	20.979	0.0477	18.9874	0.0527	9.3342
21	1.110	0.9006	22.084	0.0453	19.8880	0.0503	9.8172
22	1.116	0.8961	23.194	0.0431	20.7841	0.0481	10.2993
23	1.122	0.8916	24.310	0.0411	21.6757	0.0461	10.7806
24	1.127	0.8872	25.432	0.0393	22.5629	0.0443	11.2611
25	1.133	0.8828	26.559	0.0377	23.4456	0.0427	11.7407
26	1.138	0.8784	27.692	0.0361	24.3240	0.0411	12.2195
27	1.144	0.8740	28.830	0.0347	25.1980	0.0397	12.6975
28	1.150	0.8697	29.975	0.0334	26.0677	0.0384	13.1747
29	1.156	0.8653	31.124	0.0321	26.9330	0.0371	13.6510
30	1.161	0.8610	32.280	0.0310	27.7941	0.0360	14.1265
31	1.167	0.8568	33.441	0.0299	28.6508	0.0349	14.6012
32	1.173	0.8525	34.609	0.0289	29.5033	0.0339	15.0750
33	1.179	0.8483	35.782	0.0280	30.3515	0.0330	15.5480
34	1.185	0.8440	36.961	0.0271	31.1956	0.0321	16.0202
35	1.191	0.8398	38.145	0.0262	32.0354	0.0312	16.4915
40	1.221	0.8191	44.159	0.0227	36.1722	0.0277	18.8358
45	1.252	0.7990	50.324	0.0199	40.2072	0.0249	21.1595
50	1.283	0.7793	56.645	0.0177	44.1428	0.0227	23.4624
55	1.316	0.7601	63.126	0.0159	47.9815	0.0209	25.7447
60	1.349	0.7414	69.770	0.0143	51.7256	0.0193	28.0064
65	1.383	0.7231	76.582	0.0131	55.3775	0.0181	30.2475
70	1.418	0.7053	83.566	0.0120	58.9394	0.0170	32.4680
75	1.454	0.6879	90.727	0.0110	62.4137	0.0160	34.6679
80	1.490	0.6710	98.068	0.0102	65.8023	0.0152	36.8474
85	1.528	0.6545	105.594	0.0095	69.1075	0.0145	39.0065
90	1.567	0.6384	113.311	0.0088	72.3313	0.0138	41.1451
95	1.606	0.6226	121.222	0.0083	75.4757	0.0133	43.2633
100	1.647	0.6073	129.334	0.0077	78.5427	0.0127	45.3613

Table D.2. ¾% INTEREST FACTORS FOR ANNUAL COMPOUNDING INTEREST

	Single Payment		Equal Payment Series				Uniform gradient-series factor
	Compound-amount factor	Present-worth factor	Compound-amount factor	Sinking-fund factor	Present-worth factor	Capital-recovery factor	
n	To find F Given P F/P i,n	To find P Given F P/F i,n	To find F Given A F/A i,n	To find A Given F A/F i,n	To find P Given A P/A i,n	To find A Given P A/P i,n	To find A Given g A/G i,n
1	1.008	0.9926	1.000	1.0000	0.9926	1.0075	0.0000
2	1.015	0.9852	2.008	0.4981	1.9777	0.5056	0.4981
3	1.023	0.9778	3.023	0.3309	2.9556	0.3384	0.9950
4	1.030	0.9706	4.045	0.2472	3.9261	0.2547	1.4907
5	1.038	0.9633	5.076	0.1970	4.8894	0.2045	1.9851
6	1.046	0.9562	6.114	0.1636	5.8456	0.1711	2.4782
7	1.054	0.9491	7.159	0.1397	6.7946	0.1472	2.9701
8	1.062	0.9420	8.213	0.1218	7.7366	0.1293	3.4608
9	1.070	0.9350	9.275	0.1078	8.6716	0.1153	3.9502
10	1.078	0.9280	10.344	0.0967	9.5996	.0.1042	4.4384
11	1.086	0.9211	11.422	0.0876	10.5207	0.0951	4.9253
12	1.094	0.9142	12.508	0.0800	11.4349	0.0875	5.4110
13	1.102	0.9074	13.601	0.0735	12.3424	0.0810	5.8954
14	1.110	0.9007	14.703	0.0680	13.2430	0.0755	6.3786
15	1.119	0.8940	15.814	0.0632	14.1370	0.0707	6.8606
16	1.127	0.8873	16.932	0.0591	15.0243	0.0666	7.3413
17	1.135	0.8807	18.059	0.0554	15.9050	0.0629	7.8207
18	1.144	0.8742	19.195	0.0521	16.7792	0.0596	8.2989
19	1.153	0.8677	20.339	0.0492	17.6468	0.0567	8.7759
20	1.161	0.8612	21.491	0.0465	18.5080	0.0540	9.2517
21	1.170	0.8548	22.652	0.0442	19.3628	0.0517	9.7261
22	1.179	0.8484	23.822	0.0420	20.2112	0.0495	10.1994
23	1.188	0.8421	25.001	0.0400	21.0533	0.0475	10.6714
24	1.196	0.8358	26.188	0.0382	21.8892	0.0457	11.1422
25	1.205	0.8296	27.385	0.0365	22.7188	0.0440	11.6117
26	1.214	0.8234	28.590	0.0350	23.5422	0.0425	12.0800
27	1.224	0.8173	29.805	0.0336	24.3595	0.0411	12.5470
28	1.233	0.8112	31.028	0.0322	25.1707	0.0397	13.0128
29	1.242	0.8052	32.261	0.0310	25.9759	0.0385	13.4774
30	1.251	0.7992	33.503	0.0299	26.7751	0.0374	13.9407
31	1.261	0.7932	34.754	0.0288	27.5683	0.0363	14.4028
32	1.270	0.7873	36.015	0.0278	28.3557	0.0353	14.8636
33	1.280	0.7815	37.285	0.0268	29.1371	0.0343	15.3232
34	1.289	0.7757	38.565	0.0259	29.9128	0.0334	15.7816
35	1.299	0.7699	39.854	0.0251	30.6827	0.0326	16.2387
40	1.348	0.7417	46.446	0.0215	34.4469	0.0290	18.5058
45	1.400	0.7145	53.290	0.0188	38.0732	0.0263	20.7421
50	1.453	0.6883	60.394	0.0166	41.5665	0.0241	22.9476
55	1.508	0.6630	67.769	0.0148	44.9316	0.0223	25.1223
60	1.566	0.6387	75.424	0.0133	48.1734	0.0208	27.2665
65	1.625	0.6153	83.371	0.0120	51.2963	0.0195	29.3801
70	1.687	0.5927	91.620	0.0109	54.3046	0.0184	31.4634
75	1.751	0.5710	100.183	0.0100	57.2027	0.0175	33.5163
80	1.818	0.5501	109.073	0.0092	59.9945	0.0167	35.5391
85	1.887	0.5299	118.300	0.0085	62.6838	0.0160	37.5318
90	1.959	0.5105	127.879	0.0078	65.2746	0.0153	39.4946
95	2.034	0.4917	137.823	0.0073	67.7704	0.0148	41.4277
100	2.111	0.4737	148.145	0.0068	70.1746	0.0143	43.3311

Table D.3. 1% INTEREST FACTORS FOR ANNUAL COMPOUNDING INTEREST

	Single Payment		Equal Payment Series				Uniform gradient-series factor
	Compound-amount factor	Present-worth factor	Compound-amount factor	Sinking-fund factor	Present-worth factor	Capital-recovery factor	
n	To find F Given P F/P i, n	To find P Given F P/F i, n	To find F Given A F/A i, n	To find A Given F A/F i, n	To find P Given A P/A i, n	To find A Given P A/P i, n	To find A Given g A/G i, n
1	1.010	0.9901	1.000	1.0000	0.9901	1.0100	0.0000
2	1.020	0.9803	2.010	0.4975	1.9704	0.5075	0.4975
3	1.030	0.9706	3.030	0.3300	2.9410	0.3400	0.9934
4	1.041	0.9610	4.060	0.2463	3.9020	0.2563	1.4876
5	1.051	0.9515	5.101	0.1960	4.8534	0.2060	1.9801
6	1.062	0.9421	6.152	0.1626	5.7955	0.1726	2.4710
7	1.072	0.9327	7.214	0.1386	6.7282	0.1486	2.9602
8	1.083	0.9235	8.286	0.1207	7.6517	0.1307	3.4478
9	1.094	0.9143	9.369	0.1068	8.5660	0.1168	3.9337
10	1.105	0.9053	10.462	0.0956	9.4713	0.1056	4.4179
11	1.116	0.8963	11.567	0.0865	10.3676	0.0965	4.9005
12	1.127	0.8875	12.683	0.0789	11.2551	0.0889	5.3815
13	1.138	0.8787	13.809	0.0724	12.1338	0.0824	5.8607
14	1.149	0.8700	14.947	0.0669	13.0037	0.0769	6.3384
15	1.161	0.8614	16.097	0.0621	13.8651	0.0721	6.8143
16	1.173	0.8528	17.258	0.0580	14.7179	0.0680	7.2887
17	1.184	0.8444	18.430	0.0543	15.5623	0.0643	7.7613
18	1.196	0.8360	19.615	0.0510	16.3983	0.0610	8.2323
19	1.208	0.8277	20.811	0.0481	17.2260	0.0581	8.7017
20	1.220	0.8196	22.019	0.0454	18.0456	0.0554	9.1694
21	1.232	0.8114	23.239	0.0430	18.8570	0.0530	9.6354
22	1.245	0.8034	24.472	0.0409	19.6604	0.0509	10.0998
23	1.257	0.7955	25.716	0.0389	20.4558	0.0489	10.5626
24	1.270	0.7876	26.973	0.0371	21.2434	0.0471	11.0237
25	1.282	0.7798	28.243	0.0354	22.0232	0.0454	11.4831
26	1.295	0.7721	29.526	0.0339	22.7952	0.0439	11.9409
27	1.308	0.7644	30.821	0.0325	23.5596	0.0425	12.3971
28	1.321	0.7568	32.129	0.0311	24.3165	0.0411	12.8516
29	1.335	0.7494	33.450	0.0299	25.0658	0.0399	13.3045
30	1.348	0.7419	34.785	0.0288	25.8077	0.0388	13.7557
31	1.361	0.7346	36.133	0.0277	26.5423	0.0377	14.2052
32	1.375	0.7273	37.494	0.0267	27.2696	0.0367	14.6532
33	1.389	0.7201	38.869	0.0257	27.9897	0.0357	15.0995
34	1.403	0.7130	40.258	0.0248	28.7027	0.0348	15.5441
35	1.417	0.7059	41.660	0.0240	29.4086	0.0340	15.9871
40	1.489	0.6717	48.886	0.0205	32.8347	0.0305	18.1776
45	1.565	0.6391	56.481	0.0177	36.0945	0.0277	20.3273
50	1.645	0.6080	64.463	0.0155	39.1961	0.0255	22.4363
55	1.729	0.5785	72.852	0.0137	42.1472	0.0237	24.5049
60	1.817	0.5505	81.670	0.0123	44.9550	0.0223	26.5333
65	1.909	0.5237	90.937	0.0110	47.6266	0.0210	28.5217
70	2.007	0.4983	100.676	0.0099	50.1685	0.0199	30.4703
75	2.109	0.4741	110.913	0.0090	52.5871	0.0190	32.3793
80	2.217	0.4511	121.672	0.0082	54.8882	0.0182	34.2492
85	2.330	0.4292	132.979	0.0075	57.0777	0.0175	36.0801
90	2.449	0.4084	144.863	0.0069	59.1609	0.0169	37.8725
95	2.574	0.3886	157.354	0.0064	61.1430	0.0164	39.6265
100	2.705	0.3697	170.481	0.0059	63.0289	0.0159	41.3426

Table D.4. 1½% INTEREST FACTORS FOR ANNUAL COMPOUNDING INTEREST

n	Single Payment		Equal Payment Series				Uniform gradient-series factor
	Compound-amount factor	Present-worth factor	Compound-amount factor	Sinking-fund factor	Present-worth factor	Capital-recovery factor	
	To find F Given P F/P i,n	To find P Given F P/F i,n	To find F Given A F/A i,n	To find A Given F A/F i,n	To find P Given A P/A i,n	To find A Given P A/P i,n	To find A Given g A/G i,n
1	1.015	0.9852	1.000	1.0000	0.9852	1.0150	0.0000
2	1.030	0.9707	2.015	0.4963	1.9559	0.5113	0.4963
3	1.046	0.9563	3.045	0.3284	2.9122	0.3434	0.9901
4	1.061	0.9422	4.091	0.2445	3.8544	0.2595	1.4814
5	1.077	0.9283	5.152	0.1941	4.7827	0.2091	1.9702
6	1.093	0.9146	6.230	0.1605	5.6972	0.1755	2.4566
7	1.110	0.9010	7.323	0.1366	6.5982	0.1516	2.9405
8	1.127	0.8877	8.433	0.1186	7.4859	0.1336	3.4219
9	1.143	0.8746	9.559	0.1046	8.3605	0.1196	3.9008
10	1.161	0.8617	10.703	0.0934	9.2222	0.1084	4.3772
11	1.178	0.8489	11.863	0.0843	10.0711	0.0993	4.8512
12	1.196	0.8364	13.041	0.0767	10.9075	0.0917	5.3227
13	1.214	0.8240	14.237	0.0703	11.7315	0.0853	5.7917
14	1.232	0.8119	15.450	0.0647	12.5434	0.0797	6.2582
15	1.250	0.7999	16.682	0.0600	13.3432	0.0750	6.7223
16	1.269	0.7880	17.932	0.0558	14.1313	0.0708	7.1839
17	1.288	0.7764	19.201	0.0521	14.9077	0.0671	7.6431
18	1.307	0.7649	20.489	0.0488	15.6726	0.0638	8.0997
19	1.327	0.7536	21.797	0.0459	16.4262	0.0609	8.5539
20	1.347	0.7425	23.124	0.0433	17.1686	0.0583	9.0057
21	1.367	0.7315	24.471	0.0409	17.9001	0.0559	9.4550
22	1.388	0.7207	25.838	0.0387	18.6208	0.0537	9.9018
23	1.408	0.7100	27.225	0.0367	19.3309	0.0517	10.3462
24	1.430	0.6996	28.634	0.0349	20.0304	0.0499	10.7881
25	1.451	0.6892	30.063	0.0333	20.7196	0.0483	11.2276
26	1.473	0.6790	31.514	0.0317	21.3986	0.0467	11.6646
27	1.495	0.6690	32.987	0.0303	22.0676	0.0453	12.0992
28	1.517	0.6591	34.481	0.0290	22.7267	0.0440	12.5313
29	1.540	0.6494	35.999	0.0278	23.3761	0.0428	12.9610
30	1.563	0.6398	37.539	0.0266	24.0158	0.0416	13.3883
31	1.587	0.6303	39.102	0.0256	24.6462	0.0406	13.8131
32	1.610	0.6210	40.688	0.0246	25.2671	0.0396	14.2355
33	1.634	0.6118	42.299	0.0237	25.8790	0.0387	14.6555
34	1.659	0.6028	43.933	0.0228	26.4817	0.0378	15.0731
35	1.684	0.5939	45.592	0.0219	27.0756	0.0369	15.4882
40	1.814	0.5513	54.268	.0.0184	29.9159	0.0334	17.5277
45	1.954	0.5117	63.614	0.0157	32.5523	0.0307	19.5074
50	2.105	0.4750	73.683	0.0136	34.9997	0.0286	21.4277
55	2.268	0.4409	84.530	0.0118	37.2715	0.0268	23.2894
60	2.443	0.4093	96.215	0.0104	39.3803	0.0254	25.0930
65	2.632	0.3799	108.803	0.0092	41.3378	0.0242	26.8392
70	2.835	0.3527	122.364	0.0082	43.1549	0.0232	28.5290
75	3.055	0.3274	136.973	0.0073	44.8416	0.0223	30.1631
80	3.291	0.3039	152.711	0.0066	46.4073	0.0216	31.7423
85	3.545	0.2821	169.665	0.0059	47.8607	0.0209	33.2676
90	3.819	0.2619	187.930	0.0053	49.2099	0.0203	34.7399
95	4.114	0.2431	207.606	0.0048	50.4622	0.0198	36.1602
100	4.432	0.2256	228.803	0.0044	51.6247	0.0194	37.5295

Table D.5. 2% INTEREST FACTORS FOR ANNUAL COMPOUNDING INTEREST

	Single Payment		Equal Payment Series				Uniform gradient-series factor
	Compound-amount factor	Present-worth factor	Compound-amount factor	Sinking-fund factor	Present-worth factor	Capital-recovery factor	
n	To find F Given P F/P i,n	To find P Given F P/F i,n	To find F Given A F/A i,n	To find A Given F A/F i,n	To find P Given A P/A i,n	To find A Given P A/P i,n	To find A Given g A/G i,n
1	1.020	0.9804	1.000	1.0000	0.9804	1.0200	0.0000
2	1.040	0.9612	2.020	0.4951	1.9416	0.5151	0.4951
3	1.061	0.9423	3.060	0.3268	2.8839	0.3468	0.9868
4	1.082	0.9239	4.122	0.2426	3.8077	0.2626	1.4753
5	1.104	0.9057	5.204	0.1922	4.7135	0.2122	1.9604
6	1.126	0.8880	6.308	0.1585	5.6014	0.1785	2.4423
7	1.149	0.8706	7.434	0.1345	6.4720	0.1545	2.9208
8	1.172	0.8535	8.583	0.1165	7.3255	0.1365	3.3961
9	1.195	0.8368	9.755	0.1025	8.1622	0.1225	3.8681
10	1.219	0.8204	10.950	0.0913	8.9826	0.1113	4.3367
11	1.243	0.8043	12.169	0.0822	9.7869	0.1022	4.8021
12	1.268	0.7885	13.412	0.0746	10.5754	0.0946	5.2643
13	1.294	0.7730	14.680	0.0681	11.3484	0.0881	5.7231
14	1.319	0.7579	15.974	0.0626	12.1063	0.0826	6.1786
15	1.346	0.7430	17.293	0.0578	12.8493	0.0778	6.6309
16	1.373	0.7285	18.639	0.0537	13.5777	0.0737	7.0799
17	1.400	0.7142	20.012	0.0500	14.2919	0.0700	7.5256
18	1.428	0.7002	21.412	0.0467	14.9920	0.0667	7.9681
19	1.457	0.6864	22.841	0.0438	15.6785	0.0638	8.4073
20	1.486	0.6730	24.297	0.0412	16.3514	0.0612	8.8433
21	1.516	0.6598	25.783	0.0388	17.0112	0.0588	9.2760
22	1.546	0.6468	27.299	0.0366	17.6581	0.0566	9.7055
23	1.577	0.6342	28.845	0.0347	18.2922	0.0547	10.1317
24	1.608	0.6217	30.422	0.0329	18.9139	0.0529	10.5547
25	1.641	0.6095	32.030	0.0312	19.5235	0.0512	10.9745
26	1.673	0.5976	33.671	0.0297	20.1210	0.0497	11.3910
27	1.707	0.5859	35.344	0.0283	20.7069	0.0483	11.8043
28	1.741	0.5744	37.051	0.0270	21.2813	0.0470	12.2145
29	1.776	0.5631	38.792	0.0258	21.8444	0.0458	12.6214
30	1.811	0.5521	40.568	0.0247	22.3965	0.0447	13.0251
31	1.848	0.5413	42.379	0.0236	22.9377	0.0436	13.4257
32	1.885	0.5306	44.227	0.0226	23.4683	0.0426	13.8230
33	1.922	0.5202	46.112	0.0217	23.9886	0.0417	14.2172
34	1.961	0.5100	48.034	0.0208	24.4986	0.0408	14.6083
35	2.000	0.5000	49.994	0.0200	24.9986	0.0400	14.9961
40	2.208	0.4529	60.402	0.0166	27.3555	0.0366	16.8885
45	2.438	0.4102	71.893	0.0139	29.4902	0.0339	18.7034
50	2.692	0.3715	84.579	0.0118	31.4236	0.0318	20.4420
55	2.972	0.3365	98.587	0.0102	33.1748	0.0302	22.1057
60	3.281	0.3048	114.052	0.0088	34.7609	0.0288	23.6961
65	3.623	0.2761	131.126	0.0076	36.1975	0.0276	25.2147
70	4.000	0.2500	149.978	0.0067	37.4986	0.0267	26.6632
75	4.416	0.2265	170.792	0.0059	38.6771	0.0259	28.0434
80	4.875	0.2051	193.772	0.0052	39.7445	0.0252	29.3572
85	5.383	0.1858	219.144	0.0046	40.7113	0.0246	30.6064
90	5.943	0.1683	247.157	0.0041	41.5869	0.0241	31.7929
95	6.562	0.1524	278.085	0.0036	42.3800	0.0236	32.9189
100	7.245	0.1380	312.232	0.0032	43.0984	0.0232	33.9863

Table D.6. 3% INTEREST FACTORS FOR ANNUAL COMPOUNDING INTEREST

	Single Payment		Equal Payment Series				Uniform gradient-series factor
	Compound-amount factor	Present-worth factor	Compound-amount factor	Sinking-fund factor	Present-worth factor	Capital-recovery factor	
n	To find F Given P F/P i, n	To find P Given F P/F i, n	To find F Given A F/A i, n	To find A Given F A/F i, n	To find P Given A P/A i, n	To find A Given P A/P i, n	To find A Given g A/G i, n
1	1.030	0.9709	1.000	1.0000	0.9709	1.0300	0.0000
2	1.061	0.9426	2.030	0.4926	1.9135	0.5226	0.4926
3	1.093	0.9152	3.091	0.3235	2.8286	0.3535	0.9803
4	1.126	0.8885	4.184	0.2390	3.7171	0.2690	1.4631
5	1.159	0.8626	5.309	0.1884	4.5797	0.2184	1.9409
6	1.194	0.8375	6.468	0.1546	5.4172	0.1846	2.4138
7	1.230	0.8131	7.662	0.1305	6.2303	0.1605	2.8819
8	1.267	0.7894	8.892	0.1125	7.0197	0.1425	3.3450
9	1.305	0.7664	10.159	0.0984	7.7861	0.1284	3.8032
10	1.344	0.7441	11.464	0.0872	8.5302	0.1172	4.2565
11	1.384	0.7224	12.808	0.0781	9.2526	0.1081	4.7049
12	1.426	0.7014	14.192	0.0705	9.9540	0.1005	5.1485
13	1.469	0.6810	15.618	0.0640	10.6350	0.0940	5.5872
14	1.513	0.6611	17.086	0.0585	11.2961	0.0885	6.0211
15	1.558	0.6419	18.599	0.0538	11.9379	0.0838	6.4501
16	1.605	0.6232	20.157	0.0496	12.5611	0.0796	6.8742
17	1.653	0.6050	21.762	0.0460	13.1661	0.0760	7.2936
18	1.702	0.5874	23.414	0.0427	13.7535	0.0727	7.7081
19	1.754	0.5703	25.117	0.0398	14.3238	0.0698	8.1179
20	1.806	0.5537	26.870	0.0372	14.8775	0.0672	8.5229
21	1.860	0.5376	28.676	0.0349	15.4150	0.0649	8.9231
22	1.916	0.5219	30.537	0.0328	15.9369	0.0628	9.3186
23	1.974	0.5067	32.453	0.0308	16.4436	0.0608	9.7094
24	2.033	0.4919	34.426	0.0291	16.9356	0.0591	10.0954
25	2.094	0.4776	36.459	0.0274	17.4132	0.0574	10.4768
26	2.157	0.4637	38.553	0.0259	17.8769	0.0559	10.8535
27	2.221	0.4502	40.710	0.0246	18.3270	0.0546	11.2256
28	2.288	0.4371	42.931	0.0233	18.7641	0.0533	11.5930
29	2.357	0.4244	45.219	0.0221	19.1885	0.0521	11.9558
30	2.427	0.4120	47.575	0.0210	19.6005	0.0510	12.3141
31	2.500	0.4000	50.003	0.0200	20.0004	0.0500	12.6678
32	2.575	0.3883	52.503	0.0191	20.3888	0.0491	13.0169
33	2.652	0.3770	55.078	0.0182	20.7658	0.0482	13.3616
34	2.732	0.3661	57.730	0.0173	21.1318	0.0473	13.7018
35	2.814	0.3554	60.462	0.0165	21.4872	0.0465	14.0375
40	3.262	0.3066	75.401	0.0133	23.1148	0.0433	15.6502
▶45	3.782	0.2644	92.720	0.0108	24.5187	0.0408	17.1556
50	4.384	0.2281	112.797	0.0089	25.7298	0.0389	18.5575
55	5.082	0.1968	136.072	0.0074	26.7744	0.0374	19.8600
60	5.892	0.1697	163.053	0.0061	27.6756	0.0361	21.0674
65	6.830	0.1464	194.333	0.0052	28.4529	0.0352	22.1841
70	7.918	0.1263	230.594	0.0043	29.1234	0.0343	23.2145
75	9.179	0.1090	272.631	0.0037	29.7018	0.0337	24.1634
80	10.641	0.0940	321.363	0.0031	30.2008	0.0331	25.0354
85	12.336	0.0811	377.857	0.0027	30.6312	0.0327	25.8349
90	14.300	0.0699	443.349	0.0023	31.0024	0.0323	26.5667
95	16.578	0.0603	519.272	0.0019	31.3227	0.0319	27.2351
100	19.219	0.0520	607.288	0.0017	31.5989	0.0317	27.8445

Table D.7. 4% INTEREST FACTORS FOR ANNUAL COMPOUNDING INTEREST

	Single Payment		Equal Payment Series				Uniform gradient-series factor
	Compound-amount factor	Present-worth factor	Compound-amount factor	Sinking-fund factor	Present-worth factor	Capital-recovery factor	
n	To find F Given P F/P i, n	To find P Given F P/F i, n	To find F Given A F/A i, n	To find A Given F A/F i, n	To find P Given A P/A i, n	To find A Given P A/P i, n	To find A Given g A/G i, n
1	1.040	0.9615	1.000	1.0000	0.9615	1.0400	0.0000
2	1.082	0.9246	2.040	0.4902	1.8861	0.5302	0.4902
3	1.125	0.8890	3.122	0.3204	2.7751	0.3604	0.9739
4	1.170	0.8548	4.246	0.2355	3.6299	0.2755	1.4510
5	1.217	0.8219	5.416	0.1846	4.4518	0.2246	1.9216
6	1.265	0.7903	6.633	0.1508	5.2421	0.1908	2.3857
7	1.316	0.7599	7.898	0.1266	6.0021	0.1666	2.8433
8	1.369	0.7307	9.214	0.1085	6.7328	0.1485	3.2944
9	1.423	0.7026	10.583	0.0945	7.4353	0.1345	3.7391
10	1.480	0.6756	12.006	0.0833	8.1109	0.1233	4.1773
11	1.539	0.6496	13.486	0.0742	8.7605	0.1142	4.6090
12	1.601	0.6246	15.026	0.0666	9.3851	0.1066	5.0344
13	1.665	0.6006	16.627	0.0602	9.9857	0.1002	5.4533
14	1.732	0.5775	18.292	0.0547	10.5631	0.0947	5.8659
15	1.801	0.5553	20.024	0.0500	11.1184	0.0900	6.2721
16	1.873	0.5339	21.825	0.0458	11.6523	0.0858	6.6720
17	1.948	0.5134	23.698	0.0422	12.1657	0.0822	7.0656
18	2.026	0.4936	25.645	0.0390	12.6593	0.0790	7.4530
19	2.107	0.4747	27.671	0.0361	13.1339	0.0761	7.8342
20	2.191	0.4564	29.778	0.0336	13.5903	0.0736	8.2091
21	2.279	0.4388	31.969	0.0313	14.0292	0.0713	8.5780
22	2.370	0.4220	34.248	0.0292	14.4511	0.0692	8.9407
23	2.465	0.4057	36.618	0.0273	14.8569	0.0673	9.2973
24	2.563	0.3901	39.083	0.0256	15.2470	0.0656	9.6479
25	2.666	0.3751	41.646	0.0240	15.6221	0.0640	9.9925
26	2.772	0.3607	44.312	0.0226	15.9828	0.0626	10.3312
27	2.883	0.3468	47.084	0.0212	16.3296	0.0612	10.6640
28	2.999	0.3335	49.968	0.0200	16.6631	0.0600	10.9909
29	3.119	0.3207	52.966	0.0189	16.9837	0.0589	11.3121
30	3.243	0.3083	56.085	0.0178	17.2920	0.0578	11.6274
31	3.373	0.2965	59.328	0.0169	17.5885	0.0569	11.9371
32	3.508	0.2851	62.701	0.0160	17.8736	0.0560	12.2411
33	3.648	0.2741	66.210	0.0151	18.1477	0.0551	12.5396
34	3.794	0.2636	69.858	0.0143	18.4112	0.0543	12.8325
35	3.946	0.2534	73.652	0.0136	18.6646	0.0536	13.1199
40	4.801	0.2083	95.026	0.0105	19.7928	0.0505	14.4765
45	5.841	0.1712	121.029	0.0083	20.7200	0.0483	15.7047
50	7.107	0.1407	152.667	0.0066	21.4822	0.0466	16.8123
55	8.646	0.1157	191.159	0.0052	22.1086	0.0452	17.8070
60	10.520	0.0951	237.991	0.0042	22.6235	0.0442	18.6972
65	12.799	0.0781	294.968	0.0034	23.0467	0.0434	19.4909
70	15.572	0.0642	364.290	0.0028	23.3945	0.0428	20.1961
75	18.945	0.0528	448.631	0.0022	23.6804	0.0422	20.8206
80	23.050	0.0434	551.245	0.0018	23.9154	0.0418	21.3719
85	28.044	0.0357	676.090	0.0015	24.1085	0.0415	21.8569
90	34.119	0.0293	817.983	0.0012	24.2673	0.0412	22.2826
95	41.511	0.0241	1012.785	0.0010	24.3978	0.0410	22.6550
100	50.505	0.0198	1237.624	0.0008	24.5050	0.0408	22.9800

Table D.8. 5% INTEREST FACTORS FOR ANNUAL COMPOUNDING INTEREST

	Single Payment		Equal Payment Series				Uniform gradient-series factor
	Compound-amount factor	Present-worth factor	Compound-amount factor	Sinking-fund factor	Present-worth factor	Capital-recovery factor	
n	To find F Given P F/P i, n	To find P Given F P/F i, n	To find F Given A F/A i, n	To find A Given F A/F i, n	To find P Given A P/A i, n	To find A Given P A/P i, n	To find A Given g A/G i, n
1	1.050	0.9524	1.000	1.0000	0.9524	1.0500	0.0000
2	1.103	0.9070	2.050	0.4878	1.8594	0.5378	0.4878
3	1.158	0.8638	3.153	0.3172	2.7233	0.3672	0.9675
4	1.216	0.8227	4.310	0.2320	3.5460	0.2820	1.4391
5	1.276	0.7835	5.526	0.1810	4.3295	0.2310	1.9025
6	1.340	0.7462	6.802	0.1470	5.0757	0.1970	2.3579
7	1.407	0.7107	8.142	0.1228	5.7864	0.1728	2.8052
8	1.477	0.6768	9.549	0.1047	6.4632	0.1547	3.2445
9	1.551	0.6446	11.027	0.0907	7.1078	0.1407	3.6758
10	1.629	0.6139	12.587	0.0795	7.7217	0.1295	4.0991
11	1.710	0.5847	14.207	0.0704	8.3064	0.1204	4.5145
12	1.796	0.5568	15.917	0.0628	8.8633	0.1128	4.9219
13	1.886	0.5303	17.713	0.0565	9.3936	0.1065	5.3215
14	1.980	0.5051	19.599	0.0510	9.8987	0.1010	5.7133
15	2.079	0.4810	21.579	0.0464	10.3797	0.0964	6.0973
16	2.183	0.4581	23.658	0.0423	10.8378	0.0923	6.4736
17	2.292	0.4363	25.840	0.0387	11.2741	0.0887	6.8423
18	2.407	0.4155	28.132	0.0356	11.6896	0.0856	7.2034
19	2.527	0.3957	30.539	0.0328	12.0853	0.0828	7.5569
20	2.653	0.3769	33.066	0.0303	12.4622	0.0803	7.9030
21	2.786	0.3590	35.719	0.0280	12.8212	0.0780	8.2416
22	2.925	0.3419	38.505	0.0260	13.1630	0.0760	8.5730
23	3.072	0.3256	41.430	0.0241	13.4886	0.0741	8.8971
24	3.225	0.3101	44.502	0.0225	13.7987	0.0725	9.2140
25	3.386	0.2953	47.727	0.0210	14.0940	0.0710	9.5238
26	3.556	0.2813	51.113	0.0196	14.3752	0.0696	9.8266
27	3.733	0.2679	54.669	0.0183	14.6430	0.0683	10.1224
28	3.920	0.2551	58.403	0.0171	14.8981	0.0671	10.4114
29	4.116	0.2430	62.323	0.0161	15.1411	0.0661	10.6936
30	4.322	0.2314	66.439	0.0151	15.3725	0.0651	10.9691
31	4.538	0.2204	70.761	0.0141	15.5928	0.0641	11.2381
32	4.765	0.2099	75.299	0.0133	15.8027	0.0633	11.5005
33	5.003	0.1999	80.064	0.0125	16.0026	0.0625	11.7566
34	5.253	0.1904	85.067	0.0118	16.1929	0.0618	12.0063
35	5.516	0.1813	90.320	0.0111	16.3742	0.0611	12.2498
40	7.040	0.1421	120.800	0.0083	17.1591	0.0583	13.3775
45	8.985	0.1113	159.700	0.0063	17.7741	0.0563	14.3644
50	11.467	0.0872	209.348	0.0048	18.2559	0.0548	15.2233
55	14.636	0.0683	272.713	0.0037	18.6335	0.0537	15.9665
60	18.679	0.0535	353.584	0.0028	18.9293	0.0528	16.6062
65	23.840	0.0420	456.798	0.0022	19.1611	0.0522	17.1541
70	30.426	0.0329	588.529	0.0017	19.3427	0.0517	17.6212
75	38.833	0.0258	756.654	0.0013	19.4850	0.0513	18.0176
80	49.561	0.0202	971.229	0.0010	19.5965	0.0510	18.3526
85	63.254	0.0158	1245.087	0.0008	19.6838	0.0508	18.6346
90	80.730	0.0124	1594.607	0.0006	19.7523	0.0506	18.8712
95	103.035	0.0097	2040.694	0.0005	19.8059	0.0505	19.0689
100	131.501	0.0076	2610.025	0.0004	19.8479	0.0504	19.2337

Table D.9. 6% INTEREST FACTORS FOR ANNUAL COMPOUNDING INTEREST

	Single Payment		Equal Payment Series				Uniform gradient-series factor
	Compound-amount factor	Present-worth factor	Compound-amount factor	Sinking-fund factor	Present-worth factor	Capital-recovery factor	
n	To find F Given P F/P i, n	To find P Given F P/F i, n	To find F Given A F/A i, n	To find A Given F A/F i, n	To find P Given A P/A i, n	To find A Given P A/P i, n	To find A Given g A/G i, n
1	1.060	0.9434	1.000	1.0000	0.9434	1.0600	0.0000
2	1.124	0.8900	2.060	0.4854	1.8334	0.5454	0.4854
3	1.191	0.8396	3.184	0.3141	2.6730	0.3741	0.9612
4	1.262	0.7921	4.375	0.2286	3.4651	0.2886	1.4272
5	1.338	0.7473	5.637	0.1774	4.2124	0.2374	1.8836
6	1.419	0.7050	6.975	0.1434	4.9173	0.2034	2.3304
7	1.504	0.6651	8.394	0.1191	5.5824	0.1791	2.7676
8	1.594	0.6274	9.897	0.1010	6.2098	0.1610	3.1952
9	1.689	0.5919	11.491	0.0870	6.8017	0.1470	3.6133
10	1.791	0.5584	13.181	0.0759	7.3601	0.1359	4.0220
11	1.898	0.5268	14.972	0.0668	7.8869	0.1268	4.4213
12	2.012	0.4970	16.870	0.0593	8.3839	0.1193	4.8113
13	2.133	0.4688	18.882	0.0530	8.8527	0.1130	5.1920
14	2.261	0.4423	21.015	0.0476	9.2950	0.1076	5.5635
15	2.397	0.4173	23.276	0.0430	9.7123	0.1030	5.9260
16	2.540	0.3937	25.673	0.0390	10.1059	0.0990	6.2794
17	2.693	0.3714	28.213	0.0355	10.4773	0.0955	6.6240
18	2.854	0.3504	30.906	0.0324	10.8276	0.0924	6.9597
19	3.026	0.3305	33.760	0.0296	11.1581	0.0896	7.2867
20	3.207	0.3118	36.786	0.0272	11.4699	0.0872	7.6052
21	3.400	0.2942	39.993	0.0250	11.7641	0.0850	7.9151
22	3.604	0.2775	43.392	0.0231	12.0416	0.0831	8.2166
23	3.820	0.2618	46.996	0.0213	12.3034	0.0813	8.5099
24	4.049	0.2470	50.816	0.0197	12.5504	0.0797	8.7951
25	4.292	0.2330	54.865	0.0182	12.7834	0.0782	9.0722
26	4.549	0.2198	59.156	0.0169	13.0032	0.0769	9.3415
27	4.822	0.2074	63.706	0.0157	13.2105	0.0757	9.6030
28	5.112	0.1956	68.528	0.0146	13.4062	0.0746	9.8568
29	5.418	0.1846	73.640	0.0136	13.5907	0.0736	10.1032
30	5.744	0.1741	79.058	0.0127	13.7648	0.0727	10.3422
31	6.088	0.1643	84.802	0.0118	13.9291	0.0718	10.5740
32	6.453	0.1550	90.890	0.0110	14.0841	0.0710	10.7988
33	6.841	0.1462	97.343	0.0103	14.2302	0.0703	11.0166
34	7.251	0.1379	104.184	0.0096	14.3682	0.0696	11.2276
35	7.686	0.1301	111.435	0.0090	14.4983	0.0690	11.4319
40	10.286	0.0972	154.762	0.0065	15.0463	0.0665	12.3590
45	13.765	0.0727	212.744	0.0047	15.4558	0.0647	13.1413
50	18.420	0.0543	290.336	0.0035	15.7619	0.0635	13.7964
55	24.650	0.0406	394.172	0.0025	15.9906	0.0625	14.3411
60	32.988	0.0303	533.128	0.0019	16.1614	0.0619	14.7910
65	44.145	0.0227	719.083	0.0014	16.2891	0.0614	15.1601
70	59.076	0.0169	967.932	0.0010	16.3846	0.0610	15.4614
75	79.057	0.0127	1300.949	0.0008	16.4559	0.0608	15.7058
80	105.796	0.0095	1746.600	0.0006	16.5091	0.0606	15.9033
85	141.579	0.0071	2342.982	0.0004	16.5490	0.0604	16.0620
90	189.465	0.0053	3141.075	0.0003	16.5787	0.0603	16.1891
95	253.546	0.0040	4209.104	0.0002	16.6009	0.0602	16.2905
100	339.302	0.0030	5638.368	0.0002	16.6176	0.0602	16.3711

Table D.10. 7% INTEREST FACTORS FOR ANNUAL COMPOUNDING INTEREST

	Single Payment		Equal Payment Series				Uniform
	Compound-amount factor	Present-worth factor	Compound-amount factor	Sinking-fund factor	Present-worth factor	Capital-recovery factor	gradient-series factor
n	To find F Given P F/P i, n	To find P Given F P/F i, n	To find F Given A F/A i, n	To find A Given F A/F i, n	To find P Given A P/A i, n	To find A Given P A/P i, n	To find A Given g A/G i, n
1	1.070	0.9346	1.000	1.0000	0.9346	1.0700	0.0000
2	1.145	0.8734	2.070	0.4831	1.8080	0.5531	0.4831
3	1.225	0.8163	3.215	0.3111	2.6243	0.3811	0.9549
4	1.311	0.7629	4.440	0.2252	3.3872	0.2952	1.4155
5	1.403	0.7130	5.751	0.1739	4.1002	0.2439	1.8650
6	1.501	0.6664	7.153	0.1398	4.7665	0.2098	2.3032
7	1.606	0.6228	8.654	0.1156	5.3893	0.1856	2.7304
8	1.718	0.5820	10.260	0.0975	5.9713	0.1675	3.1466
9	1.838	0.5439	11.978	0.0835	6.5152	0.1535	3.5517
10	1.967	0.5084	13.816	0.0724	7.0236	0.1424	3.9461
11	2.105	0.4751	15.784	0.0634	7.4987	0.1334	4.3296
12	2.252	0.4440	17.888	0.0559	7.9427	0.1259	4.7025
13	2.410	0.4150	20.141	0.0497	8.3577	0.1197	5.0649
14	2.579	0.3878	22.550	0.0444	8.7455	0.1144	5.4167
15	2.759	0.3625	25.129	0.0398	9.1079	0.1098	5.7583
16	2.952	0.3387	27.888	0.0359	9.4467	0.1059	6.0897
17	3.159	0.3166	30.840	0.0324	9.7632	0.1024	6.4110
18	3.380	0.2959	33.999	0.0294	10.0591	0.0994	6.7225
19	3.617	0.2765	37.379	0.0268	10.3356	0.0968	7.0242
20	3.870	0.2584	40.996	0.0244	10.5940	0.0944	7.3163
21	4.141	0.2415	44.865	0.0223	10.8355	0.0923	7.5990
22	4.430	0.2257	49.006	0.0204	11.0613	0.0904	7.8725
23	4.741	0.2110	53.436	0.0187	11.2722	0.0887	8.1369
24	5.072	0.1972	58.177	0.0172	11.4693	o.0872	8.3923
25	5.427	0.1843	63.249	0.0158	11.6536	0.0858	8.6391
26	5.807	0.1722	68.676	0.0146	11.8258	0.0846	8.8773
27	6.214	0.1609	74.484	0.0134	11.9867	0.0834	9.1072
28	6.649	0.1504	80.698	0.0124	12.1371	0.0824	9.3290
29	7.114	0.1406	87.347	0.0115	12.2777	0.0815	9.5427
30	7.612	0.1314	94.461	0.0106	12.4091	0.0806	9.7487
31	8.145	0.1228	102.073	0.0098	12.5318	0.0798	9.9471
32	8.715	0.1148	110.218	0.0091	12.6466	0.0791	10.1381
33	9.325	0.1072	118.933	0.0084	12.7538	0.0784	10.3219
34	9.978	0.1002	128.259	0.0078	12.8540	0.0778	10.4987
35	10.677	0.0937	138.237	0.0072	12.9477	0.0772	10.6687
40	14.974	0.0668	199.635	0.0050	13.3317	0.0750	11.4234
45	21.002	0.0476	285.749	0.0035	13.6055	0.0735	12.0360
50	29.457	0.0340	406.529	0.0025	13.8008	0 0725	12.5287
55	41.315	0.0242	575.929	0.0017	13.9399	0.0717	12.9215
60	57.946	0.0173	813.520	0.0012	14.0392	0.0712	13.2321
65	81.273	0.0123	1146.755	0.0009	14.1099	0.0709	13.4760
70	113.989	0.0088	1614.134	0.0006	14.1604	0.0706	13.6662
75	159.876	0.0063	2269.657	0.0005	14.1964	0.0705	13.8137
80	224.234	0.0045	3189.063	0.0003	14.2220	0.0703	13.9274
85	314.500	0.0032	4478.576	0.0002	14.2403	0.0702	14.0146
90	441.103	0.0023	6287.185	0.0002	14.2533	0.0702	14.0812
95	618.670	0.0016	8823.854	0.0001	14.2626	0.0701	14.1319
100	867.716	0.0012	12381.662	0.0001	14.2693	0.0701	14.1703

Table D.11. 8% INTEREST FACTORS FOR ANNUAL COMPOUNDING INTEREST

	Single Payment		Equal Payment Series				Uniform gradient-series factor
	Compound-amount factor	Present-worth factor	Compound-amount factor	Sinking-fund factor	Present-worth factor	Capital-recovery factor	
n	To find F Given P F/P i,n	To find P Given F P/F i,n	To find F Given A F/A i,n	To find A Given F A/F i,n	To find P Given A P/A i,n	To find A Given P A/P i,n	To find A Given g A/G i,n
1	1.080	0.9259	1.000	1.0000	0.9259	1.0800	0.0000
2	1.166	0.8573	2.080	0.4808	1.7833	0.5608	0.4808
3	1.260	0.7938	3.246	0.3080	2.5771	0.3880	0.9488
4	1.360	0.7350	4.506	0.2219	3.3121	0.3019	1.4040
5	1.469	0.6806	5.867	0.1705	3.9927	0.2505	1.8465
6	1.587	0.6302	7.336	0.1363	4.6229	0.2163	2.2764
7	1.714	0.5835	8.923	0.1121	5.2064	0.1921	2.6937
8	1.851	0.5403	10.637	0.0940	5.7466	0.1740	2.0985
9	1.999	0.5003	12.488	0.0801	6.2469	0.1601	3.4910
10	2.159	0.4632	14.487	0.0690	6.7101	0.1490	3.8713
11	2.332	0.4289	16.645	0.0601	7.1390	0.1401	4.2395
12	2.518	0.3971	18.977	0.0527	7.5361	0.1327	4.5958
13	2.720	0.3677	21.495	0.0465	7.9038	0.1265	4.9402
14	2.937	0.3405	24.215	0.0413	8.2442	0.1213	5.2731
15	3.172	0.3153	27.152	0.0368	8.5595	0.1168	5.5945
16	3.426	0.2919	30.324	0.0330	8.8514	0.1130	5.9046
17	3.700	0.2703	33.750	0.0296	9.1216	0.1096	6.2038
18	3.996	0.2503	37.450	0.0267	9.3719	0.1067	6.4920
19	4.316	0.2317	41.446	0.0241	9.6036	0.1041	6.7697
20	4.661	0.2146	45.762	0.0219	9.8182	0.1019	7.0370
21	5.034	0.1987	50.423	0.0198	10.0168	0.0998	7.2940
22	5.437	0.1840	55.457	0.0180	10.2008	0.0980	7.5412
23	5.871	0.1703	60.893	0.0164	10.3711	0.0964	7.7786
24	6.341	0.1577	66.765	0.0150	10.5288	0.0950	8.0066
25	6.848	0.1460	73.106	0.0137	10.6748	0.0937	8.2254
26	7.396	0.1352	79.954	0.0125	10.8100	0.0925	8.4352
27	7.988	0.1252	87.351	0.0115	10.9352	0.0915	8.6363
28	8.627	0.1159	95.339	0.0105	11.0511	0.0905	8.8289
29	9.317	0.1073	103.966	0.0096	11.1584	0.0896	9.0133
30	10.063	0.0994	113.283	0.0088	11.2578	0.0888	9.1897
31	10.868	0.0920	123.346	0.0081	11.3498	0.0881	9.3584
32	11.737	0.0852	134.214	0.0075	11.4350	0.0875	9.5197
33	12.676	0.0789	145.951	0.0069	11.5139	0.0869	9.6737
34	13.690	0.0731	158.627	0.0063	11.5869	0.0863	9.8208
35	14.785	0.0676	172.317	0.0058	11.6546	0.0858	9.9611
40	21.725	0.0460	259.057	0.0039	11.9246	0.0839	10.5699
45	31.920	0.0313	386.506	0.0026	12.1084	0.0826	11.0447
50	46.902	0.0213	573.770	0.0018	12.2335	0.0818	11.4107
55	68.914	0.0145	848.923	0.0012	12.3186	0.0812	11.6902
60	101.257	0.0099	1253.213	0.0008	12.3766	0.0808	11.9015
65	148.780	0.0067	1847.248	0.0006	12.4160	0.0806	12.0602
70	218.606	0.0046	2720.080	0.0004	12.4428	0.0804	12.1783
75	321.205	0.0031	4002.557	0.0003	12.4611	0.0803	12.2658
80	471.955	0.0021	5886.935	0.0002	12.4735	0.0802	12.3301
85	693.456	0.0015	8655.706	0.0001	12.4820	0.0801	12.3773
90	1018.915	0.0010	12723.939	0.0001	12.4877	0.0801	12.4116
95	1497.121	0.0007	18701.507	0.0001	12.4917	0.0801	12.4365
100	2199.761	0.0005	27484.516	0.0001	12.4943	0.0800	12.4545

Table D.12. 9% INTEREST FACTORS FOR ANNUAL COMPOUNDING INTEREST

	Single Payment		Equal Payment Series				Uniform gradient-series factor
	Compound-amount factor	Present-worth factor	Compound-amount factor	Sinking-fund factor	Present-worth factor	Capital-recovery factor	
n	To find F Given P F/P i, n	To find P Given F P/F i, n	To find F Given A F/A i, n	To find A Given F A/F i, n	To find P Given A P/A i, n	To find A Given P A/P i, n	To find A Given g A/G i, n
1	1.090	0.9174	1.000	1.0000	0.9174	1.0900	0.0000
2	1.188	0.8417	2.090	0.4785	1.7591	0.5685	0.4785
3	1.295	0.7722	3.278	0.3051	2.5313	0.3951	0.9426
4	1.412	0.7084	4.573	0.2187	3.2397	0.3087	1.3925
5	1.539	0.6499	5.985	0.1671	3.8897	0.2571	1.8282
6	1.677	0.5963	7.523	0.1329	4.4859	0.2229	2.2498
7	1.828	0.5470	9.200	0.1087	5.0330	0.1987	2.6574
8	1.993	0.5019	11.028	0.0907	5.5348	0.1807	3.0512
9	2.172	0.4604	13.021	0.0768	5.9953	0.1668	3.4312
10	2.367	0.4224	15.193	0.0658	6.4177	0.1558	3.7978
11	2.580	0.3875	17.560	0.0570	6.8052	0.1470	4.1510
12	2.813	0.3555	20.141	0.0497	7.1607	0.1397	4.4910
13	3.066	0.3262	22.953	0.0436	7.4869	0.1336	4.8182
14	3.342	0.2993	26.019	0.0384	7.7862	0.1284	5.1326
15	3.642	0.2745	29.361	0.0341	8.0607	0.1241	5.4346
16	3.970	0.2519	33.003	0.0303	8.3126	0.1203	5.7245
17	4.328	0.2311	36.974	0.0271	8.5436	0.1171	6.0024
18	4.717	0.2120	41.301	0.0242	8.7556	0.1142	6.2687
19	5.142	0.1945	46.018	0.0217	8.9501	0.1117	6.5236
20	5.604	0.1784	51.160	0.0196	9.1286	0.1096	6.7675
21	6.109	0.1637	56.765	0.0176	9.2923	0.1076	7.0006
22	6.659	0.1502	62.873	0.0159	9.4424	0.1059	7.2232
23	7.258	0.1378	69.532	0.0144	9.5802	0.1044	7.4358
24	7.911	0.1264	76.790	0.0130	9.7066	0.1030	7.6384
25	8.623	0.1160	84.701	0.0118	9.8226	0.1018	7.8316
26	9.399	0.1064	93.324	0.0107	9.9290	0.1007	8.0156
27	10.245	0.0976	102.723	0.0097	10.0266	0.0997	8.1906
28	11.167	0.0896	112.968	0.0089	10.1161	0.0989	8.3572
29	12.172	0.0822	124.135	0.0081	10.1983	0.0981	8.5154
30	13.268	0.0754	136.308	0.0073	10.2737	0.0973	8.6657
31	14.462	0.0692	149.575	0.0067	10.3428	0.0967	8.8083
32	15.763	0.0634	164.037	0.0061	10.4063	0.0961	8.9436
33	17.182	0.0582	179.800	0.0056	10.4645	0.0956	9.0718
34	18.728	0.0534	196.982	0.0051	10.5178	0.0951	9.1933
35	20.414	0.0490	215.711	0.0046	10.5668	0.0946	9.3083
40	31.409	0.0318	337.882	0.0030	10.7574	0.0930	9.7957
45	48.327	0.0207	525.859	0.0019	10.8812	0.0919	10.1603
50	74.358	0.0135	815.084	0.0012	10.9617	0.0912	10.4295
55	114.408	0.0088	1260.092	0.0008	11.0140	0.0908	10.6261
60	176.031	0.0057	1944.792	0.0005	11.0480	0.0905	10.7683
65	270.846	0.0037	2998.288	0.0003	11.0701	0.0903	10.8702
70	416.730	0.0024	4619.223	0.0002	11.0845	0.0902	10.9427
75	641.191	0.0016	7113.232	0.0002	11.0938	0.0902	10.9940
80	986.552	0.0010	10950.574	0.0001	11.0999	0.0901	11.0299
85	1517.932	0.0007	16854.800	0.0001	11.1038	0.0901	11.0551
90	2335.527	0.0004	25939.184	0.0001	11.1064	0.0900	11.0726
95	3593.497	0.0003	39916.635	0.0000	11.1080	0.0900	11.0847
100	5529.041	0.0002	61422.675	0.0000	11.1091	0.0900	11.0930

Table D.13. 10% INTEREST FACTORS FOR ANNUAL COMPOUNDING INTEREST

	Single Payment		Equal Payment Series				Uniform gradient-series factor
	Compound-amount factor	Present-worth factor	Compound-amount factor	Sinking-fund factor	Present-worth factor	Capital-recovery factor	
n	To find F Given P F/P i,n	To find P Given F P/F i,n	To find F Given A F/A i,n	To find A Given F A/F i,n	To find P Given A P/A i,n	To find A Given P A/P i,n	To find A Given g A/G i,n
1	1.100	0.9091	1.000	1.0000	0.9091	1.1000	0.0000
2	1.210	0.8265	2.100	0.4762	1.7355	0.5762	0.4762
3	1.331	0.7513	3.310	0.3021	2.4869	0.4021	0.9366
4	1.464	0.6830	4.641	0.2155	3.1699	0.3155	1.3812
5	1.611	0.6209	6.105	0.1638	3.7908	0.2638	1.8101
6	1.772	0.5645	7.716	0.1296	4.3553	0.2296	2.2236
7	1.949	0.5132	9.487	0.1054	4.8684	0.2054	2.6216
8	2.144	0.4665	11.436	0.0875	5.3349	0.1875	3.0045
9	2.358	0.4241	13.579	0.0737	5.7590	0.1737	3.3724
10	2.594	0.3856	15.937	0.0628	6.1446	0.1628	3.7255
11	2.853	0.3505	18.531	0.0540	6.4951	0.1540	4.0641
12	3.138	0.3186	21.384	0.0468	6.8137	0.1468	4.3884
13	3.452	0.2897	24.523	0.0408	7.1034	0.1408	4.6988
14	3.798	0.2633	27.975	0.0358	7.3667	0.1358	4.9955
15	4.177	0.2394	31.772	0.0315	7.6061	0.1315	5.2789
16	4.595	0.2176	35.950	0.0278	7.8237	0.1278	5.5493
17	5.054	0.1979	40.545	0.0247	8.0216	0.1247	5.8071
18	5.560	0.1799	45.599	0.0219	8.2014	0.1219	6.0526
19	6.116	0.1635	51.159	0.0196	8.3649	0.1196	6.2861
20	6.728	0.1487	57.275	0.0175	8.5136	0.1175	6.5081
21	7.400	0.1351	64.003	0.0156	8.6487	0.1156	6.7189
22	8.140	0.1229	71.403	0.0140	8.7716	0.1140	6.9189
23	8.954	0.1117	79.543	0.0126	8.8832	0.1126	7.1085
24	9.850	0.1015	88.497	0.0113	8.9848	0.1113	7.2881
25	10.835	0.0923	98.347	0.0102	9.0771	0.1102	7.4580
26	11.918	0.0839	109.182	0.0092	9.1610	0.1092	7.6187
27	13.110	0.0763	121.100	0.0083	9.2372	0.1083	7.7704
28	14.421	0.0694	134.210	0.0075	9.3066	0.1075	7.9137
29	15.863	0.0630	148.631	0.0067	9.3696	0.1067	8.0489
30	17.449	0.0573	164.494	0.0061	9.4269	0.1061	8.1762
31	19.194	0.0521	181.943	0.0055	9.4790	0.1055	8.2962
32	21.114	0.0474	201.138	0.0050	9.5264	0.1050	8.4091
33	23.225	0.0431	222.252	0.0045	9.5694	0.1045	8.5152
34	25.548	0.0392	245.477	0.0041	9.6086	0.1041	8.6149
35	28.102	0.0356	271.024	0.0037	9.6442	0.1037	8.7086
40	45.259	0.0221	442.593	0.0023	9.7791	0:1023	9.0962
45	72.890	0.0137	718.905	0.0014	9.8628	0.1014	9.3741
50	117.391	0.0085	1163.909	0.0009	9.9148	0.1009	9.5704
55	189.059	0.0053	1880.591	0.0005	9.9471	0.1005	9.7075
60	304.482	0.0033	3034.816	0.0003	9.9672	0.1003	9.8023
65	490.371	0.0020	4893.707	0.0002	9.9796	0.1002	9.8672
70	789.747	0.0013	7887.470	0.0001	9.9873	0.1001	9.9113
75	1271.895	0.0008	12708.954	0.0001	9.9921	0.1001	9.9410
80	2048.400	0.0005	20474.002	0.0001	9.9951	0.1001	9.9609
85	3298.969	0.0003	32979.690	0.0000	9.9970	0.1000	9.9742
90	5313.023	0.0002	53120.226	0.0000	9.9981	0.1000	9.9831
95	8556.676	0.0001	85556.760	0.0000	9.9988	0.1000	9.9889
100	13780.612	0.0001	137796.123	0.0000	9.9993	0.1000	9.9928

Table D.14. 12% INTEREST FACTORS FOR ANNUAL COMPOUNDING INTEREST

	Single Payment		Equal Payment Series				Uniform gradient-series factor
	Compound-amount factor	Present-worth factor	Compound-amount factor	Sinking-fund factor	Present-worth factor	Capital-recovery factor	
n	To find F Given P F/P i, n	To find P Given F P/F i, n	To find F Given A F/A i, n	To find A Given F A/F i, n	To find P Given A P/A i, n	To find A Given P A/P i, n	To find A Given g A/G i, n
1	1.120	0.8929	1.000	1.0000	0.8929	1.1200	0.0000
2	1.254	0.7972	2.120	0.4717	1.6901	0.5917	0.4717
3	1.405	0.7118	3.374	0.2964	2.4018	0.4164	0.9246
4	1.574	0.6355	4.779	0.2092	3.0374	0.3292	1.3589
5	1.762	0.5674	6.353	0.1574	3.6048	0.2774	1.7746
6	1.974	0.5066	8.115	0.1232	4.1114	0.2432	2.1721
7	2.211	0.4524	10.089	0.0991	4.5638	0.2191	2.5515
8	2.476	0.4039	12.300	0.0813	4.9676	0.2013	2.9132
9	2.773	0.3606	14.776	0.0677	5.3283	0.1877	3.2574
10	3.106	0.3220	17.549	0.0570	5.6502	0.1770	3.5847
11	3.479	0.2875	20.655	0.0484	5.9377	0.1684	3.8953
12	3.896	0.2567	24.133	0.0414	6.1944	0.1614	4.1897
13	4.364	0.2292	28.029	0.0357	6.4236	0.1557	4.4683
14	4.887	0.2046	32.393	0.0309	6.6282	0.1509	4.7317
15	5.474	0.1827	37.280	0.0268	6.8109	0.1468	4.9803
16	6.130	0.1631	42.753	0.0234	6.9740	0.1434	5.2147
17	6.866	0.1457	48.884	0.0205	7.1196	0.1405	5.4353
18	7.690	0.1300	55.750	0.0179	7.2497	0.1379	5.6427
19	8.613	0.1161	63.440	0.0158	7.3658	0.1358	5.8375
20	9.646	0.1037	72.052	0.0139	7.4695	0.1339	6.0202
21	10.804	0.0926	81.699	0.0123	7.5620	0.1323	6.1913
22	12.100	0.0827	92.503	0.0108	7.6447	0.1308	6.3514
23	13.552	0.0738	104.603	0.0096	7.7184	0.1296	6.5010
24	15.179	0.0659	118.155	0.0085	7.7843	0.1285	6.6407
25	17.000	0.0588	133.334	0.0075	7.8431	0.1275	6.7708
26	19.040	0.0525	150.334	0.0067	7.8957	0.1267	6.8921
27	21.325	0.0469	169.374	0.0059	7.9426	0.1259	7.0049
28	23.884	0.0419	190.699	0.0053	7.9844	0.1253	7.1098
29	26.750	0.0374	214.583	0.0047	8.0218	0.1247	7.2071
30	29.960	0.0334	241.333	0.0042	8.0552	0.1242	7.2974
31	33.555	0.0298	271.293	0.0037	8.0850	0.1237	7.3811
32	37.582	0.0266	304.848	0.0033	8.1116	0.1233	7.4586
33	42.092	0.0238	342.429	0.0029	8.1354	0.1229	7.5303
34	47.143	0.0212	384.521	0.0026	8.1566	0.1226	7.5965
35	52.800	0.0189	431.664	0.0023	8.1755	0.1223	7.6577
40	93.051	0.0108	767.091	0.0013	8.2438	0.1213	7.8988
45	163.988	0.0061	1358.230	0.0007	8.2825	0.1207	8.0572
50	289.002	0.0035	2400.018	0.0004	8.3045	0.1204	8.1597

Table D.15. 15% INTEREST FACTORS FOR ANNUAL COMPOUNDING INTEREST

	Single Payment		Equal Payment Series				Uniform gradient-series factor
	Compound-amount factor	Present-worth factor	Compound-amount factor	Sinking-fund factor	Present-worth factor	Capital-recovery factor	
n	To find F Given P F/P i, n	To find P Given F P/F i, n	To find F Given A F/A i, n	To find A Given F A/F i, n	To find P Given A P/A i, n	To find A Given P A/P i, n	To find A Given g A/G i, n
1	1.150	0.8696	1.000	1.0000	0.8696	1.1500	0.0000
2	1.323	0.7562	2.150	0.4651	1.6257	0.6151	0.4651
3	1.521	0.6575	3.473	0.2880	2.2832	0.4380	0.9071
4	1.749	0.5718	4.993	0.2003	2.8550	0.3503	1.3263
5	2.011	0.4972	6.742	0.1483	3.3522	0.2983	1.7228
6	2.313	0.4323	8.754	0.1142	3.7845	0.2642	2.0972
7	2.660	0.3759	11.067	0.0904	4.1604	0.2404	2.4499
8	3.059	0.3269	13.727	0.0729	4.4873	0.2229	2.7813
9	3.518	0.2843	16.786	0.0596	4.7716	0.2096	3.0922
10	4.046	0.2472	20.304	0.0493	5.0188	0.1993	3.3832
11	4.652	0.2150	24.349	0.0411	5.2337	0.1911	3.6550
12	5.350	0.1869	29.002	0.0345	5.4206	0.1845	3.9082
13	6.153	0.1625	34.352	0.0291	5.5832	0.1791	4.1438
14	7.076	0.1413	40.505	0.0247	5.7245	0.1747	4.3624
15	8.137	0.1229	47.580	0.0210	5.8474	0.1710	4.5650
16	9.358	0.1069	55.717	0.0180	5.9542	0.1680	4.7523
17	10.761	0.0929	65.075	0.0154	6.0472	0.1654	4.9251
18	12.375	0.0808	75.836	0.0132	6.1280	0.1632	5.0843
19	14.232	0.0703	88.212	0.0113	6.1982	0.1613	5.2307
20	16.367	0.0611	102.444	0.0098	6.2593	0.1598	5.3651
21	18.822	0.0531	118.810	0.0084	6.3125	0.1584	5.4883
22	21.645	0.0462	137.632	0.0073	6.3587	0.1573	5.6010
23	24.891	0.0402	159.276	0.0063	6.3988	0.1563	5.7040
24	28.625	0.0349	184.168	0.0054	6.4338	0.1554	5.7979
25	32.919	0.0304	212.793	0.0047	6.4642	0.1547	5.8834
26	37.857	0.0264	245.712	0.0041	6.4906	0.1541	5.9612
27	43.535	0.0230	283.569	0.0035	6.5135	0.1535	6.0319
28	50.066	0.0200	327.104	0.0031	6.5335	0.1531	6.0960
29	57.575	0.0174	377.170	0.0027	6.5509	0.1527	6.1541
30	66.212	0.0151	434.745	0.0023	6.5660	0.1523	6.2066
31	76.144	0.0131	500.957	0.0020	6.5791	0.1520	6.2541
32	87.565	0.0114	577.100	0.0017	6.5905	0.1517	6.2970
33	100.700	0.0099	664.666	0.0015	6.6005	0.1515	6.3357
34	115.805	0.0086	765.365	0.0013	6.6091	0.1513	6.3705
35	133.176	0.0075	881.170	0.0011	6.6166	0.1511	6.4019
40	267.864	0.0037	1779.090	0.0006	6.6418	0.1506	6.5168
45	538.769	0.0019	3585.128	0.0003	6.6543	0.1503	6.5830
50	1083.657	0.0009	7217.716	0.0002	6.6605	0.1501	6.6205

Table D.16. 20% INTEREST FACTORS FOR ANNUAL COMPOUNDING INTEREST

	Single Payment		Equal Payment Series				Uniform gradient-series factor
	Compound-amount factor	Present-worth factor	Compound-amount factor	Sinking-fund factor	Present-worth factor	Capital-recovery factor	
n	To find F Given P F/P i, n	To find P Given F P/F i, n	To find F Given A F/A i, n	To find A Given F A/F i, n	To find P Given A P/A i, n	To find A Given P A/P i, n	To find A Given g A/G i, n
1	1.200	0.8333	1.000	1.0000	0.8333	1.2000	0.0000
2	1.440	0.6945	2.200	0.4546	1.5278	0.6546	0.4546
3	1.728	0.5787	3.640	0.2747	2.1065	0.4747	0.8791
4	2.074	0.4823	5.368	0.1863	2.5887	0.3863	1.2742
5	2.488	0.4019	7.442	0.1344	2.9906	0.3344	1.6405
6	2.986	0.3349	9.930	0.1007	3.3255	0.3007	1.9788
7	3.583	0.2791	12.916	0.0774	3.6046	0.2774	2.2902
8	4.300	0.2326	16.499	0.0606	3.8372	0.2606	2.5756
9	5.160	0.1938	20.799	0.0481	4.0310	0.2481	2.8364
10	6.192	0.1615	25.959	0.0385	4.1925	0.2385	3.0739
11	7.430	0.1346	32.150	0.0311	4.3271	0.2311	3.2893
12	8.916	0.1122	39.581	0.0253	4.4392	0.2253	3.4841
13	10.699	0.0935	48.497	0.0206	4.5327	0.2206	3.6597
14	12.839	0.0779	59.196	0.0169	4.6106	0.2169	3.8175
15	15.407	0.0649	72.035	0.0139	4.6755	0.2139	3.9589
16	18.488	0.0541	87.442	0.0114	4.7296	0.2114	4.0851
17	22.186	0.0451	105.931	0.0095	4.7746	0.2095	4.1976
18	26.623	0.0376	128.117	0.0078	4.8122	0.2078	4.2975
19	31.948	0.0313	154.740	0.0065	4.8435	0.2065	4.3861
20	38.338	0.0261	186.688	0.0054	4.8696	0.2054	4.4644
21	46.005	0.0217	225.026	0.0045	4.8913	0.2045	4.5334
22	55.206	0.0181	271.031	0.0037	4.9094	0.2037	4.5942
23	66.247	0.0151	326.237	0.0031	4.9245	0.2031	4.6475
24	79.497	0.0126	392.484	0.0026	4.9371	0.2026	4.6943
25	95.396	0.0105	471.981	0.0021	4.9476	0.2021	4.7352
26	114.475	0.0087	567.377	0.0018	4.9563	0.2018	4.7709
27	137.371	0.0073	681.853	0.0015	4.9636	0.2015	4.8020
28	164.845	0.0061	819.223	0.0012	4.9697	0.2012	4.8291
29	197.814	0.0051	984.068	0.0010	4.9747	0.2010	4.8527
30	237.376	0.0042	1181.882	0.0009	4.9789	0.2009	4.8731
31	284.852	0.0035	1419.258	0.0007	4.9825	0.2007	4.8908
32	341.822	0.0029	1704.109	0.0006	4.9854	0.2006	4.9061
33	410.186	0.0024	2045.931	0.0005	4.9878	0.2005	4.9194
34	492.224	0.0020	2456.118	0.0004	4.9899	0.2004	4.9308
35	590.668	0.0017	2948.341	0.0003	4.9915	0.2003	4.9407
40	1469.772	0.0007	7343.858	0.0002	4.9966	0.2001	4.9728
45	3657.262	0.0003	18281.310	0.0001	4.9986	0.2001	4.9877
50	9100.438	0.0001	45497.191	0.0000	4.9995	0.2000	4.9945

Table D.17. 25% INTEREST FACTORS FOR ANNUAL COMPOUNDING INTEREST

	Single Payment		Equal Payment Series				Uniform gradient-series factor
	Compound-amount factor	Present-worth factor	Compound-amount factor	Sinking-fund factor	Present-worth factor	Capital-recovery factor	
n	To find F Given P F/P i,n	To find P Given F P/F i,n	To find F Given A F/A i,n	To find A Given F A/F i,n	To find P Given A P/A i,n	To find A Given P A/P i,n	To find A Given g A/G i,n
1	1.250	0.8000	1.000	1.0000	0.8000	1.2500	0.0000
2	1.563	0.6400	2.250	0.4445	1.4400	0.6945	0.4445
3	1.953	0.5120	3.813	0.2623	1.9520	0.5123	0.8525
4	2.441	0.4096	5.766	0.1735	2.3616	0.4235	1.2249
5	3.052	0.3277	8.207	0.1219	2.6893	0.3719	1.5631
6	3.815	0.2622	11.259	0.0888	2.9514	0.3388	1.8683
7	4.768	0.2097	15.073	0.0664	3.1611	0.3164	2.1424
8	5.960	0.1678	19.842	0.0504	3.3289	0.3004	2.3873
9	7.451	0.1342	25.802	0.0388	3.4631	0.2888	2.6048
10	9.313	0.1074	33.253	0.0301	3.5705	0.2801	2.7971
11	11.642	0.0859	42.566	0.0235	3.6564	0.2735	2.9663
12	14.552	0.0687	54.208	0.0185	3.7251	0.2685	3.1145
13	18.190	0.0550	68.760	0.0146	3.7801	0.2646	3.2438
14	22.737	0.0440	86.949	0.0115	3.8241	0.2615	3.3560
15	28.422	0.0352	109.687	0.0091	3.8593	0.2591	3.4530
16	35.527	0.0282	138.109	0.0073	3.8874	0.2573	3.5366
17	44.409	0.0225	173.636	0.0058	3.9099	0.2558	3.6084
18	55.511	0.0180	218.045	0.0046	3.9280	0.2546	3.6698
19	69.389	0.0144	273.556	0.0037	3.9424	0.2537	3.7222
20	86.736	0.0115	342.945	0.0029	3.9539	0.2529	3.7667
21	108.420	0.0092	429.681	0.0023	3.9631	0.2523	3.8045
22	135.525	0.0074	538.101	0.0019	3.9705	0.2519	3.8365
23	169.407	0.0059	673.626	0.0015	3.9764	0.2515	3.8634
24	211.758	0.0047	843.033	0.0012	3.9811	0.2512	3.8861
25	264.698	0.0038	1054.791	0.0010	3.9849	0.2510	3.9052
26	330.872	0.0030	1319.489	0.0008	3.9879	0.2508	3.9212
27	413.590	0.0024	1650.361	0.0006	3.9903	0.2506	3.9346
28	516.988	0.0019	2063.952	0.0005	3.9923	0.2505	3.9457
29	646.235	0.0016	2580.939	0.0004	3.9938	0.2504	3.9551
30	807.794	0.0012	3227.174	0.0003	3.9951	0.2503	3.9628
31	1009.742	0.0010	4034.968	0.0003	3.9960	0.2503	3.9693
32	1262.177	0.0008	5044.710	0.0002	3.9968	0.2502	3.9746
33	1577.722	0.0006	6306.887	0.0002	3.9975	0.2502	3.9791
34	1972.152	0.0005	7884.609	0.0001	3.9980	0.2501	3.9828
35	2465.190	0.0004	9856.761	0.0001	3.9984	0.2501	3.9858

Table D.18. 30% INTEREST FACTORS FOR ANNUAL COMPOUNDING INTEREST

	Single Payment		Equal Payment Series				Uniform gradient-series factor
	Compound-amount factor	Present-worth factor	Compound-amount factor	Sinking-fund factor	Present-worth factor	Capital-recovery factor	
n	To find F Given P F/P i, n	To find P Given F P/F i, n	To find F Given A F/A i, n	To find A Given F A/F i, n	To find P Given A P/A i, n	To find A Given P A/P i, n	To find A Given g A/G i, n
1	1.300	0.7692	1.000	1.0000	0.7692	1.3000	0.0000
2	1.690	0.5917	2.300	0.4348	1.3610	0.7348	0.4348
3	2.197	0.4552	3.990	0.2506	1.8161	0.5506	0.8271
4	2.856	0.3501	6.187	0.1616	2.1663	0.4616	1.1783
5	3.713	0.2693	9.043	0.1106	2.4356	0.4106	1.4903
6	4.827	0.2072	12.756	0.0784	2.6428	0.3784	1.7655
7	6.275	0.1594	17.583	0.0569	2.8021	0.3569	2.0063
8	8.157	0.1226	23.858	0.0419	2.9247	0.3419	2.2156
9	10.605	0.0943	32.015	0.0312	3.0190	0.3312	2.3963
10	13.786	0.0725	42.620	0.0235	3.0915	0.3235	2.5512
11	17.922	0.0558	56.405	0.0177	3.1473	0.3177	2.6833
12	23.298	0.0429	74.327	0.0135	3.1903	0.3135	2.7952
13	30.288	0.0330	97.625	0.0103	3.2233	0.3103	2.8895
14	39.374	0.0254	127.913	0.0078	3.2487	0.3078	2.9685
15	51.186	0.0195	167.286	0.0060	3.2682	0.3060	3.0345
16	66.542	0.0150	218.472	0.0046	3.2832	0.3046	3.0892
17	86.504	0.0116	285.014	0.0035	3.2948	0.3035	3.1345
18	112.455	0.0089	371.518	0.0027	3.3037	0.3027	3.1718
19	146.192	0.0069	483.973	0.0021	3.3105	0.3021	3.2025
20	190.050	0.0053	630.165	0.0016	3.3158	0.3016	3.2276
21	247.065	0.0041	820.215	0.0012	3.3199	0.3012	3.2480
22	321.184	0.0031	1067.280	0.0009	3.3230	0.3009	3.2646
23	417.539	0.0024	1388.464	0.0007	3 3254	0.3007	3.2781
24	542.801	0.0019	1806.003	0.0006	3.3272	0.3006	3.2890
25	705.641	0.0014	2348.803	0.0004	3.3286	0.3004	3.2979
26	917.333	0.0011	3054.444	0.0003	3.3297	0.3003	3.3050
27	1192.533	0.0008	3971.778	0.0003	3.3305	0.3003	3.3107
28	1550.293	0.0007	5164.311	0.0002	3.3312	0.3002	3.3153
29	2015.381	0.0005	6714.604	0.0002	3.3317	0.3002	3.3189
30	2619.996	0.0004	8729.985	0.0001	3.3321	0.3001	3.3219
31	3405.994	0.0003	11349.981	0.0001	3.3324	0.3001	3.3242
32	4427.793	0.0002	14755.975	0.0001	3.3326	0.3001	3.3261
33	5756.130	0.0002	19183.768	0.0001	3.3328	0.3001	3.3276
34	7482.970	0.0001	24939.899	0.0001	3.3329	0.3001	3.3288
35	9727.860	0.0001	32422.868	0.0000	3.3330	0.3000	3.3297

Table D.19. EFFECTIVE INTEREST RATES CORRESPONDING TO NOMINAL RATE r

r	Compounding Frequency					
	Semi-annually $\left(1+\dfrac{r}{2}\right)^2 - 1$	Quarterly $\left(1+\dfrac{r}{4}\right)^4 - 1$	Monthly $\left(1+\dfrac{r}{12}\right)^{12} - 1$	Weekly $\left(1+\dfrac{r}{52}\right)^{52} - 1$	Daily $\left(1+\dfrac{r}{365}\right)^{365} - 1$	Continuously $\left(1+\dfrac{r}{\infty}\right)^{\infty} - 1$
1	1.0025	1.0038	1.0046	1.0049	1.0050	1.0050
2	2.0100	2.0151	2.0184	2.0197	2.0200	2.0201
3	3.0225	3.0339	3.0416	3.0444	3.0451	3.0455
4	4.0400	4.0604	4.0741	4.0793	4.0805	4.0811
5	5.0625	5.0945	5.1161	5.1244	5.1261	5.1271
6	6.0900	6.1364	6.1678	6.1797	6.1799	6.1837
7	7.1225	7.1859	7.2290	7.2455	7.2469	7.2508
8	8.1600	8.2432	8.2999	8.3217	8.3246	8.3287
9	9.2025	9.3083	9.3807	9.4085	9.4132	9.4174
10	10.2500	10.3813	10.4713	10.5060	10.5126	10.5171
11	11.3025	11.4621	11.5718	11.6144	11.6231	11.6278
12	12.3600	12.5509	12.6825	12.7336	12.7447	12.7497
13	13.4225	13.6476	13.8032	13.8644	13.8775	13.8828
14	14.4900	14.7523	14.9341	15.0057	15.0217	15.0274
15	15.5625	15.8650	16.0755	16.1582	16.1773	16.1834
16	16.6400	16.9859	17.2270	17.3221	17.3446	17.3511
17	17.7225	18.1148	18.3891	18.4974	18.5235	18.5305
18	18.8100	19.2517	19.5618	19.6843	19.7142	19.7217
19	19.9025	20.3971	20.7451	20.8828	20.9169	20.9250
20	21.0000	21.5506	21.9390	22.0931	22.1316	22.1403
21	22.1025	22.7124	23.1439	23.3153	23.3584	23.3678
22	23.2100	23.8825	24.3596	24.5494	24.5976	24.6077
23	24.3225	25.0609	25.5863	25.7957	25.8492	25.8600
24	25.4400	26.2477	26.8242	27.0542	27.1133	27.1249
25	26.5625	27.4429	28.0731	28.3250	28.3901	28.4025
26	27.6900	28.6466	29.3333	29.6090	29.6796	29.6930
27	28.8225	29.8588	30.6050	30.9049	30.9821	30.9964
28	29.9600	31.0796	31.8880	32.2135	32.2976	32.3130
29	31.1025	32.3089	33.1826	33.5350	33.6264	33.6428
30	32.2500	33.5469	34.4889	34.8693	34.9684	34.9859
31	33.4025	34.7936	35.8068	36.2168	36.3238	36.3425
32	34.5600	36.0489	37.1366	37.5775	37.6928	37.7128
33	35.7225	37.3130	38.4784	38.9515	39.0756	39.0968
34	36.8900	38.5859	39.8321	40.3389	40.4722	40.4948
35	38.0625	39.8676	41.1979	41.7399	41.8827	41.9068

Table D.20. 1% INTEREST FACTORS FOR CONTINUOUS COMPOUNDING INTEREST

	Single Payment		Equal Payment Series				Uniform gradient-series factor
	Compound-amount factor	Present-worth factor	Compound-amount factor	Sinking-fund factor	Present-worth factor	Capital-recovery factor	
n	To find F Given P F/P r, n	To find P Given F P/F r, n	To find F Given A F/A r, n	To find A Given F A/F r, n	To find P Given A P/A r, n	To find A Given P A/P r, n	To find A Given g A/G r, n
1	1.010	0.9901	1.000	1.0000	0.9901	1.0101	0.0000
2	1.020	0.9802	2.010	0.4975	1.9703	0.5076	0.4975
3	1.030	0.9705	3.030	0.3300	2.9407	0.3401	0.9933
4	1.041	0.9608	4.061	0.2463	3.9015	0.2563	1.4875
5	1.051	0.9512	5.102	0.1960	4.8527	0.2061	1.9800
6	1.062	0.9418	6.153	0.1625	5.7945	0.1726	2.4708
7	1.073	0.9324	7.215	0.1386	6.7269	0.1487	2.9600
8	1.083	0.9231	8.287	0.1207	7.6500	0.1307	3.4475
9	1.094	0.9139	9.370	0.1067	8.5639	0.1168	3.9334
10	1.105	0.9048	10.465	0.0956	9.4688	0.1056	4.4175
11	1.116	0.8958	11.570	0.0864	10.3646	0.0965	4.9000
12	1.128	0.8869	12.686	0.0788	11.2515	0.0889	5.3809
13	1.139	0.8781	13.814	0.0724	12.1296	0.0825	5.8600
14	1.150	0.8694	14.952	0.0669	12.9990	0.0769	6.3376
15	1.162	0.8607	16.103	0.0621	13.8597	0.0722	6.8134
16	1.174	0.8522	17.264	0.0579	14.7118	0.0680	7.2876
17	1.185	0.8437	18.438	0.0542	15.5555	0.0643	7.7601
18	1.197	0.8353	19.623	0.0510	16.3908	0.0610	8.2310
19	1.209	0.8270	20.821	0.0480	17.2177	0.0581	8.7002
20	1.221	0.8187	22.030	0.0454	18.0365	0.0555	9.1677
21	1.234	0.8106	23.251	0.0430	18.8470	0.0531	9.6336
22	1.246	0.8025	24.485	0.0409	19.6496	0.0509	10.0978
23	1.259	0.7945	25.731	0.0389	20.4441	0.0489	10.5604
24	1.271	0.7866	26.990	0.0371	21.2307	0.0471	11.0213
25	1.284	0.7788	28.261	0.0354	22.0095	0.0454	11.4806
26	1.297	0.7711	29.545	0.0339	22.7806	0.0439	11.9381
27	1.310	0.7634	30.842	0.0324	23.5439	0.0425	12.3941
28	1.323	0.7558	32.152	0.0311	24.2997	0.0412	12.8484
29	1.336	0.7483	33.475	0.0299	25.0480	0.0399	13.3010
30	1.350	0.7408	34.811	0.0287	25.7888	0.0388	13.7520
31	1.363	0.7335	36.161	0.0277	26.5223	0.0377	14.2013
32	1.377	0.7262	37.525	0.0267	27.2484	0.0367	14.6490
33	1.391	0.7189	38.902	0.0257	27.9673	0.0358	15.0950
34	1.405	0.7118	40.293	0.0248	28.6791	0.0349	15.5394
35	1.419	0.7047	41.698	0.0240	29.3838	0.0340	15.9821
40	1.492	0.6703	48.937	0.0204	32.8034	0.0305	18.1711
45	1.568	0.6376	56.548	0.0177	36.0563	0.0277	20.3190
50	1.649	0.6065	64.548	0.0155	39.1505	0.0256	22.4261
55	1.733	0.5770	72.959	0.0137	42.0939	0.0238	24.4926
60	1.822	0.5488	81.802	0.0122	44.8936	0.0223	26.5187
65	1.916	0.5221	91.097	0.0110	47.5569	0.0210	28.5045
70	2.014	0.4966	100.869	0.0099	50.0902	0.0200	30.4505
75	2.117	0.4724	111.142	0.0090	52.5000	0.0191	32.3567
80	2.226	0.4493	121.942	0.0082	54.7922	0.0183	34.2235
85	2.340	0.4274	133.296	0.0075	56.9727	0.0176	36.0513
90	2.460	0.4066	145.232	0.0069	59.0468	0.0169	37.8402
95	2.586	0.3868	157.779	0.0063	61.0198	0.0164	39.5907
100	2.718	0.3679	170.970	0.0059	62.8965	0.0159	41.3032

Table D.21. 2% INTEREST FACTORS FOR CONTINUOUS COMPOUNDING INTEREST

	Single Payment		Equal Payment Series				Uniform gradient-series factor
	Compound-amount factor	Present-worth factor	Compound-amount factor	Sinking-fund factor	Present-worth factor	Capital-recovery factor	
n	To find F Given P $F/P \quad r, n$	To find P Given F $P/F \quad r, n$	To find F Given A $F/A \quad r, n$	To find A Given F $A/F \quad r, n$	To find P Given A $P/A \quad r, n$	To find A Given P $A/P \quad r, n$	To find A Given g $A/G \quad r, n$
1	1.020	0.9802	1.000	1.0000	0.9802	1.0202	0.0000
2	1.041	0.9608	2.020	0.4950	1.9410	0.5152	0.4950
3	1.062	0.9418	3.061	0.3267	2.8828	0.3469	0.9867
4	1.083	0.9231	4.123	0.2426	3.8059	0.2628	1.4750
5	1.105	0.9048	5.206	0.1921	4.7107	0.2123	1.9600
6	1.128	0.8869	6.311	0.1585	5.5976	0.1787	2.4417
7	1.150	0.8694	7.439	0.1344	6.4670	0.1546	2.9200
8	1.174	0.8522	8.589	0.1164	7.3191	0.1366	3.3951
9	1.197	0.8353	9.763	0.1024	8.1544	0.1226	3.8667
10	1.221	0.8187	10.960	0.0913	8.9731	0.1115	4.3351
11	1.246	0.8025	12.181	0.0821	9.7757	0.1023	4.8002
12	1.271	0.7866	13.427	0.0745	10.5623	0.0947	5.2619
13	1.297	0.7711	14.699	0.0680	11.3333	0.0882	5.7203
14	1.323	0.7558	15.995	0.0625	12.0891	0.0827	6.1754
15	1.350	0.7408	17.319	0.0578	12.8299	0.0780	6.6272
16	1.377	0.7262	18.668	0.0536	13.5561	0.0738	7.0757
17	1.405	0.7118	20.046	0.0499	14.2679	0.0701	7.5209
18	1.433	0.6977	21.451	0.0466	14.9655	0.0668	7.9628
19	1.462	0.6839	22.884	0.0437	15.6494	0.0639	8.4015
20	1.492	0.6703	24.346	0.0411	16.3197	0.0613	8.8368
21	1.522	0.6571	25.838	0.0387	16.9768	0.0589	9.2688
22	1.553	0.6440	27.360	0.0366	17.6208	0.0568	9.6976
23	1.584	0.6313	28.913	0.0346	18.2521	0.0548	10.1231
24	1.616	0.6188	30.497	0.0328	18.8709	0.0530	10.5453
25	1.649	0.6065	32.113	0.0312	19.4774	0.0514	10.9643
26	1.682	0.5945	33.762	0.0296	20.0719	0.0498	11.3801
27	1.716	0.5828	35.444	0.0282	20.6547	0.0484	11.7925
28	1.751	0.5712	37.160	0.0269	21.2259	0.0471	12.2018
29	1.786	0.5599	38.910	0.0257	21.7858	0.0459	12.6078
30	1.822	0.5488	40.696	0.0246	22.3346	0.0448	13.0106
31	1.859	0.5380	42.518	0.0235	22.8725	0.0437	13.4102
32	1.896	0.5273	44.377	0.0225	23.3998	0.0427	13.8065
33	1.935	0.5169	46.274	0.0216	23.9167	0.0418	14.1997
34	1.974	0.5066	48.209	0.0208	24.4233	0.0410	14.5897
35	2.014	0.4966	50.182	0.0199	24.9199	0.0401	14.9765
40	2.226	0.4493	60.666	0.0165	27.2591	0.0367	16.8630
45	2.460	0.4066	72.253	0.0139	29.3758	0.0341	18.6714
50	2.718	0.3679	85.058	0.0118	31.2910	0.0320	20.4028
55	3.004	0.3329	99.210	0.0101	33.0240	0.0303	22.0588
60	3.320	0.3012	114.850	0.0087	34.5921	0.0289	23.6409
65	3.669	0.2725	132.135	0.0076	36.0109	0.0278	25.1507
70	4.055	0.2466	151.238	0.0066	37.2947	0.0268	26.5899
75	4.482	0.2231	172.349	0.0058	38.4564	0.0260	27.9604
80	4.953	0.2019	195.682	0.0051	39.5075	0.0253	29.2640
85	5.474	0.1827	221.468	0.0045	40.4585	0.0247	30.5028
90	6.050	0.1653	249.966	0.0040	41.3191	0.0242	31.6786
95	6.686	0.1496	281.461	0.0036	42.0978	0.0238	32.7937
100	7.389	0.1353	316.269	0.0032	42.8024	0.0234	33.8499

Table D.22. 3% INTEREST FACTORS FOR CONTINUOUS COMPOUNDING INTEREST

	Single Payment		Equal Payment Series				Uniform gradient-series factor
	Compound-amount factor	Present-worth factor	Compound-amount factor	Sinking-fund factor	Present-worth factor	Capital-recovery factor	
n	To find F Given P F/P r, n	To find P Given F P/F r, n	To find F Given A F/A r, n	To find A Given F A/F r, n	To find P Given A P/A r, n	To find A Given P A/P r, n	To find A Given g A/G r, n
1	1.030	0.9705	1.000	1.0000	0.9705	1.0305	0.0000
2	1.062	0.9418	2.030	0.4925	1.9122	0.5230	0.4925
3	1.094	0.9139	3.092	0.3234	2.8262	0.3538	0.9800
4	1.128	0.8869	4.186	0.2389	3.7131	0.2693	1.4625
5	1.162	0.8607	5.314	0.1882	4.5738	0.2186	1.9400
6	1.197	0.8353	6.476	0.1544	5.4090	0.1849	2.4126
7	1.234	0.8106	7.673	0.1303	6.2196	0.1608	2.8801
8	1.271	0.7866	8.907	0.1123	7.0063	0.1427	3.3427
9	1.310	0.7634	10.178	0.0983	7.7696	0.1287	3.8003
10	1.350	0.7408	11.488	0.0871	8.5105	0.1175	4.2529
11	1.391	0.7189	12.838	0.0779	9.2294	0.1084	4.7006
12	1.433	0.6977	14.229	0.0703	9.9271	0.1007	5.1433
13	1.477	0.6771	15.662	0.0639	10.6041	0.0943	5.5811
14	1.522	0.6571	17.139	0.0584	11.2612	0.0888	6.0139
15	1.568	0.6376	18.661	0.0536	11.8988	0.0841	6.4419
16	1.616	0.6188	20.229	0.0494	12.5176	0.0799	6.8650
17	1.665	0.6005	21.845	0.0458	13.1181	0.0762	7.2831
18	1.716	0.5828	23.511	0.0425	13.7008	0.0730	7.6964
19	1.768	0.5655	25.227	0.0397	14.2663	0.0701	8.1049
20	1.822	0.5488	26.995	0.0371	14.8152	0.0675	8.5085
21	1.878	0.5326	28.817	0.0347	15.3477	0.0652	8.9072
22	1.935	0.5169	30.695	0.0326	15.8646	0.0630	9.3012
23	1.994	0.5016	32.629	0.0307	16.3662	0.0611	9.6904
24	2.054	0.4868	34.623	0.0289	16.8529	0.0593	10.0748
25	2.117	0.4724	36.678	0.0273	17.3253	0.0577	10.4545
26	2.181	0.4584	38.795	0.0258	17.7837	0.0562	10.8294
27	2.248	0.4449	40.976	0.0244	18.2286	0.0549	11.1996
28	2.316	0.4317	43.224	0.0231	18.6603	0.0536	11.5652
29	2.387	0.4190	45.540	0.0220	19.0792	0.0524	11.9261
30	2.460	0.4066	47.927	0.0209	19.4858	0.0513	12.2823
31	2.535	0.3946	50.387	0.0199	19.8803	0.0503	12.6339
32	2.612	0.3829	52.921	0.0189	20.2632	0.0494	12.9810
33	2.691	0.3716	55.533	0.0180	20.6348	0.0485	13.3235
34	2.773	0.3606	58.224	0.0172	20.9954	0.0476	13.6614
35	2.858	0.3499	60.998	0.0164	21.3453	0.0469	13.9948
40	3.320	0.3012	76.183	0.0131	22.9459	0.0436	15.5953
45	3.857	0.2593	93.826	0.0107	24.3235	0.0411	17.0874
50	4.482	0.2231	114.324	0.0088	25.5092	0.0392	18.4750
55	5.207	0.1921	138.140	0.0072	26.5297	0.0377	19.7623
60	6.050	0.1653	165.809	0.0060	27.4081	0.0365	20.9538
65	7.029	0.1423	197.957	0.0051	28.1642	0.0355	22.0540
70	8.166	0.1225	235.307	0.0043	28.8149	0.0347	23.0677
75	9.488	0.1054	278.702	0.0036	29.3750	0.0341	23.9996
80	11.023	0.0907	329.119	0.0030	29.8570	0.0335	24.8543
85	12.807	0.0781	387.696	0.0026	30.2720	0.0330	25.6368
90	14.880	0.0672	455.753	0.0022	30.6291	0.0327	26.3516
95	17.288	0.0579	534.823	0.0019	30.9365	0.0323	27.0033
100	20.086	0.0498	626.690	0.0016	31.2010	0.0321	27.5963

Table D.23. 4% INTEREST FACTORS FOR CONTINUOUS COMPOUNDING INTEREST

	Single Payment		Equal Payment Series				Uniform gradient-series factor
	Compound-amount factor	Present-worth factor	Compound-amount factor	Sinking-fund factor	Present-worth factor	Capital-recovery factor	
n	To find F Given P F/P r, n	To find P Given F P/F r, n	To find F Given A F/A r, n	To find A Given F A/F r, n	To find P Given A P/A r, n	To find A Given P A/P r, n	To find A Given g A/G r, n
1	1.041	0.9608	1.000	1.0000	0.9608	1.0408	0.0000
2	1.083	0.9231	2.041	0.4900	1.8839	0.5308	0.4900
3	1.128	0.8869	3.124	0.3201	2.7708	0.3609	0.9734
4	1.174	0.8522	4.252	0.2352	3.6230	0.2760	1.4500
5	1.221	0.8187	5.425	0.1843	4.4417	0.2251	1.9201
6	1.271	0.7866	6.647	0.1505	5.2283	0.1913	2.3835
7	1.323	0.7558	7.918	0.1263	5.9841	0.1671	2.8402
8	1.377	0.7262	9.241	0.1082	6.7103	0.1490	3.2904
9	1.433	0.6977	10.618	0.0942	7.4079	0.1350	3.7339
10	1.492	0.6703	12.051	0.0830	8.0783	0.1238	4.1709
11	1.553	0.6440	13.543	0.0738	8.7223	0.1147	4.6013
12	1.616	0.6188	15.096	0.0663	9.3411	0.1071	5.0252
13	1.682	0.5945	16.712	0.0598	9.9356	0.1007	5.4425
14	1.751	0.5712	18.394	0.0544	10.5068	0.0952	5.8534
15	1.822	0.5488	20.145	0.0497	11.0556	0.0905	6.2578
16	1.896	0.5273	21.967	0.0455	11.5829	0.0863	6.6558
17	1.974	0.5066	23.863	0.0419	12.0895	0.0827	7.0474
18	2.054	0.4868	25.837	0.0387	12.5763	0.0795	7.4326
19	2.138	0.4677	27.892	0.0359	13.0440	0.0767	7.8114
20	2.226	0.4493	30.030	0.0333	13.4933	0.0741	8.1840
21	2.316	0.4317	32.255	0.0310	13.9250	0.0718	8.5503
22	2.411	0.4148	34.572	0.0289	14.3398	0.0697	8.9105
23	2.509	0.3985	36.983	0.0270	14.7383	0.0679	9.2644
24	2.612	0.3829	39.492	0.0253	15.1212	0.0661	9.6122
25	2.718	0.3679	42.104	0.0238	15.4891	0.0646	9.9539
26	2.829	0.3535	44.822	0.0223	15.8425	0.0631	10.2896
27	2.945	0.3396	47.651	0.0210	16.1821	0.0618	10.6193
28	3.065	0.3263	50.596	0.0198	16.5084	0.0606	10.9431
29	3.190	0.3135	53.661	0.0186	16.8219	0.0595	11.2609
30	3.320	0.3012	56.851	0.0176	17.1231	0.0584	11.5730
31	3.456	0.2894	60.171	0.0166	17.4125	0.0574	11.8792
32	3.597	0.2780	63.626	0.0157	17.6905	0.0565	12.1797
33	3.743	0.2671	67.223	0.0149	17.9576	0.0557	12.4746
34	3.896	0.2567	70.966	0.0141	18.2143	0.0549	12.7638
35	4.055	0.2466	74.863	0.0134	18.4609	0.0542	13.0475
40	4.953	0.2019	96.862	0.0103	19.5562	0.0511	14.3845
45	6.050	0.1653	123.733	0.0081	20.4530	0.0489	15.5918
50	7.389	0.1353	156.553	0.0064	21.1872	0.0472	16.6775
55	9.025	0.1108	196.640	0.0051	21.7883	0.0459	17.6498
60	11.023	0.0907	245.601	0.0041	22.2805	0.0449	18.5172
65	13.464	0.0743	305.403	0.0033	22.6834	0.0441	19.2882
70	16.445	0.0608	378.445	0.0027	23.0133	0.0435	19.9710
75	20.086	0.0498	467.659	0.0021	23.2834	0.0430	20.5737
80	24.533	0.0408	576.625	0.0017	23.5045	0.0426	21.1038
85	29.964	0.0334	709.717	0.0014	23.6856	0.0422	21.5687
90	36.598	0.0273	872.275	0.0012	23.8338	0.0420	21.9751
95	44.701	0.0224	1070.825	0.0009	23.9552	0.0418	22.3295
100	54.598	0.0183	1313.333	0.0008	24.0545	0.0416	22.6376

Table D.24. 5% INTEREST FACTORS FOR CONTINUOUS COMPOUNDING INTEREST

	Single Payment		Equal Payment Series				Uniform gradient-series factor
	Compound-amount factor	Present-worth factor	Compound-amount factor	Sinking-fund factor	Present-worth factor	Capital-recovery factor	
n	To find F Given P $F/P\ \ r, n$	To find P Given F $P/F\ \ r, n$	To find F Given A $F/A\ \ r, n$	To find A Given F $A/F\ \ r, n$	To find P Given A $P/A\ \ r, n$	To find A Given P $A/P\ \ r, n$	To find A Given g $A/G\ \ r, n$
1	1.051	0.9512	1.000	1.0000	0.9512	1.0513	0.0000
2	1.105	0.9048	2.051	0.4875	1.8561	0.5388	0.4875
3	1.162	0.8607	3.156	0.3168	2.7168	0.3681	0.9667
4	1.221	0.8187	4.318	0.2316	3.5355	0.2829	1.4376
5	1.284	0.7788	5.540	0.1805	4.3143	0.2318	1.9001
6	1.350	0.7408	6.824	0.1466	5.0551	0.1978	2.3544
7	1.419	0.7047	8.174	0.1224	5.7598	0.1736	2.8004
8	1.492	0.6703	9.593	0.1043	6.4301	0.1555	3.2382
9	1.568	0.6376	11.084	0.0902	7.0678	0.1415	3.6678
10	1.649	0.6065	12.653	0.0790	7.6743	0.1303	4.0892
11	1.733	0.5770	14.301	0.0699	8.2513	0.1212	4.5025
12	1.822	0.5488	16.035	0.0624	8.8001	0.1136	4.9077
13	1.916	0.5221	17.857	0.0560	9.3221	0.1073	5.3049
14	2.014	0.4966	19.772	0.0506	9.8187	0.1019	5.6941
15	2.117	0.4724	21.786	0.0459	10.2911	0.0972	6.0753
16	2.226	0.4493	23.903	0.0418	10.7404	0.0931	6.4487
17	2.340	0.4274	26.129	0.0383	11.1678	0.0896	6.8143
18	2.460	0.4066	28.468	0.0351	11.5744	0.0864	7.1721
19	2.586	0.3868	30.928	0.0323	11.9611	0.0836	7.5222
20	2.718	0.3679	33.514	0.0298	12.3290	0.0811	7.8646
21	2.858	0.3499	36.232	0.0276	12.6789	0.0789	8.1996
22	3.004	0.3329	39.090	0.0256	13.0118	0.0769	8.5270
23	3.158	0.3166	42.094	0.0238	13.3284	0.0750	8.8471
24	3.320	0.3012	45.252	0.0221	13.6296	0.0734	9.1599
25	3.490	0.2865	48.572	0.0206	13.9161	0.0719	9.4654
26	3.669	0.2725	52.062	0.0192	14.1887	0.0705	9.7638
27	3.857	0.2593	55.732	0.0180	14.4479	0.0692	10.0551
28	4.055	0.2466	59.589	0.0168	14.6945	0.0681	10.3395
29	4.263	0.2346	63.644	0.0157	14.9291	0.0670	10.6170
30	4.482	0.2231	67.907	0.0147	15.1522	0.0660	10.8877
31	4.711	0.2123	72.389	0.0138	15.3645	0.0651	11.1517
32	4.953	0.2019	77.101	0.0130	15.5664	0.0643	11.4091
33	5.207	0.1921	82.054	0.0122	15.7584	0.0635	11.6601
34	5.474	0.1827	87.261	0.0115	15.9411	0.0627	11.9046
35	5.755	0.1738	92.735	0.0108	16.1149	0.0621	12.1429
40	7.389	0.1353	124.613	0.0080	16.8646	0.0593	13.2435
45	9.488	0.1054	165.546	0.0061	17.4485	0.0573	14.2024
50	12.183	0.0821	218.105	0.0046	17.9032	0.0559	15.0329
55	15.643	0.0639	285.592	0.0035	18.2573	0.0548	15.7480
60	20.086	0.0498	372.247	0.0027	18.5331	0.0540	16.3604
65	25.790	0.0388	483.515	0.0021	18.7479	0.0533	16.8822
70	33.115	0.0302	626.385	0.0016	18.9152	0.0529	17.3245
75	42.521	0.0235	809.834	0.0012	19.0455	0.0525	17.6979
80	54.598	0.0183	1045.387	0.0010	19.1469	0.0522	18.0116
85	70.105	0.0143	1347.843	0.0008	19.2260	0.0520	18.2742
90	90.017	0.0111	1736.205	0.0006	19.2875	0.0519	18.4931
95	115.584	0.0087	2234.871	0.0005	19.3354	0.0517	18.6751
100	148.413	0.0067	2875.171	0.0004	19.3728	0.0516	18.8258

Table D.25. 6% INTEREST FACTORS FOR CONTINUOUS COMPOUNDING INTEREST

	Single Payment		Equal Payment Series				Uniform gradient-series factor
	Compound-amount factor	Present-worth factor	Compound-amount factor	Sinking-fund factor	Present-worth factor	Capital-recovery factor	
n	To find F Given P F/P r, n	To find P Given F P/F r, n	To find F Given A F/A r, n	To find A Given F A/F r, n	To find P Given A P/A r, n	To find A Given P A/P r, n	To find A Given g A/G r, n
1	1.062	0.9418	1.000	1.0000	0.9418	1.0618	0.0000
2	1.128	0.8869	2.062	0.4850	1.8287	0.5469	0.4850
3	1.197	0.8353	3.189	0.3136	2.6640	0.3754	0.9600
4	1.271	0.7866	4.387	0.2280	3.4506	0.2898	1.4251
5	1.350	0.7408	5.658	0.1768	4.1914	0.2386	1.8802
6	1.433	0.6977	7.008	0.1427	4.8891	0.2045	2.3254
7	1.522	0.6571	8.441	0.1185	5.5461	0.1803	2.7607
8	1.616	0.6188	9.963	0.1004	6.1649	0.1622	3.1862
9	1.716	0.5828	11.579	0.0864	6.7477	0.1482	3.6020
10	1.822	0.5488	13.295	0.0752	7.2965	0.1371	4.0080
11	1.935	0.5169	15.117	0.0662	7.8133	0.1280	4.4044
12	2.054	0.4868	17.052	0.0587	8.3001	0.1205	4.7912
13	2.181	0.4584	19.106	0.0523	8.7585	0.1142	5.1685
14	2.316	0.4317	21.288	0.0470	9.1902	0.1088	5.5363
15	2.460	0.4066	23.604	0.0424	9.5968	0.1042	5.8949
16	2.612	0.3829	26.064	0.0384	9.9797	0.1002	6.2442
17	2.773	0.3606	28.676	0.0349	10.3403	0.0967	6.5845
18	2.945	0.3396	31.449	0.0318	10.6799	0.0936	6.9157
19	3.127	0.3198	34.393	0.0291	10.9997	0.0909	7.2379
20	3.320	0.3012	37.520	0.0267	11.3009	0.0885	7.5514
21	3.525	0.2837	40.840	0.0245	11.5845	0.0863	7.8562
22	3.743	0.2671	44.366	0.0225	11.8517	0.0844	8.1525
23	3.975	0.2516	48.109	0.0208	12.1032	0.0826	8.4403
24	4.221	0.2369	52.084	0.0192	12.3402	0.0810	8.7199
25	4.482	0.2231	56.305	0.0178	12.5633	0.0796	8.9913
26	4.759	0.2101	60.786	0.0165	12.7734	0.0783	9.2546
27	5.053	0.1979	65.545	0.0153	12.9713	0.0771	9.5101
28	5.366	0.1864	70.598	0.0142	13.1577	0.0760	9.7578
29	5.697	0.1755	75.964	0.0132	13.3332	0.0750	9.9980
30	6.050	0.1653	81.661	0.0123	13.4985	0.0741	10.2307
31	6.424	0.1557	87.711	0.0114	13.6542	0.0732	10.4561
32	6.821	0.1466	94.135	0.0106	13.8008	0.0725	10.6743
33	7.243	0.1381	100.956	0.0099	13.9389	0.0718	10.8855
34	7.691	0.1300	108.198	0.0093	14.0689	0.0711	11.0899
35	8.166	0.1225	115.889	0.0086	14.1914	0.0705	11.2876
40	11.023	0.0907	162.091	0.0062	14.7046	0.0680	12.1809
45	14.880	0.0672	224.458	0.0045	15.0849	0.0663	12.9295
50	20.086	0.0498	308.645	0.0032	15.3665	0.0651	13.5519
55	27.113	0.0369	422.285	0.0024	15.5752	0.0642	14.0654
60	36.598	0.0273	575.683	0.0017	15.7298	0.0636	14.4862
65	49.402	0.0203	782.748	0.0013	15.8443	0.0631	14.8288
70	66.686	0.0150	1062.257	0.0010	15.9292	0.0628	15.1060
75	90.017	0.0111	1439.555	0.0007	15.9920	0.0625	15.3291
80	121.510	0.0082	1948.854	0.0005	16.0386	0.0624	15.5078
85	164.022	0.0061	2636.336	0.0004	16.0731	0.0622	15.6503
90	221.406	0.0045	3564.339	0.0003	16.0986	0.0621	15.7633
95	298.867	0.0034	4817.012	0.0002	16.1176	0.0621	15.8527
100	403.429	0.0025	6507.944	0.0002	16.1316	0.0620	15.9232

Table D.26. 7% INTEREST FACTORS FOR CONTINUOUS COMPOUNDING INTEREST

	Single Payment		Equal Payment Series				Uniform gradient-series factor
	Compound-amount factor	Present-worth factor	Compound-amount factor	Sinking-fund factor	Present-worth factor	Capital-recovery factor	
n	To find F Given P F/P r,n	To find P Given F P/F r,n	To find F Given A F/A r,n	To find A Given F A/F r,n	To find P Given A P/A r,n	To find A Given P A/P r,n	To find A Given g A/G r,n
1	1.073	0.9324	1.000	1.0000	0.9324	1.0725	0.0000
2	1.150	0.8694	2.073	0.4825	1.8018	0.5550	0.4825
3	1.234	0.8106	3.223	0.3103	2.6123	0.3828	0.9534
4	1.323	0.7558	4.456	0.2244	3.3681	0.2969	1.4126
5	1.419	0.7047	5.780	0.1730	4.0728	0.2455	1.8603
6	1.522	0.6571	7.199	0.1389	4.7299	0.2114	2.2965
7	1.632	0.6126	8.721	0.1147	5.3425	0.1872	2.7211
8	1.751	0.5712	10.353	0.0966	5.9137	0.1691	3.1344
9	1.878	0.5326	12.104	0.0826	6.4463	0.1551	3.5364
10	2.014	0.4966	13.981	0.0715	6.9429	0.1440	3.9272
11	2.160	0.4630	15.995	0.0625	7.4059	0.1350	4.3069
12	2.316	0.4317	18.155	0.0551	7.8376	0.1276	4.6756
13	2.484	0.4025	20.471	0.0489	8.2401	0.1214	5.0334
14	2.664	0.3753	22.955	0.0436	8.6154	0.1161	5.3804
15	2.858	0.3499	25.620	0.0390	8.9654	0.1161	5.7168
16	3.065	0.3263	28.478	0.0351	9.2917	0.1076	6.0428
17	3.287	0.3042	31.542	0.0317	9.5959	0.1042	6.3585
18	3.525	0.2837	34.829	0.0287	9.8795	0.1012	6.6640
19	3.781	0.2645	38.355	0.0261	10.1440	0.0986	6.9596
20	4.055	0.2466 −	42.136	0.0237	10.3906 −	0.0963	7.2453
21	4.349	0.2299	46.191	0.0217	10.6205	0.0942	7.5215
22	4.665	0.2144	50.540	0.0198	10.8349	0.0923	7.7882
23	5.003	0.1999	55.205	0.0181	11.0348	0.0906	8.0456
24	5.366	0.1864	60.208	0.0166	11.2212	0.0891	8.2940
25	5.755	0.1738	65.573	0.0153	13.3949	0.0878	8.5335
26	6.172	0.1620	71.328	0.0140	11.5570	0.0865	8.7643
27	6.619	0.1511	77.500	0.0129	11.7080	0.0854	8.9867
28	7.099	0.1409	84.119	0.0119	11.8489	0.0844	9.2009
29	7.614	0.1313	91.218	0.0110	11.9802	0.0835	9.4070
30	8.166	0.1225	98.833	0.0101	12.1027	0.0826	9.6052
31	8.758	0.1142	106.999	0.0094	12.2169	0.0819	9.7958
32	9.393	0.1065	115.757	0.0086	12.3233	0.0812	9.9790
33	10.047	0.0993	125.150	0.0080	12.4226	0.0805	10.1550
34	10.805	0.0926	135.225	0.0074	12.5151	0.0799	10.3239
35	11.588	0.0863	146.030	0.0069	12.6014	0.0794	10.4860
40	16.445	0.0608	213.006	0.0047	12.9529	0.0772	11.2017
45	23.336	0.0429	308.049	0.0033	13.2006	0.0758	11.7769
50	33.115	0.0302	442.922	0.0023	13.3751	0.0748	12.2347
55	46.993	0.0213	634.316	0.0016	13.4981	0.0741	12.5957
60	66.686	0.0150	905.916	0.0011	13.5847	0.0736	12.8781
65	94.632	0.0106	1291.336	0.0008	13.6458	0.0733	13.0974
70	134.290	0.0075	1838.272	0.0006	13.6889	0.0731	13.2664
75	190.566	0.0053	2614.412	0.0004	13.7192	0.0729	13.3959
80	270.426	0.0037	3715.807	0.0003	13.7406	0.0728	13.4946
85	383.753	0.0026	5278.761	0.0002	13.7556	0.0727	13.5695
90	544.572	0.0019	7496.698	0.0001	13.7662	0.0727	13.6260
95	772.784	0.0013	10644.100	0.0001	13.7737	0.0726	13.6685
100	1096.633	0.0009	15110.476	0.0001	13.7790	0.0726	13.7003

Table D.27. 8% INTEREST FACTORS FOR CONTINUOUS COMPOUNDING INTEREST

	Single Payment		Equal Payment Series				Uniform gradient-series factor
	Compound-amount factor	Present-worth factor	Compound-amount factor	Sinking-fund factor	Present-worth factor	Capital-recovery factor	
n	To find F Given P F/P r, n	To find P Given F P/F r, n	To find F Given A F/A r, n	To find A Given F A/F r, n	To find P Given A P/A r, n	To find A Given P A/P r, n	To find A Given g A/G r, n
1	1.083	0.9231	1.000	1.0000	0.9231	1.0833	0.0000
2	1.174	0.8522	2.083	0.4800	1.7753	0.5633	0.4800
3	1.271	0.7866	3.257	0.3071	2.5619	0.3903	0.9467
4	1.377	0.7262	4.528	0.2209	3.2880	0.3041	1.4002
5	1.492	0.6703	5.905	0.1694	3.9584	0.2526	1.8405
6	1.616	0.6188	7.397	0.1352	4.5772	0.2185	2.2676
7	1.751	0.5712	9.013	0.1110	5.1484	0.1942	2.6817
8	1.896	0.5273	10.764	0.0929	5.6757	0.1762	3.0829
9	2.054	0.4868	12.660	0.0790	6.1624	0.1623	3.4713
10	2.226	0.4493	14.715	0.0680	6.6117	0.1513	3.8470
11	2.411	0.4148	16.940	0.0590	7.0265	0.1423	4.2102
12	2.612	0.3829	19.351	0.0517	7.4094	0.1350	4.5611
13	2.829	0.3535	21.963	0.0455	7.7629	0.1288	4.8998
14	3.065	0.3263	24.792	0.0403	8.0891	0.1236	5.2265
15	3.320	0.3012	27.857	0.0359	8.3903	0.1192	5.5415
16	3.597	0.2780	31.177	0.0321	8.6684	0.1154	5.8449
17	3.896	0.2567	34.774	0.0288	8.9250	0.1121	6.1369
18	4.221	0.2369	38.670	0.0259	9.1620	0.1092	6.4178
19	4.572	0.2187	42.891	0.0233	9.3807	0.1066	6.6879
20	4.953	0.2019	47.463	0.0211	9.5826	0.1044	6.9473
21	5.366	0.1864	52.416	0.0191	9.7689	0.1024	7.1963
22	5.812	0.1721	57.781	0.0173	9.9410	0.1006	7.4352
23	6.297	0.1588	63.594	0.0157	10.0998	0.0990	7.6642
24	6.821	0.1466	69.890	0.0143	10.2464	0.0976	7.8836
25	7.389	0.1353	76.711	0.0130	10.3818	0.0963	8.0937
26	8.004	0.1249	84.100	0.0119	10.5067	0.0952	8.2948
27	8.671	0.1153	92.105	0.0109	10.6220	0.0942	8.4870
28	9.393	0.1065	100.776	0.0099	10.7285	0.0932	8.6707
29	10.176	0.0983	110.169	0.0091	10.8267	0.0924	8.8461
30	11.023	0.0907	120.345	0.0083	10.9175	0.0916	9.0136
31	11.941	0.0838	131.368	0.0076	11.0012	0.0909	9.1734
32	12.936	0.0773	143.309	0.0070	11.0785	0.0903	9.3257
33	14.013	0.0714	156.245	0.0064	11.1499	0.0897	9.4708
34	15.180	0.0659	170.258	0.0059	11.2157	0.0892	9.6090
35	16.445	0.0608	185.439	0.0054	11.2765	0.0887	9.7405
40	24.533	0.0408	282.547	0.0035	11.5173	0.0868	10.3069
45	36.598	0.0273	427.416	0.0023	11.6786	0.0856	10.7426
50	54.598	0.0183	643.535	0.0016	11.7868	0.0849	11.0738
55	81.451	0.0123	965.947	0.0010	11.8593	0.0843	11.3230
60	121.510	0.0082	1446.928	0.0007	11.9079	0.0840	11.5088
65	181.272	0.0055	2164.469	0.0005	11.9404	0.0838	11.6461
70	270.426	0.0037	3234.913	0.0003	11.9623	0.0836	11.7469
75	403.429	0.0025	4831.828	0.0002	11.9769	0.0835	11.8203
80	601.845	0.0017	7214.146	0.0002	11.9867	0.0834	11.8735
85	897.847	0.0011	10768.146	0.0001	11.9933	0.0834	11.9119
90	1339.431	0.0008	16070.091	0.0001	11.9977	0.0834	11.9394
95	1998.196	0.0005	23979.664	0.0001	12.0007	0.0833	11.9591
100	2980.958	0.0004	35779.360	0.0000	12.0026	0.0833	11.9731

Table D.28. 9% INTEREST FACTORS FOR CONTINUOUS COMPOUNDING INTEREST

	Single Payment		Equal Payment Series				Uniform gradient-series factor
	Compound-amount factor	Present-worth factor	Compound-amount factor	Sinking-fund factor	Present-worth factor	Capital-recovery factor	
n	To find F Given P F/P r,n	To find P Given F P/F r,n	To find F Given A F/A r,n	To find A Given F A/F r,n	To find P Given A P/A r,n	To find A Given P A/P r,n	To find A Given g A/G r,n
1	1.094	0.9139	1.000	1.0000	0.9139	1.0942	0.0000
2	1.197	0.8353	2.094	0.4775	1.7492	0.5717	0.4775
3	1.310	0.7634	3.291	0.3038	2.5126	0.3980	0.9401
4	1.433	0.6977	4.601	0.2173	3.2103	0.3115	1.3878
5	1.568	0.6376	6.035	0.1657	3.8479	0.2599	1.8206
6	1.716	0.5828	7.603	0.1315	4.4306	0.2257	2.2388
7	1.878	0.5326	9.319	0.1073	4.9632	0.2015	2.6424
8	2.054	0.4868	11.197	0.0893	5.4500	0.1835	3.0316
9	2.248	0.4449	13.251	0.0755	5.8948	0.1697	3.4065
10	2.460	0.4066	15.499	0.0645	6.3014	0.1587	3.7674
11	2.691	0.3716	17.959	0.0557	6.6730	0.1499	4.1145
12	2.945	0.3396	20.650	0.0484	7.0126	0.1426	4.4479
13	3.222	0.3104	23.594	0.0424	7.3230	0.1366	4.7680
14	3.525	0.2837	26.816	0.0373	7.6066	0.1315	5.0750
15	3.857	0.2593	30.342	0.0330	7.8658	0.1271	5.3691
16	4.221	0.2369	34.199	0.0293	8.1028	0.1234	5.6507
17	4.618	0.2165	38.420	0.0260	8.3193	0.1202	5.9201
18	5.053	0.1979	43.038	0.0232	8.5172	0.1174	6.1776
19	5.529	0.1809	48.091	0.0208	8.6981	0.1150	6.4234
20	6.050	0.1653	53.620	0.0187	8.8634	0.1128	6.6579
21	6.619	0.1511	59.670	0.0168	9.0144	0.1109	6.8815
22	7.243	0.1381	66.289	0.0151	9.1525	0.1093	7.0945
23	7.925	0.1262	73.532	0.0136	9.2787	0.1078	7.2972
24	8.671	0.1153	81.457	0.0123	9.3940	0.1065	7.4900
25	9.488	0.1054	90.128	0.0111	9.4994	0.1053	7.6732
26	10.381	0.0963	99.616	0.0100	9.5958	0.1042	7.8471
27	11.359	0.0880	109.997	0.0091	9.6838	0.1033	8.0122
28	12.429	0.0805	121.356	0.0083	9.7643	0.1024	8.1686
29	13.599	0.0735	133.784	0.0075	9.8378	0.1017	8.3169
30	14.880	0.0672	147.383	0.0068	9.9050	0.1010	8.4572
31	16.281	0.0614	162.263	0.0062	9.9664	0.1003	8.5900
32	17.814	0.0561	178.544	0.0056	10.0225	0.0998	8.7155
33	19.492	0.0513	196.358	0.0051	10.0739	0.0993	8.8341
34	21.328	0.0469	215.850	0.0046	10.1207	0.0988	8.9460
35	23.336	0.0429	237.178	0.0042	10.1636	0.0984	9.0516
40	36.598	0.0273	378.004	0.0027	10.3285	0.0968	9.4950
45	57.397	0.0174	598.863	0.0017	10.4336	0.0959	9.8207
50	90.017	0.0111	945.238	0.0011	10.5007	0.0952	10.0569
55	141.175	0.0071	1488.463	0.0007	10.5434	0.0949	10.2263
60	221.406	0.0045	2340.410	0.0004	10.5707	0.0946	10.3464
65	347.234	0.0029	3676.528	0.0003	10.5880	0.0945	10.4309
70	544.572	0.0019	5771.978	0.0002	10.5991	0.0944	10.4898
75	854.059	0.0012	9058.298	0.0001	10.6062	0.0943	10.5307
80	1339.431	0.0008	14212.274	0.0001	10.6107	0.0943	10.5588
85	2100.646	0.0005	22295.318	0.0001	10.6136	0.0942	10.5781
90	3294.468	0.0003	34972.053	0.0000	10.6154	0.0942	10.5913
95	5166.754	0.0002	54853.132	0.0000	10.6166	0.0942	10.6002
100	8103.084	0.0001	86032.870	0.0000	10.6173	0.0942	10.6063

Table D.29. 10% INTEREST FACTORS FOR CONTINUOUS COMPOUNDING INTEREST

	Single Payment		Equal Payment Series				Uniform
	Compound-amount factor	Present-worth factor	Compound-amount factor	Sinking-fund factor	Present-worth factor	Capital-recovery factor	gradient-series factor
n	To find F Given P F/P r,n	To find P Given F P/F r,n	To find F Given A F/A r,n	To find A Given F A/F r,n	To find P Given A P/A r,n	To find A Given P A/P r,n	To find A Given g A/G r,n
1	1.105	0.9048	1.000	1.0000	0.9048	1.1052	0.0000
2	1.221	0.8187	2.105	0.4750	1.7236	0.5802	0.4750
3	1.350	0.7408	3.327	0.3006	2.4644	0.4058	0.9335
4	1.492	0.6703	4.676	0.2138	3.1347	0.3190	1.3754
5	1.649	0.6065	6.168	0.1621	3.7412	0.2673	1.8009
6	1.822	0.5488	7.817	0.1279	4.2901	0.2331	2.2101
7	2.014	0.4966	9.639	0.1038	4.7866	0.2089	2.6033
8	2.226	0.4493	11.653	0.0858	5.2360	0.1910	2.9806
9	2.460	0.4066	13.878	0.0721	5.6425	0.1772	3.3423
10	2.718	0.3679	16.338	0.0612	6.0104	0.1664	3.6886
11	3.004	0.3329	19.056	0.0525	6.3433	0.1577	4.0198
12	3.320	0.3012	22.060	0.0453	6.6445	0.1505	4.3362
13	3.669	0.2725	25.381	0.0394	6.9170	0.1446	4.6381
14	4.055	0.2466	29.050	0.0344	7.1636	0.1396	4.9260
15	4.482	0.2231	33.105	0.0302	7.3867	0.1354	5.2001
16	4.953	0.2019	37.587	0.0266	7.5886	0.1318	5.4608
17	5.474	0.1827	42.540	0.0235	7.7713	0.1287	5.7086
18	6.050	0.1653	48.014	0.0208	7.9366	0.1260	5.9437
19	6.686	0.1496	54.063	0.0185	8.0862	0.1237	6.1667
20	7.389	0.1353	60.749	0.0165	8.2215	0.1216	6.3780
21	8.166	0.1225	68.138	0.0147	8.3440	0.1199	6.5779
22	9.025	0.1108	76.305	0.0131	8.4548	0.1183	6.7669
23	9.974	0.1003	85.330	0.0117	8.5550	0.1169	6.9454
24	11.023	0.0907	95.304	0.0105	8.6458	0.1157	7.1139
25	12.183	0.0821	106.327	0.0094	8.7279	0.1146	7.2727
26	13.464	0.0743	118.509	0.0084	8.8021	0.1136	7.4223
27	14.880	0.0672	131.973	0.0076	8.8693	0.1128	7.5631
28	16.445	0.0608	146.853	0.0068	8.9301	0.1120	7.6954
29	18.174	0.0550	163.297	0.0061	8.9852	0.1113	7.8198
30	20.086	0.0498	181.472	0.0055	9.0349	0.1107	7.9365
31	22.198	0.0451	201.557	0.0050	9.0800	0.1101	8.0459
32	24.533	0.0408	223.755	0.0045	9.1208	0.1097	8.1485
33	27.113	0.0369	248.288	0.0040	9.1576	0.1092	8.2446
34	29.964	0.0334	275.400	0.0036	9.1910	0.1088	8.3345
35	33.115	0.0302	305.364	0.0033	9.2212	0.1085	8.4185
40	54.598	0.0183	509.629	0.0020	9.3342	0.1071	8.7620
45	90.017	0.0111	846.404	0.0012	9.4027	0.1064	9.0028
50	148.413	0.0067	1401.653	0.0007	9.4443	0.1059	9.1692
55	244.692	0.0041	2317.104	0.0004	9.4695	0.1056	9.2826
60	403.429	0.0025	3826.427	0.0003	9.4848	0.1054	9.3592
65	665.142	0.0015	6314.879	0.0002	9.4940	0.1053	9.4105
70	1096.633	0.0009	10417.644	0.0001	9.4997	0.1053	9.4445
75	1808.042	0.0006	17181.959	0.0001	9.5031	0.1052	9.4668
80	2980.958	0.0004	28334.430	0.0001	9.5052	0.1052	9.4815
85	4914.769	0.0002	46721.745	0.0000	9.5064	0.1052	9.4910
90	8103.084	0.0001	77037.303	0.0000	9.5072	0.1052	9.4972
95	13359.727	0.0001	127019.209	0.0000	9.5076	0.1052	9.5012
100	22026.466	0.0001	209425.440	0.0000	9.5079	0.1052	9.5038

Table D.30. 12% INTEREST FACTORS FOR CONTINUOUS COMPOUNDING INTEREST

	Single Payment		Equal Payment Series				Uniform gradient-series factor
	Compound-amount factor	Present-worth factor	Compound-amount factor	Sinking-fund factor	Present-worth factor	Capital-recovery factor	
n	To find F Given P $F/P \quad r, n$	To find P Given F $P/F \quad r, n$	To find F Given A $F/A \quad r, n$	To find A Given F $A/F \quad r, n$	To find P Given A $P/A \quad r, n$	To find A Given P $A/P \quad r, n$	To find A Given g $A/G \quad r, n$
1	1.128	0.8869	1.000	1.0000	0.8869	1.1275	0.0000
2	1.271	0.7866	2.128	0.4700	1.6736	0.5975	0.4700
3	1.433	0.6977	3.399	0.2942	2.3712	0.4217	0.9202
4	1.616	0.6188	4.832	0.2070	2.9900	0.3345	1.3506
5	1.822	0.5488	6.448	0.1551	3.5388	0.2826	1.7615
6	2.054	0.4868	8.270	0.1209	4.0256	0.2484	2.1531
7	2.316	0.4317	10.325	0.0969	4.4573	0.2244	2.5257
8	2.612	0.3829	12.641	0.0791	4.8402	0.2066	2.8796
9	2.945	0.3396	15.253	0.0656	5.1798	0.1931	3.2153
10	3.320	0.3012	18.197	0.0550	5.4810	0.1825	3.5332
11	3.743	0.2671	21.518	0.0465	5.7481	0.1740	3.8337
12	4.221	0.2369	25.261	0.0396	5.9850	0.1671	4.1174
13	4.759	0.2101	29.482	0.0339	6.1952	0.1614	4.3848
14	5.366	0.1864	34.241	0.0292	6.3815	0.1567	4.6364
15	6.050	0.1653	39.606	0.0253	6.5468	0.1528	4.8728
16	6.821	0.1466	45.656	0.0219	6.6935	0.1494	5.0947
17	7.691	0.1300	52.477	0.0191	6.8235	0.1466	5.3025
18	8.671	0.1153	60.167	0.0166	6.9388	0.1441	5.4969
19	9.777	0.1023	68.838	0.0145	7.0411	0.1420	5.6785
20	11.023	0.0907	78.615	0.0127	7.1318	0.1402	5.8480
21	12.429	0.0805	89.638	0.0112	7.2123	0.1387	6.0058
22	14.013	0.0714	102.067	0.0098	7.2836	0.1373	6.1528
23	15.800	0.0633	116.080	0.0086	7.3469	0.1361	6.2893
24	17.814	0.0561	131.880	0.0076	7.4031	0.1351	6.4160
25	20.086	0.0498	149.694	0.0067	7.4528	0.1342	6.5334
26	22.646	0.0442	169.780	0.0059	7.4970	0.1334	6.6422
27	25.534	0.0392	192.426	0.0052	7.5362	0.1327	6.7428
28	28.789	0.0347	217.960	0.0046	7.5709	0.1321	6.8358
29	32.460	0.0308	246.749	0.0041	7.6017	0.1316	6.9215
30	36.598	0.0273	279.209	0.0036	7.6290	0.1311	7.0006
31	41.264	0.0242	315.807	0.0032	7.6533	0.1307	7.0734
32	46.525	0.0215	357.071	0.0028	7.6748	0.1303	7.1404
33	52.457	0.0191	403.597	0.0025	7.6938	0.1300	7.2020
34	59.145	0.0169	456.054	0.0022	7.7107	0.1297	7.2586
35	66.686	0.0150	515.200	0.0020	7.7257	0.1294	7.3105
40	121.510	0.0082	945.203	0.0011	7.7788	0.1286	7.5114
45	221.406	0.0045	1728.720	0.0006	7.8079	0.1281	7.6392
50	403.429	0.0025	3156.382	0.0003	7.8239	0.1278	7.7191

Table D.31. 15% INTEREST FACTORS FOR CONTINUOUS COMPOUNDING INTEREST

	Single Payment		Equal Payment Series				Uniform gradient-series factor
	Compound-amount factor	Present-worth factor	Compound-amount factor	Sinking-fund factor	Present-worth factor	Capital-recovery factor	
n	To find F Given P F/P r,n	To find P Given F P/F r,n	To find F Given A F/A r,n	To find A Given F A/F r,n	To find P Given A P/A r,n	To find A Given P A/P r,n	To find A Given g A/G r,n
1	1.162	0.8607	1.000	1.0000	0.8607	1.1618	0.0000
2	1.350	0.7408	2.162	0.4626	1.6015	0.6244	0.4626
3	1.568	0.6376	3.512	0.2848	2.2392	0.4466	0.9004
4	1.822	0.5488	5.080	0.1969	2.7880	0.3587	1.3137
5	2.117	0.4724	6.902	0.1449	3.2603	0.3067	1.7029
6	2.460	0.4066	9.019	0.1109	3.6669	0.2727	2.0685
7	2.858	0.3499	11.479	0.0871	4.0168	0.2490	2.4110
8	3.320	0.3012	14.336	0.0698	4.3180	0.2316	2.7311
9	3.857	0.2593	17.657	0.0566	4.5773	0.2185	3.0295
10	4.482	0.2231	21.514	0.0465	4.8004	0.2083	3.3070
11	5.207	0.1921	25.996	0.0385	4.9925	0.2003	3.5645
12	6.050	0.1653	31.203	0.0321	5.1578	0.1939	3.8028
13	7.029	0.1423	37.252	0.0269	5.3000	0.1887	4.0228
14	8.166	0.1225	44.281	0.0226	5.4225	0.1844	4.2255
15	9.488	0.1054	52.447	0.0191	5.5279	0.1809	4.4119
16	11.023	0.0907	61.935	0.0162	5.6186	0.1780	4.5829
17	12.807	0.0781	72.958	0.0137	5.6967	0.1756	4.7394
18	14.880	0.0672	85.765	0.0117	5.7639	0.1735	4.8823
19	17.288	0.0579	100.645	0.0099	5.8217	0.1718	5.0127
20	20.086	0.0498	117.933	0.0085	5.8715	0.1703	5.1313
21	23.336	0.0429	138.018	0.0073	5.9144	0.1691	5.2390
22	27.113	0.0369	161.354	0.0062	5.9513	0.1680	5.3367
23	31.500	0.0318	188.467	0.0053	5.9830	0.1672	5.4251
24	36.598	0.0273	219.967	0.0046	6.0103	0.1664	5.5050
25	42.521	0.0235	256.566	0.0039	6.0339	0.1657	5.5771
26	49.402	0.0203	299.087	0.0034	6.0541	0.1652	5.6420
27	57.397	0.0174	348.489	0.0029	6.0715	0.1647	5.7004
28	66.686	0.0150	405.886	0.0025	6.0865	0.1643	5.7529
29	77.478	0.0129	472.573	0.0021	6.0994	0.1640	5.8000
30	90.017	0.0111	550.051	0.0018	6.1105	0.1637	5.8422
31	104.585	0.0096	640.068	0.0016	6.1201	0.1634	5.8799
32	121.510	0.0082	744.653	0.0014	6.1283	0.1632	5.9136
33	141.175	0.0071	866.164	0.0012	6.1354	0.1630	5.9438
34	164.022	0.0061	1007.339	0.0010	6.1415	0.1628	5.9706
35	190.566	0.0053	1171.361	0.0009	6.1467	0.1627	5.9945
40	403.429	0.0025	2486.673	0.0004	6.1639	0.1622	6.0798
45	854.059	0.0012	5271.188	0.0002	6.1719	0.1620	6.1264
50	1808.042	0.0006	11166.008	0.0001	6.1758	0.1619	6.1515

Table D.32. 20% INTEREST FACTORS FOR CONTINUOUS COMPOUNDING INTEREST

	Single Payment		Equal Payment Series				Uniform gradient-series factor
	Compound-amount factor	Present-worth factor	Compound-amount factor	Sinking-fund factor	Present-worth factor	Capital-recovery factor	
n	To find F Given P F/P r,n	To find P Given F P/F r,n	To find F Given A F/A r,n	To find A Given F A/F r,n	To find P Given A P/A r,n	To find A Given P A/P r,n	To find A Given g A/G r,n
1	1.221	0.8187	1.000	1.0000	0.8187	1.2214	0.0000
2	1.492	0.6703	2.221	0.4502	1.4891	0.6716	0.4502
3	1.822	0.5488	3.713	0.2693	2.0379	0.4907	0.8676
4	2.226	0.4493	5.535	0.1807	2.4872	0.4021	1.2528
5	2.718	0.3679	7.761	0.1289	2.8551	0.3503	1.6068
6	3.320	0.3012	10.479	0.0954	3.1563	0.3168	1.9306
7	4.055	0.2466	13.799	0.0725	3.4029	0.2939	2.2255
8	4.953	0.2019	17.854	0.0560	3.6048	0.2774	2.4929
9	6.050	0.1653	22.808	0.0439	3.7701	0.2653	2.7344
10	7.389	0.1353	28.857	0.0347	3.9054	0.2561	2.9515
11	9.025	0.1108	36.246	0.0276	4.0162	0.2490	3.1460
12	11.023	0.0907	45.271	0.0221	4.1069	0.2435	3.3194
13	13.464	0.0743	56.294	0.0178	4.1812	0.2392	3.4736
14	16.445	0.0608	69.758	0.0143	4.2420	0.2357	3.6102
15	20.086	0.0498	86.203	0.0116	4.2918	0.2330	3.7307
16	24.533	0.0408	106.288	0.0094	4.3326	0.2308	3.8368
17	29.964	0.0334	130.821	0.0077	4.3659	0.2291	3.9297
18	36.598	0.0273	160.785	0.0062	4.3933	0.2276	4.0110
19	44.701	0.0224	197.383	0.0051	4.4156	0.2265	4.0819
20	54.598	0.0183	242.084	0.0041	4.4339	0.2255	4.1435
21	66.686	0.0150	296.683	0.0034	4.4489	0.2248	4.1970
22	81.451	0.0123	363.369	0.0028	4.4612	0.2242	4.2432
23	99.484	0.0101	444.820	0.0023	4.4713	0.2237	4.2831
24	121.510	0.0082	544.304	0.0018	4.4795	0.2232	4.3175
25	148.413	0.0067	665.814	0.0015	4.4862	0.2229	4.3471
26	181.272	0.0055	814.228	0.0012	4.4917	0.2226	4.3724
27	221.406	0.0045	995.500	0.0010	4.4963	0.2224	4.3942
28	270.426	0.0037	1216.906	0.0008	4.5000	0.2222	4.4127
29	330.300	0.0030	1487.333	0.0007	4.5030	0.2221	4.4286
30	403.429	0.0025	1817.632	0.0006	4.5055	0.2220	4.4421
31	492.749	0.0020	2221.061	0.0005	4.5075	0.2219	4.4536
32	601.845	0.0017	2713.810	0.0004	4.5092	0.2218	4.4634
33	735.095	0.0014	3315.655	0.0003	4.5105	0.2217	4.4717
34	897.847	0.0011	4050.750	0.0003	4.5116	0.2217	4.4788
35	1096.633	0.0009	4948.598	0.0002	4.5125	0.2216	4.4847
40	2980.958	0.0004	13459.444	0.0001	4.5152	0.2215	4.5032
45	8103.084	0.0001	36594.322	0.0000	4.5161	0.2214	4.5111
50	22026.466	0.0001	99481.443	0.0000	4.5165	0.2214	4.5144

Table D.33. 25% INTEREST FACTORS FOR CONTINUOUS COMPOUNDING INTEREST

	Single Payment		Equal Payment Series				Uniform
	Compound-amount factor	Present-worth factor	Compound-amount factor	Sinking-fund factor	Present-worth factor	Capital-recovery factor	gradient-series factor
n	To find F Given P $F/P \ \ r, n$	To find P Given F $P/F \ \ r, n$	To find F Given A $F/A \ \ r, n$	To find A Given F $A/F \ \ r, n$	To find P Given A $P/A \ \ r, n$	To find A Given P $A/P \ \ r, n$	To find A Given g $A/G \ \ r, n$
1	1.284	0.7788	1.000	1.0000	0.7788	1.2840	0.0000
2	1.649	0.6065	2.284	0.4378	1.3853	0.7219	0.4378
3	2.117	0.4724	3.933	0.2543	1.8577	0.5383	0.8351
4	2.718	0.3679	6.050	0.1653	2.2256	0.4493	1.1929
5	3.490	0.2865	8.768	0.1141	2.5121	0.3981	1.5131
6	4.482	0.2231	12.258	0.0816	2.7352	0.3656	1.7975
7	5.755	0.1738	16.740	0.0597	2.9090	0.3438	2.0486
8	7.389	0.1353	22.495	0.0445	3.0443	0.3285	2.2687
9	9.488	0.1054	29.884	0.0335	3.1497	0.3175	2.4605
10	12.183	0.0821	39.371	0.0254	3.2318	0.3094	2.6266
11	15.643	0.0639	51.554	0.0194	3.2957	0.3034	2.7696
12	20.086	0.0498	67.197	0.0149	3.3455	0.2989	2.8921
13	25.790	0.0388	87.282	0.0115	3.3843	0.2955	2.9964
14	33.115	0.0302	113.072	0.0089	3.4145	0.2929	3.0849
15	42.521	0.0235	146.188	0.0069	3.4380	0.2909	3.1596
16	54.598	0.0183	188.709	0.0053	3.4563	0.2893	3.2223
17	70.105	0.0143	243.307	0.0041	3.4706	0.2881	3.2748
18	90.017	0.0111	313.413	0.0032	3.4817	0.2872	3.3186
19	115.584	0.0087	403.430	0.0025	3.4904	0.2865	3.3550
20	148.413	0.0067	519.014	0.0019	3.4971	0.2860	3.3851
21	190.566	0.0053	667.427	0.0015	3.5023	0.2855	3.4100
22	244.692	0.0041	857.993	0.0012	3.5064	0.2852	3.4305
23	314.191	0.0032	1102.685	0.0009	3.5096	0.2849	3.4474
24	403.429	0.0025	1416.876	0.0007	3.5121	0.2847	3.4612
25	518.013	0.0019	1820.305	0.0006	3.5140	0.2846	3.4725
26	665.142	0.0015	2338.318	0.0004	3.5155	0.2845	3.4817
27	854.059	0.0012	3003.459	0.0003	3.5167	0.2844	3.4892
28	1096.633	0.0009	3857.518	0.0003	3.5176	0.2843	3.4953
29	1408.105	0.0007	4954.151	0.0002	3.5183	0.2842	3.5002
30	1808.042	0.0006	6362.256	0.0002	3.5189	0.2842	3.5042
31	2321.572	0.0004	8170.298	0.0001	3.5193	0.2842	3.5075
32	2980.958	0.0004	10491.871	0.0001	3.5196	0.2841	3.5101
33	3827.626	0.0003	13472.829	0.0001	3.5199	0.2841	3.5122
34	4914.769	0.0002	17300.455	0.0001	3.5201	0.2841	3.5139
35	6310.688	0.0002	22215.223	0.0001	3.5203	0.2841	3.5153

Table D.34. 30% INTEREST FACTORS FOR CONTINUOUS COMPOUNDING INTEREST

	Single Payment		Equal Payment Series				Uniform gradient-series factor
	Compound-amount factor	Present-worth factor	Compound-amount factor	Sinking-fund factor	Present-worth factor	Capital-recovery factor	
n	To find F Given P $F/P \quad r,n$	To find P Given F $P/F \quad r,n$	To find F Given A $F/A \quad r,n$	To find A Given F $A/F \quad r,n$	To find P Given A $P/A \quad r,n$	To find A Given P $A/P \quad r,n$	To find A Given g $A/G \quad r,n$
1	1.350	0.7408	1.000	1.0000	0.7408	1.3499	0.0000
2	1.822	0.5488	2.350	0.4256	1.2896	0.7754	0.4256
3	2.460	0.4066	4.172	0.2397	1.6962	0.5896	0.8030
4	3.320	0.3012	6.632	0.1508	1.9974	0.5007	1.1343
5	4.482	0.2231	9.952	0.1005	2.2205	0.4504	1.4222
6	6.050	0.1653	14.433	0.0693	2.3858	0.4192	1.6701
7	8.166	0.1225	20.483	0.0488	2.5083	0.3987	1.8815
8	11.023	0.0907	28.649	0.0349	2.5990	0.3848	2.0602
9	14.880	0.0672	39.672	0.0252	2.6662	0.3751	2.2099
10	20.086	0.0498	54.552	0.0183	2.7160	0.3682	2.3343
11	27.113	0.0369	74.638	0.0134	2.7529	0.3633	2.4371
12	36.598	0.0273	101.750	0.0098	2.7802	0.3597	2.5212
13	49.402	0.0203	138.349	0.0072	2.8004	0.3571	2.5897
14	66.686	0.0150	187.751	0.0053	2.8154	0.3552	2.6452
15	90.017	0.0111	254.437	0.0039	2.8266	0.3538	2.6898
16	121.510	0.0082	344.454	0.0029	2.8348	0.3528	2.7255
17	164.022	0.0061	465.965	0.0022	2.8409	0.3520	2.7540
18	221.406	0.0045	629.987	0.0016	2.8454	0.3515	2.7766
19	298.867	0.0034	851.393	0.0012	2.8487	0.3510	2.7945
20	403.429	0.0025	1150.261	0.0009	2.8512	0.3507	2.8086
21	544.572	0.0018	1553.689	0.0007	2.8531	0.3505	2.8197
22	735.095	0.0014	2098.261	0.0005	2.8544	0.3503	2.8283
23	992.275	0.0010	2833.356	0.0004	2.8554	0.3502	2.8351
24	1339.431	0.0008	3825.631	0.0003	2.8562	0.3501	2.8404
25	1808.042	0.0006	5165.062	0.0002	2.8567	0.3501	2.8445
26	2440.602	0.0004	6973.104	0.0002	2.8571	0.3500	2.8476
27	3294.468	0.0003	9413.706	0.0001	2.8574	0.3500	2.8501
28	4447.067	0.0002	12708.174	0.0001	2.8577	0.3499	2.8520
29	6002.912	0.0002	17155.241	0.0001	2.8578	0.3499	2.8535
30	8103.084	0.0001	23158.153	0.0001	2.8580	0.3499	2.8546
31	10938.019	0.0001	31261.237	0.0000	2.8580	0.3499	2.8555
32	14764.782	0.0001	42199.257	0.0000	2.8581	0.3499	2.8561
33	19930.370	0.0001	56964.038	0.0000	2.8582	0.3499	2.8566
34	26903.186	0.0001	76894.409	0.0000	2.8582	0.3499	2.8570
35	36315.503	0.0000	103797.595	0.0000	2.8582	0.3499	2.8573

Table D.35. FUNDS FLOW CONVERSION FACTOR

r	$\dfrac{e^r - 1}{r}$ $(A/\overline{A}\ r)$
1	1.005020
2	1.010065
3	1.015150
4	1.020270
5	1.025422
6	1.030608
7	1.035831
8	1.041088
9	1.046381
10	1.051709
11	1.057073
12	1.062474
13	1.067910
14	1.073384
15	1.078894
16	1.084443
17	1.090028
18	1.095652
19	1.101313
20	1.107014
21	1.112752
22	1.118530
23	1.124347
24	1.130204
25	1.136101
26	1.142038
27	1.148016
28	1.154035
29	1.160094
30	1.166196
31	1.172339
32	1.178524
33	1.184751
34	1.191022
35	1.197335
36	1.203692
37	1.210093
38	1.216538
39	1.223027
40	1.229561

E

Selected References

Barish, N. N., *Economic Analysis* (New York : McGraw-Hill Book Company, Inc., 1962).

Barnard, Chester I., *The Functions of the Executive* (Cambridge, Mass. : Harvard University Press, 1938).

Bierman, H. J., and S. Smidt, *The Capital Budgeting Decision* (New York : The Macmillan Company, 1966).

Buffa, E. S., *Models for Production and Operations Management* (New York : John Wiley & Sons, Inc., 1963).

Bullinger, C. E., *Engineering Economy* (New York : McGraw-Hill Book Company, Inc., 1958).

Dean, J., *Managerial Economics* (Englewood Cliffs, N.J. : Prentice-Hall, Inc., 1951).

DeGarmo, E. P., *Engineering Economy* (New York : The Macmillan Company, 1967).

English, J. M., Editor, *Cost Effectiveness* (New York : John Wiley and Sons, Inc., 1968).

Fabrycky, W. J., and P. E. Torgersen, *Operations Economy: Industrial Applications of Operations Research* (Englewood Cliffs, N. J. : Prentice-Hall, Inc. 1966).

Fleischer, G. A., *Capital Allocation Theory* (New York : Appleton-Century-Crofts, 1969).

Grant, E. L., *Basic Accounting and Cost Accounting* (New York : McGraw-Hill Book Company, Inc., 1956).

Grant, E. L. and W. G. Ireson, *Principles of Engineering Economy* (New York : The Ronald Press Company, 1970).

Marston, A., R. Winfrey, and J. C. Hempstead, *Engineering Valuation and Depreciation* (New York : McGraw-Hill Book Company, Inc., 1953).

Miller, D. W. and M. K. Starr, *Executive Decisions and Operations Research* (Englewood Cliffs, N. J. : Prentice-Hall, Inc., 1960).

Morris, W. T., *The Analysis of Management Decisions* (Homewood, Ill. : Richard D. Irwin, Inc., 1964).

Riggs, J. L., *Economic Decision Models* (New York : McGraw-Hill Book Company, Inc., 1968).

Roscoe, E. S., *Project Economy* (Homewood, Ill. : Richard D. Irwin, Inc., 1961).

Schweyer, H. E., *Analytical Models for Managerial and Engineering Economics* (New York: Reinhold Publishing Corporation, 1964).

Smith, G. W., *Engineering Economy* (Ames, Iowa: Iowa State University Press, 1969).

Specthrie, S. W., *Industrial Accounting* (Englewood Cliffs, N.J.: Prentice-Hall, Inc., 1959).

Spencer, M. H. and L. Siegelman, *Managerial Economics* (Homewood, Ill.: Richard D. Irwin, Inc., 1959).

Taylor, G. A., *Managerial and Engineering Economy* (Princeton, N.J.: D. van Nostrand Company, Inc., 1964).

Terborgh, G., *Business Investment Policy* (Washington, D.C.: Machinery and Allied Products Institute, 1958).

Terborgh, G., *Dynamic Equipment Policy* (New York: McGraw-Hill Book Company, Inc. 1949).

Tucker, H. and M. C. Leager, *Highway Economics* (Scranton, Pa.: International Textbook Company, 1942).

Tyler, C. and C. H. Winter, Jr., *Chemical Engineering Economics* (New York: McGraw-Hill Book Company, Inc., 1959).

Winfrey, R., *Economic Analysis for Highways* (Scranton, Pennsylvania: International Textbook Company, 1969).

INDEX

Index